# AVR WORKSHOP

# AVR WORKSHOP

## A Hands-on Introduction with 60 Projects

by John Boxall

no starch press

San Francisco

Printed in the United States of America

First printing

26 25 24 23 22    1 2 3 4 5

ISBN-13: 978-1-7185-0258-1 (print)
ISBN-13: 978-1-7185-0259-8 (ebook)

Publisher: William Pollock
Managing Editor: Jill Franklin
Production Manager: Rachel Monaghan
Developmental Editor: Abigail Schott-Rosenfield
Production Editor: Rachel Monaghan
Cover and Interior Design: Octopod Studios
Technical Reviewer: Xander Soldaat
Copyeditor: Rachel Head
Production Services: Scribe Inc.

For information on distribution, bulk sales, corporate sales, or translations, please contact No Starch Press, Inc. directly at info@nostarch.com or:

No Starch Press, Inc.
245 8th Street, San Francisco, CA 94103
phone: 1.415.863.9900
www.nostarch.com

Library of Congress Control Number 2022020563



[S]

To Cyril, John, David, and all
the people who enjoy making
something out of nothing.

## About the Author

John Boxall has been an electronics enthusiast for over 30 years, spending much of that time in the electronics retail, wholesale, and consulting fields. He also spent several years writing Arduino tutorials, projects, and reviews of kits and accessories at *https://tronixstuff.com*. *Arduino Workshop* (No Starch Press, 2013) was his first book.

## About the Technical Reviewer

Xander Soldaat is a former Mindstorms Community Partner for LEGO® MINDSTORMS. He has a deep background in IT infrastructure architecture, as well as embedded systems, compiler, and STEM curriculum development. He is currently an OpenShift Cloud Success Architect at Red Hat. In his spare time he likes to tinker with robots, electronics, retro computers, and table-top games.

# BRIEF CONTENTS

Acknowledgments . . . . . xvii

Introduction . . . . . xix

Chapter 1: Getting Started . . . . . 1

Chapter 2: First Steps . . . . . 29

Chapter 3: Getting and Displaying Input. . . . . 57

Chapter 4: Communicating with the Outside World using the USART . . . . . 91

Chapter 5: Taking Control with Hardware Interrupts . . . . . 115

Chapter 6: Using Hardware Timers . . . . . 137

Chapter 7: Using Pulse-Width Modulation. . . . . 153

Chapter 8: Controlling Motors with Mosfets. . . . . 169

Chapter 9: Using the Internal EEPROM . . . . . 189

Chapter 10: Writing Your Own AVR Libraries . . . . . 205

Chapter 11: AVR and the SPI Bus. . . . . 219

Chapter 12: AVR and the $I^2C$ Bus. . . . . 249

Chapter 13: AVR and Character Liquid Crystal Displays . . . . . 281

Chapter 14: Controlling Servos . . . . . 313

Epilogue . . . . . 333

Index . . . . . 335

# CONTENTS IN DETAIL

**ACKNOWLEDGMENTS** **xvii**

**INTRODUCTION** **xix**

**1**
**GETTING STARTED** **1**

The Possibilities Are Endless . . . . . 2
The Microchip AVR Microcontrollers . . . . . 4
Required Parts and Accessories . . . . . 6
Electronic Components . . . . . 7
Choosing an AVR Programmer . . . . . 7
Required Software . . . . . 8
macOS 10.6 or Later . . . . . 9
Ubuntu Linux 20.04 LTS or Later . . . . . 10
Windows 7 Through 11 . . . . . 11
Current, Voltage, and Power . . . . . 15
Electronic Components . . . . . 15
Resistors . . . . . 16
Light-Emitting Diodes . . . . . 19
Power Diodes . . . . . 20
Capacitors . . . . . 21
Integrated Circuits . . . . . 23
Solderless Breadboards . . . . . 24
Powering Your Projects . . . . . 25

**2**
**FIRST STEPS** **29**

Testing the Hardware and Toolchain . . . . . 30
Build the Circuit . . . . . 30
Connect and Run the Programmer . . . . . 33
What If It Didn't Work? . . . . . 35
**Project 0: Blinking an LED** . . . . . 36
Uploading Your First AVR Code . . . . . 36
What If It Didn't Work? . . . . . 38
Controlling Digital Outputs . . . . . 39
Hardware Registers . . . . . 39
**Project 1: Experimenting with ATtiny85 Digital Outputs** . . . . . 41
The Hardware . . . . . 42
The Code . . . . . 42
Using Schematic Diagrams . . . . . 43
Components in Schematics . . . . . 44
Wires in Schematics . . . . . 45
Dissecting a Schematic . . . . . 46

**Project 2: Experimenting with ATmega328P-PU Digital Outputs** . . . . . . . . . 47
The Hardware . . . . . . . . . . . . 47
The Code . . . . . . . . . . . . 49
**Project 3: Bit-Shifting Digital Outputs** . . . . . . . . . . . . 50
**Project 4: Experimenting with NOT** . . . . . . . . . . . . 53
**Project 5: Experimenting with AND** . . . . . . . . . . . . 54
**Project 6: Experimenting with OR** . . . . . . . . . . . . 55
**Project 7: Experimenting with XOR** . . . . . . . . . . . . 56

**3**
**GETTING AND DISPLAYING INPUT** **57**

Digital Inputs . . . . . . . . . . . . 58
Introducing the Pushbutton . . . . . . . . . . . . 58
Reading the Status of Digital Input Pins . . . . . . . . . . . . 59
**Project 8: Blinking an LED on Command** . . . . . . . . . . . . 59
The Hardware . . . . . . . . . . . . 59
The Code . . . . . . . . . . . . 60
Making Decisions in Code . . . . . . . . . . . . 61
if Statements . . . . . . . . . . . . 61
if . . . else Statements . . . . . . . . . . . . 62
Making Two or More Comparisons . . . . . . . . . . . . 62
switch . . . case Statements . . . . . . . . . . . . 63
Creating Your Own Functions . . . . . . . . . . . . 63
**Project 9: A Simple Custom Function** . . . . . . . . . . . . 64
**Project 10: Custom Functions with Internal Variables** . . . . . . . . . . . . 65
**Project 11: Custom Functions That Return Values** . . . . . . . . . . . . 66
Switch Bounce . . . . . . . . . . . . 68
Protecting Your AVR from Fluctuating Voltages . . . . . . . . . . . . 69
Pullup Resistors . . . . . . . . . . . . 69
Pulldown Resistors . . . . . . . . . . . . 70
Introducing Seven-Segment LED Displays . . . . . . . . . . . . 70
**Project 12: Building a Single-Digit Numerical Counter** . . . . . . . . . . . . 71
The Hardware . . . . . . . . . . . . 72
The Code . . . . . . . . . . . . 72
Analog Inputs . . . . . . . . . . . . 74
Using ATtiny85 ADCs . . . . . . . . . . . . 75
**Project 13: Making a Single-Cell Battery Tester** . . . . . . . . . . . . 76
The Hardware . . . . . . . . . . . . 76
The Code . . . . . . . . . . . . 77
Using the ATmega328P-PU ADCs . . . . . . . . . . . . 79
Introducing the Variable Resistor . . . . . . . . . . . . 80
**Project 14: Experimenting with an ATmega328P-PU ADC** . . . . . . . . . . . . 82
The Hardware . . . . . . . . . . . . 82
The Code . . . . . . . . . . . . 83
Doing Arithmetic with an AVR . . . . . . . . . . . . 84
Using External Power . . . . . . . . . . . . 85
The TMP36 Temperature Sensor . . . . . . . . . . . . 86
**Project 15: Creating a Digital Thermometer** . . . . . . . . . . . . 87
The Hardware . . . . . . . . . . . . 87
The Code . . . . . . . . . . . . 88

## 4
## COMMUNICATING WITH THE OUTSIDE WORLD USING THE USART 91


Introducing the USART . . . 92
Hardware and Software for USART Communication . . . 93
**Project 16: Testing the USART** . . . 95
The Hardware . . . 95
The Code . . . 96
**Project 17: Sending Text with the USART** . . . 98
**Project 18: Sending Numbers with the USART** . . . 100
**Project 19: Creating a Temperature Data Logger** . . . 102
The Hardware . . . 102
The Code . . . 104
**Project 20: Receiving Data from Your Computer** . . . 108
**Project 21: Building a Four-Function Calculator** . . . 111


## 5
## TAKING CONTROL WITH HARDWARE INTERRUPTS 115


External Interrupts . . . 116
Setting Up Interrupts in Code . . . 116
**Project 22: Experimenting with Rising Edge Interrupts** . . . 118
The Hardware . . . 118
The Code . . . 119
**Project 23: Experimenting with Falling Edge Interrupts** . . . 121
The Hardware . . . 121
The Code . . . 122
**Project 24: Experimenting with Two Interrupts** . . . 124
The Hardware . . . 124
The Code . . . 125
Pin-Change Interrupts . . . 126
**Project 25: Experimenting with Pin-Change Interrupts** . . . 129
The Hardware . . . 129
The Code . . . 130
**Project 26: Creating an Up/Down Counter Using Interrupts** . . . 131
The Hardware . . . 131
The Code . . . 132
Final Notes on Interrupts . . . 135


## 6
## USING HARDWARE TIMERS 137


Introducing Timers . . . 138
**Project 27: Experimenting with Timer Overflow and Interrupts** . . . 139
The Hardware . . . 139
The Code . . . 140
Examining the Accuracy of the Internal Timer . . . 146
**Project 28: Using a CTC Timer for Repetitive Actions** . . . 142
**Project 29: Using CTC Timers for Repetitive Actions with Longer Delays** . . . 143
Addressing Registers with Bitwise Operations . . . 147
Addressing Individual Bits in a Register . . . 147
Addressing Multiple Bits in a Register . . . 149

**Project 30: Experimenting with Overflow Timers Using Bitwise Operations** . . . . . . 150

**7**
**USING PULSE-WIDTH MODULATION** **153**

Pulse-Width Modulation and Duty Cycles . . . . . . 154
**Project 31: Demonstrating PWM with the ATtiny85** . . . . . . 155
The Hardware . . . . . . 155
The Code . . . . . . 155
Individual PWM Pin Control for the ATtiny85 . . . . . . 157
**Project 32: Experimenting with Piezo and PWM** . . . . . . 159
The Hardware . . . . . . 159
The Code . . . . . . 160
Individual PWM Pin Control for the ATmega328P-PU . . . . . . 161
The RGB LED . . . . . . 163
**Project 33: Experimenting with RGB LEDs and PWM** . . . . . . 164
The Hardware . . . . . . 164

**8**
**CONTROLLING MOTORS WITH MOSFETS** **169**

The MOSFET . . . . . . 170
**Project 34: DC Motor Control with PWM and MOSFET** . . . . . . 171
The Hardware . . . . . . 171
The Code . . . . . . 172
**Project 35: Temperature-Controlled Fan** . . . . . . 174
The Hardware . . . . . . 174
The Code . . . . . . 176
The L293D Motor Driver IC . . . . . . 178
**Project 36: DC Motor Control with L293D** . . . . . . 180
The Hardware . . . . . . 181
The Code . . . . . . 181
**Project 37: Controlling a Two-Wheel-Drive Robot Vehicle** . . . . . . 183
The Hardware . . . . . . 184
The Code . . . . . . 185

**9**
**USING THE INTERNAL EEPROM** **189**

Storing Bytes in EEPROM . . . . . . 190
**Project 38: Experimenting with the ATtiny85's EEPROM** . . . . . . 191
Storing Words . . . . . . 192
**Project 39: A Simple EEPROM Datalogger** . . . . . . 193
The Hardware . . . . . . 193
The Code . . . . . . 194
Storing Floating-Point Variables . . . . . . 198
**Project 40: Temperature Logger with EEPROM** . . . . . . 199
The Hardware . . . . . . 199

## 10
## WRITING YOUR OWN AVR LIBRARIES 205

Creating Your First Library . . . 206
Anatomy of a Library . . . 207
Installing the Library . . . 208
**Project 41: Your First Library** . . . 208
Creating a Library That Accepts Values to Perform a Function . . . 210
**Project 42: Using the blinko2.c Library** . . . 212
Creating a Library That Processes Data and Returns Values . . . 212
**Project 43: Creating a Digital Thermometer with the thermometer.c Library** . . . 215

## 11
## AVR AND THE SPI BUS 219

How Buses Work . . . 220
Pin Connections and Voltages . . . 221
Implementing the SPI Bus . . . 222
Sending Data . . . 223
**Project 44: Using the 74HC595 Shift Register** . . . 223
The Hardware . . . 225
The Code . . . 226
**Project 45: Using Two 74HC595 Shift Registers** . . . 229
The Hardware . . . 229
The Code . . . 230
**Project 46: Using the MAX7219 LED Driver IC** . . . 231
The Hardware . . . 231
The Code . . . 234
**Project 47: Adding a Reset Button** . . . 237
Multiple SPI Devices on the Same Bus . . . 239
Receiving Data from the SPI Bus . . . 239
**Project 48: Using the MCP3008 ADC IC** . . . 240
The Hardware . . . 242
The Code . . . 243

## 12
## AVR AND THE I²C BUS 249

Increasing AVR Speed . . . 250
Introducing the I²C Bus . . . 252
Pin Connections and Voltages . . . 253
Writing to I²C Devices . . . 254
**Project 49: Using the MCP23017 16-Bit I/O Expander** . . . 256
The Hardware . . . 258
The Code . . . 259
Reading Data from I²C Devices . . . 262
**Project 50: Using an External IC EEPROM** . . . 263
The Hardware . . . 264
The Code . . . 265
**Project 51: Using the DS3231 Real-Time Clock** . . . 270
The Hardware . . . 272
The Code . . . 273

## 13
## AVR AND CHARACTER LIQUID CRYSTAL DISPLAYS 281


Introducing LCDs . . . 282
Send Commands to the LCD . . . 285
Initialize the LCD for Use . . . 286
Clear the LCD . . . 286
Set the Cursor . . . 286
Print to the LCD . . . 287
**Project 52: Using a Character LCD with Your AVR** . . . 288
The Hardware . . . 288
The Code . . . 289
**Project 53: Building an AVR-Based LCD Digital Clock** . . . 292
The Hardware . . . 292
The Code . . . 293
Displaying Floating-Point Numbers on the LCD . . . 300
**Project 54: LCD Digital Thermometer with Min/Max Display** . . . 301
The Hardware . . . 301
The Code . . . 302
Displaying Custom Characters on the LCD . . . 306
Write Data to CGRAM . . . 308
Send Custom Character Data to LCD . . . 308
Display Custom Characters on LCD . . . 308
**Project 55: Displaying Custom LCD Characters** . . . 309


## 14
## CONTROLLING SERVOS 313


Setting Up Your Servo . . . 314
Connecting a Servo . . . 315
Controlling a Servo . . . 315
**Project 56: Experimenting with Servos** . . . 317
The Hardware . . . 317
The Code . . . 318
**Project 57: Creating an Analog Thermometer** . . . 320
The Hardware . . . 320
The Code . . . 322
**Project 58: Controlling Two Servos** . . . 323
The Hardware . . . 323
The Code . . . 324
**Project 59: Building an Analog Clock with Servo Hands** . . . 326
The Hardware . . . 326
The Code . . . 327


## EPILOGUE 333

## INDEX 335

# ACKNOWLEDGMENTS

Thank you to Brian S. Dean for starting the AVRDUDE project, followed by Jörg Wunsch and the various contributors. Kudos and thanks to the KiCad team for their open source electronics design automation suite, which I've used throughout this book for circuit schematics.

Many thanks to my technical reviewer, Xander Soldaat, for his contributions and for once again having the tenacity to follow through with such a large project.

Thanks also to the following people (in no particular order) from whom I've received encouragement, inspiration, and support: Elizabeth Pryce, Mr. Richard Smith AC, the late Sir Clive Sinclair, and my wife, Kathleen, for her endless patience.

Finally, thank you to everyone at No Starch Press for their efforts, including Abigail Schott-Rosenfield for her editorial input; Rachel Monaghan for guiding the book through the production process; Rachel Head for copyediting; Scribe Inc. for composition, proofreading, and indexing; and of course Bill Pollock for his support and guidance and for convincing me that sometimes there is a better way to explain something.

# INTRODUCTION

A *microcontroller* (or *MCU* for short) is a small, complete computer that fits on a single integrated circuit. Just like your desktop computer, a microcontroller contains a processor, memory, devices to receive input from various sources, and outputs that can be used to control or communicate with external devices.

Thanks to the success of development platforms like Arduino and PICAXE, microcontrollers are being used increasingly often in the electronics field and among hobbyists and hackers. Such platforms simplify projects for beginners, but they can be costly; they also put a layer of abstraction between the user and the microcontroller, which decreases the microcontroller's performance and often prevents the user from accessing its full set of features. More experienced users may want to control the microcontroller directly or use less expensive parts in their projects. Or, if you're a beginner, you may want to start your microcontroller journey without any artificially imposed overheads.

That's where this book comes in. Whether you're an absolute beginner or a longtime electronics enthusiast, *AVR Workshop* shows you how to take advantage of two chips from the inexpensive range of Microchip AVR 8-bit microcontrollers made famous in the Arduino and compatible boards. Once you master these chips, you'll be able to maximize their performance to create powerful projects with cheaper hardware. Along the way you'll learn about electronics, C programming, and much more.

I'll walk you through over 55 projects of increasing difficulty based around the ATtiny85 and ATmega328P-PU microcontrollers from Microchip, and I'll explain and demonstrate everything you need to know for each project. You'll start off blinking a small light, then move on to keeping time, capturing and analyzing real-world data such as temperatures, and even controlling small motorized devices. This book doesn't cover AVR for IoT, as that's a more advanced topic, but after completing the projects here, you'll be able to harness a wide variety of devices, sensors, motors, displays, and more with your AVR microcontroller to bring your dreams and ideas to life.

I've written this book for a wide variety of people. You might be a student wanting to get a head start with microcontrollers, an electronics hobbyist with no prior experience in digital or microcontroller circuitry, an employee who wants to increase their knowledge base for work, or just someone who enjoys making things. This book is for anyone interested in learning more about AVR technology and harnessing it to create their own projects.

My goal is that you'll leave this book with lasting knowledge and the confidence to keep learning and making. Chapter 1 will get you started by introducing a few cool real-world projects that use AVR microcontrollers, then showing you how to set up your own workstation.

# GETTING STARTED

Welcome to the beginning of your AVR microcontroller journey! In this chapter I'll introduce you to the microcontrollers used in this book, as well as a few exciting examples of real-world AVR-based projects, then teach you some fundamentals about electronics.

You will learn:

- Where to get the required parts for the projects in this book
- How to install the required software for Windows, macOS, and Linux
- The basic properties of electricity and electronic components
- About electronic components, including resistors, light-emitting diodes (LEDs), power diodes, capacitors, and more
- How to use a solderless breadboard to construct circuits
- Ways to safely power your experiments

By the end of the chapter, you'll be ready to use your AVR workstation to build your first project.

## The Possibilities Are Endless

A quick scan through this book will show you that you can use AVR microcontrollers as the heart of an incredibly wide range of devices. You'll go from blinking LEDs to creating a thermostat, a GPS logger, and more—but don't limit yourself to the range of projects covered here! After working through this book, you'll be well prepared to explore more advanced projects like those I'll describe in this section.

For example, computer scientist Vassilis Serasidis built a piece of electronics test equipment called a *logic analyzer*, which can measure the values of four electrical currents at the same time and display the results. His design uses an inexpensive LCD typically found in cheap cellular phones to show the signals in a graphical form, as shown in Figure 1-1.

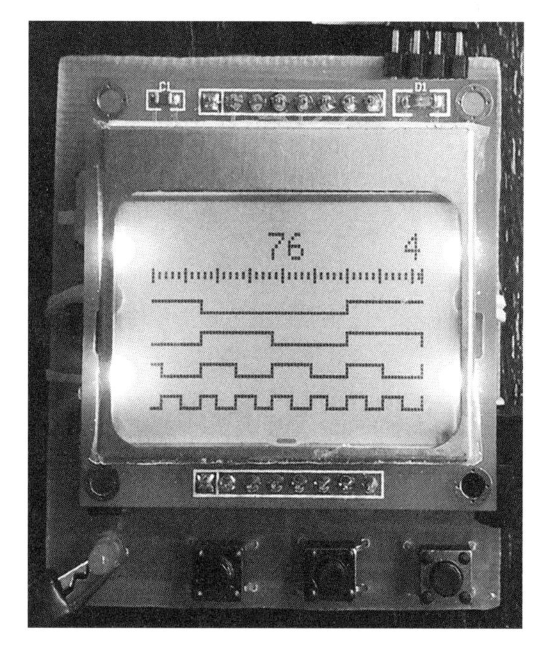


*Figure 1-1: An inexpensive logic analyzer*

You can use logic analyzers to run a huge variety of displays with an AVR microcontroller, from inexpensive black and white versions like the one in Figure 1-1 to realistic color displays. For more information on the project, visit *https://www.serasidis.gr/circuits/mini_logic_analyzer/miniLogicAnalyzer.htm.*

You can also use AVRs to build tiny projects like wearable electronics, but you need an incredibly small development board to do so. Electronics enthusiast Erik Kettenburg dreamed of such a board based on the smallest

AVR in existence. He translated this idea into the Digispark board, shown in Figure 1-2, which measures only 17.5 × 19 mm, and built a thriving business via a successful Kickstarter promotion in 2017.

*Figure 1-2: An example of the Digispark board*

The size of the Digispark board means the AVR microcontroller it uses doesn't have quite as many features as some larger chips—for example, it has less program memory space. However, the Digispark allows you to program the microcontroller directly via USB, whereas you'd normally have to purchase a separate programming device. For more information on the Digispark board, visit *http://digistump.com/.*

In addition to using AVRs professionally, plenty of people build AVR-based projects purely for fun! One example is the AVR TV Game software engineer Ben Ryves built in 2009, shown in Figure 1-3. Ben used very basic electronics and an AVR to make a device that plugs into a television and lets you play the classic games *Snake* and *Tetris.*

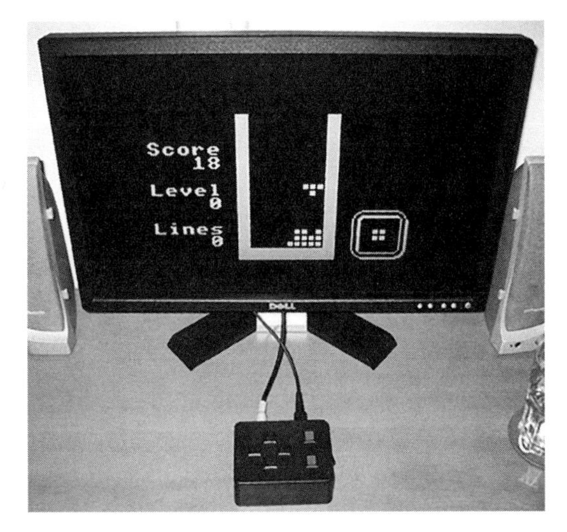


*Figure 1-3: Playing* Tetris *on the AVR TV Game*

The AVR can generate the video signals to communicate with a television without any extra hardware, and with some imagination, you can program your own games into the microcontroller. For more information, visit *http://benryves.com/products/avrtvgame/*.

Engineer Adam Heinrich even built his own cellular telephone in 2017 based on an AVR, complete with a color touchscreen interface. Adam's "AvrPhone," which is portable for mobile use, is shown in Figure 1-4. For more information, visit *http://projects.adamh.cz/avrphone/*.

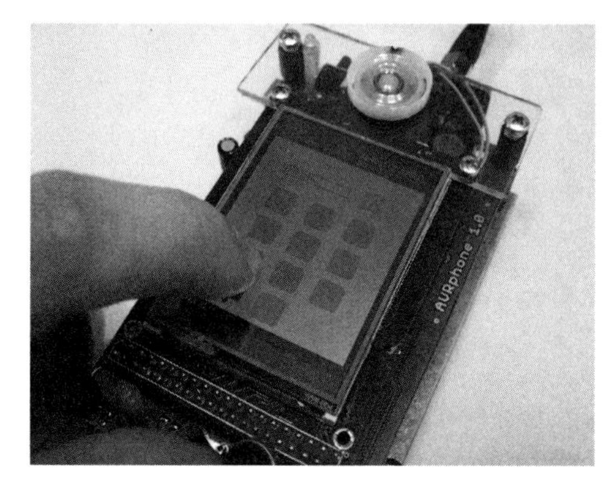


*Figure 1-4: The AvrPhone*

Just like these makers, with some effort, you can bridge the gap between hobbyist tinkering and full product development! But for now, let's start with a more detailed discussion of the parts you'll use in this book.

## The Microchip AVR Microcontrollers

Throughout this book, you'll use the two microcontrollers shown in Figure 1-5, which Microchip Technology produces as part of its AVR product line. The smaller one, called an ATtiny85, has 8 *pins*, which are the pieces of metal sticking out on the sides of the black chips that allow you to send and receive data and power from and to the microcontroller. The larger AVR is an ATmega328P-PU, which has 28 pins.

*Figure 1-5: Our AVR microcontrollers, the ATtiny85 and ATmega328P-PU*

**NOTE** *In this book and when purchasing your own parts, you may see microcontrollers labeled "Atmel." Microchip acquired Atmel in 2016, but at the time of writing some suppliers still have Atmel-branded units; either label is fine.*

Apart from size, there are several important differences between the ATtiny85 and ATmega328P-PU microcontrollers, as listed in Table 1-1.

**Table 1-1**: Specifications for the ATtiny85 and ATmega328P-PU Microcontrollers

| | **ATtiny85** | **ATmega328P-PU** |
|---|---|---|
| Schematic | PB5 1 – 8 $V_{CC}$<br>PB3 2 – 7 PB2<br>PB4 3 – 6 PB1<br>GND 4 – 5 PB0 | PC6 1 – 28 PC5<br>PD0 2 – 27 PC4<br>PD1 3 – 26 PC3<br>PD2 4 – 25 PC2<br>PD3 5 – 24 PC1<br>PD4 6 – 23 PC0<br>VCC 7 – 22 GND<br>GND 8 – 21 AREF<br>PB6 9 – 20 AVCC<br>PB7 10 – 19 PB5<br>PD5 11 – 18 PB4<br>PD6 12 – 17 PB3<br>PD7 13 – 16 PB2<br>PB0 14 – 15 PB1 |
| Maximum processing speed | 20 MHz | 20 MHz |
| Operating voltage (volts) | 1.8–5.5 V | 1.8–5.5 V |
| Number of digital pins | Up to 6 | 14 |
| Number of analog input pins | Up to 4 | 6 |
| Flash memory | 8KB | 32KB |
| EEPROM | 512 bytes | 1KB |
| SRAM | 512 bytes | 2KB |

The specifications in Table 1-1 describe each chip's physical limitations, and they will help you determine which other electronic components you can realistically use with your microcontroller. Anytime you start a new project, you need to consider this information carefully, so here's an overview of what each specification means:

**Schematic** This is a drawing that represents the connections to an electronic component, such as the microcontrollers in this table. You'll learn more about schematic symbols in Chapter 2.

**Maximum processing speed** This row tells you how fast the microcontroller can process data, measured in cycles per second. Note that the clock speed isn't always equal to processing speed, as some instructions can take multiple cycles to complete.

**Operating voltage** This row shows the range of voltages that you can safely use to power the microcontroller. If you don't supply at least 1.8 V, the chip won't turn on, but if you try to supply more than 5.5 V, it may melt!

**Digital pins** This row shows the number of pins that can send or receive digital data. Digital data is represented by voltage signals; the numbers one and zero, respectively, represent an "on" or "off" voltage. These voltage signals are then combined to represent various forms of data. All microcontrollers have pins that you can set up as digital inputs or outputs and use to control external devices.

**Analog input pins** This row shows the number of physical inputs that are available to measure voltage levels. Analog input pins let you read information from devices like *sensors*, which output different voltages based on what's happening in their surroundings.

**Flash memory** This row shows how much flash memory is available on the chip. To tell your microcontroller what to do, you'll have to write programs, which are stored in flash memory and retained even after you turn off power. If your program's file size exceeds your AVR's memory capacity, it won't fit on the microcontroller!

**EEPROM** This row tells you how much *electrically erasable programmable read-only memory (EEPROM)* is available on the chip. EEPROM can hold data your program creates even when the microcontroller is turned off. For example, if you want to display a certain image on an LCD whenever you power on your project, you can store that image in EEPROM for future use.

**SRAM** This row tells you how much *Static Random Access Memory (SRAM)* is available on the chip. This is the amount of memory available to store temporary data created by your programs. Just like the RAM in your desktop computer, SRAM is where all the information your program generates as it runs is stored until it gets deleted when the power is turned off. This could include sensor data, the results of calculations, and so on.

I'll cover these and other important features in detail in later chapters. For now, let's start setting up your AVR microcontroller lab.

## Required Parts and Accessories

You won't need to buy a ton of expensive parts to get started with the projects in this book; assuming you already have a modern personal computer, you can have a lot of fun with microcontrollers for around $50. Each project I'll walk you through includes a list of the parts you need to complete it, and you can download a list of all the parts used in this book from *https://nostarch.com/avr-workshop/*. I'd recommend you order the parts you'll need for the projects in the first few chapters now so you don't have to wait too long for delivery.

### Electronic Components

AVR microcontrollers and electronic components are available from many retailers that offer a range of products and accessories. When you're shopping, be sure to purchase the original parts I've listed and not knock-offs, or you run the risk of receiving faulty or poorly performing goods. Don't take chances buying an inferior product, as it could end up costing you more in the long run!

Always read the hardware list at the beginning of every project and be sure to buy the correct components before you start. Here are some recommended suppliers for AVR-related parts and accessories. The first five supply worldwide, while the last four are country-specific, as noted:

- DigiKey: *https://www.digikey.com/*
- element14/Farnell: *http://farnell.com/*
- PMD Way: *http://pmdway.com/*
- SparkFun Electronics: *https://www.sparkfun.com/*
- Mouser: *https://www.mouser.com/*
- Freetronics Australia: *https://www.freetronics.com/* (for Australia)
- Altronics: *https://www.altronics.com.au/* (for Australia)
- Newark: *https://www.newark.com/* (for the United States)
- MindKits: *https://www.mindkits.co.nz/* (for New Zealand)

I can vouch for these suppliers from personal experience, but there are many more across the globe. As a general rule, try to deal with organizations that offer technical and sales support and are more than just simple sales agents or listings on mega retail sites.

### Choosing an AVR Programmer

You need to connect a *programmer* like the ones shown in Figure 1-6 from your computer to your microcontroller circuit so that you can load programs and data onto the chip. Finding a good programmer is crucial for success in the AVR world, and a quick web search for "AVR programmer" will present you with many options. The projects in this book use a *USBasp programmer*, a device that interfaces between your PC and your AVR project to send your code to your project's microcontroller. You can buy one from any of the stores listed in the previous section for (at the time of printing) less than $20.

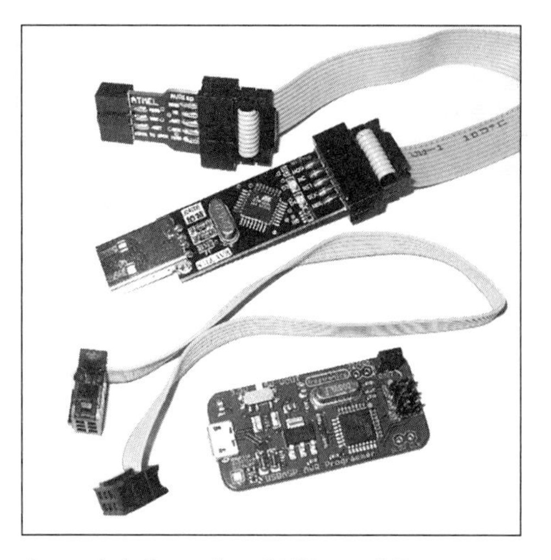

*Figure 1-6: Examples of USBasp AVR programmers*

The USBasp should work with the open source software required for AVR programming. It's an open source device created by Thomas Fischl, and only requires USB drivers when used with a Windows-based PC. When shopping for your USBasp programmer, be sure it comes with a 6-pin (not 10-pin) ribbon cable so you can use it for the projects in this book.

## Required Software

In this section, you'll set up a *toolchain* that will let you program your AVR microcontroller. This toolchain consists of a series of software that takes your typed program of instructions, translates it into a form the AVR can understand, and places it in flash memory. There are three stages to the programming process, each using a different piece of software (a different "tool" in the "chain"):

1. Using text editor software, you enter, edit, and save the code containing your instructions for the AVR.
2. Compiler software then converts the code you created into machine code that the microcontroller can understand.
3. Next, programming software takes the machine code file and uploads it to the microcontroller. At this point, the AVR should start doing what you told it to do.

You can write your programs in any basic text editor, such as Notepad on Windows or TextEdit on macOS. You'll also need a package containing the compiler and programming software. In the following three sections, I explain how to acquire and install that software for computers running MacOS, Ubuntu Linux, and Windows 7 or later.

## macOS 10.6 or Later

To install the required software on your Mac, open the macOS Terminal app and enter the following command:

```
/bin/bash -c "$(curl -fsSL https://raw.githubusercontent.com/Homebrew/install/
HEAD/install.sh)"
```

This will install software that allows you to run 32-bit programs on macOS. You may be prompted to enter your password, and then the software will start, as shown in Figure 1-7.

```
johnboxall — -zsh — 148×22
Last login: Tue Apr 12 08:08:14 on console
johnboxall@Johns-Mac-mini ~ % /bin/bash -c "$(curl -fsSL https://raw.githubusercontent.com/Homebrew/install/HEAD/install.sh)"
==> Checking for `sudo` access (which may request your password)...
Password:
==> This script will install:
/usr/local/bin/brew
/usr/local/share/doc/homebrew
/usr/local/share/man/man1/brew.1
/usr/local/share/zsh/site-functions/_brew
/usr/local/etc/bash_completion.d/brew
/usr/local/Homebrew
==> The following existing directories will be made group writable:
/usr/local/share/man/man8
==> The following existing directories will have their owner set to johnboxall:
/usr/local/share/man/man8

Press RETURN/ENTER to continue or any other key to abort:
==> /usr/bin/sudo /bin/chmod u+rwx /usr/local/share/man/man8
==> /usr/bin/sudo /bin/chmod g+rwx /usr/local/share/man/man8
==> /usr/bin/sudo /usr/sbin/chown johnboxall /usr/local/share/man/man8
==> /usr/bin/sudo /usr/sbin/chown -R johnboxall:admin /usr/local/Homebrew
==> Downloading and installing Homebrew...
```

*Figure 1-7: The toolchain installation begins in macOS.*

Once that has completed, enter the following into the command line:

```
brew tap osx-cross/avr
```

And once that has completed, enter the following command:

```
brew install avr-gcc avrdude
```

This is the final step, which installs the required AVR software. It will return you to the command prompt when finished, as shown in Figure 1-8; you can then close the Terminal app.

```
johnboxall — -zsh — 148×22
==> Installing avrdude dependency: libusb
==> Pouring libusb--1.0.26.catalina.bottle.tar.gz
🍺 /usr/local/Cellar/libusb/1.0.26: 22 files, 532.0KB
==> Installing avrdude dependency: libusb-compat
==> Pouring libusb-compat--0.1.5_1.catalina.bottle.tar.gz
🍺 /usr/local/Cellar/libusb-compat/0.1.5_1: 14 files, 98.3KB
==> Installing avrdude dependency: libftdi0
==> Pouring libftdi0--0.20.catalina.bottle.2.tar.gz
🍺 /usr/local/Cellar/libftdi0/0.20: 20 files, 226.2KB
==> Installing avrdude dependency: libhid
==> Pouring libhid--0.2.16.catalina.bottle.1.tar.gz
🍺 /usr/local/Cellar/libhid/0.2.16: 18 files, 189.9KB
==> Installing avrdude dependency: libelf
==> Pouring libelf--0.8.13_1.catalina.bottle.tar.gz
🍺 /usr/local/Cellar/libelf/0.8.13_1: 12 files, 218.2KB
==> Installing avrdude
==> Pouring avrdude--6.4_1.catalina.bottle.tar.gz
🍺 /usr/local/Cellar/avrdude/6.4_1: 14 files, 2.1MB
==> Running `brew cleanup avrdude`...
Disable this behaviour by setting HOMEBREW_NO_INSTALL_CLEANUP.
Hide these hints with HOMEBREW_NO_ENV_HINTS (see `man brew`).
johnboxall@Johns-Mac-mini ~ %
```

*Figure 1-8: Installation is complete.*

Now that you've installed your toolchain software, skip ahead to the section "Current, Voltage, and Power."

### Ubuntu Linux 20.04 LTS or Later

To download and install the AVR toolchain in Ubuntu Linux 20.04 LTS, first ensure your system has completed all the latest updates. Next, open a terminal window and enter the following command:

```
sudo apt-get install gcc build-essential
```

Enter your password if requested. You may then be asked if you want to continue: enter `Y` and press ENTER. A lot of text describing various packages should scroll by, and then the command prompt should once again appear. Enter the following command:

```
sudo apt-get install gcc-avr binutils-avr avr-libc gdb-avr
```

Again, enter your password and give permission to continue if required by the installation process. Then enter the following command:

```
sudo apt-get install avrdude
```

After a short period (depending on the speed of your computer and internet connection), all the required software should finish installing. Now you need to install the USB driver for the USBasp programmer. To do so, enter the following command:

```
sudo apt-get install libusb-dev
```

If prompted, provide your password and enter `Y` to continue. Finally, it's time to check the driver is working. Plug your USBasp into a USB port on the computer, then enter the following command:

```
lsusb
```

After a moment a list of devices connected to the machine via USB should appear, looking something like this:

```
Bus 002 Device 003: ID 045e:00cb Microsoft Corp. Basic Optical Mouse v2.0
Bus 002 Device 004: ID 16c0:05dc Van Ooijen Technische Informatica
Bus 002 Device 002: ID 8087:0020 Intel Corp. Integrated Rate Matching Hub
Bus 002 Device 001: ID 1d6b:0002 Linux Foundation 2.0 root hub
Bus 001 Device 004: ID 04f2:b1d6 Chicony Electronics Co., Ltd CNF9055
Toshiba Webcam
Bus 001 Device 003: ID 0bda:0138 Realtek Semiconductor Corp. RTS5138
Card Reader Controller
Bus 001 Device 002: ID 8087:0020 Intel Corp. Integrated Rate Matching Hub
Bus 001 Device 001: ID 1d6b:0002 Linux Foundation 2.0 root hub
```

In this example, the USBasp is the second item in the list. If your device doesn't appear, check the connection to the PC, restart the machine, and try the `lsusb` command one more time.

**NOTE** *Due to the open source nature of Linux, there are many variations of the OS that cannot be accounted for or documented in this book. If you're still having problems, or would like details on installing the toolchain on other flavors of Linux, please visit* http://www.nongnu.org/avr-libc/user-manual/install_tools.html.

Once installation is complete, move forward to the section titled "Current, Voltage, and Power."

## Windows 7 Through 11

Setting up your toolchain in Microsoft Windows requires an extra step: you'll first download and install the software, then install the appropriate drivers for your programmer.

### Installing the Toolchain

To install the required software for Windows, follow these instructions:

1. Open your web browser and visit the software download page located at *https://sourceforge.net/projects/winavr/files/WinAVR/*, as shown in Figure 1-9.

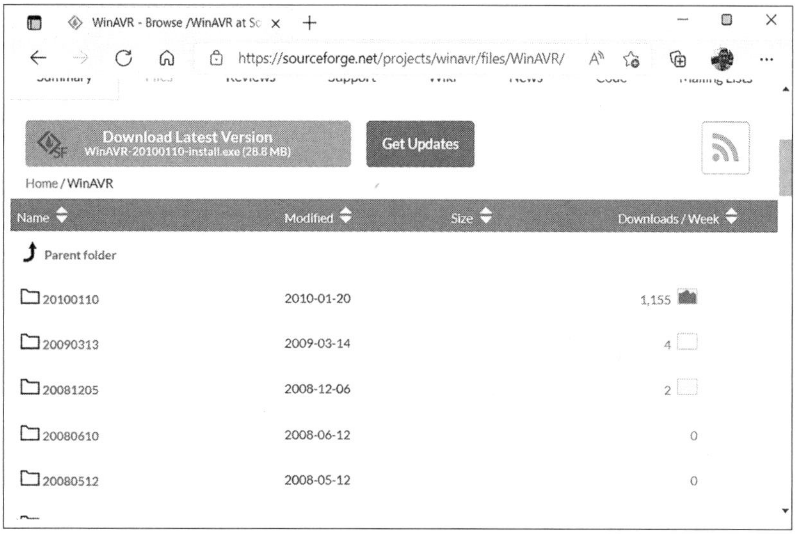

*Figure 1-9: The WinAVR download page for Windows*

2. Click the **Download Latest Version** button to start the software download. After a short period of time, the toolchain installer should finish downloading. Open the *Downloads* folder in Explorer, as shown in Figure 1-10, and you should see the WinAVR install file.

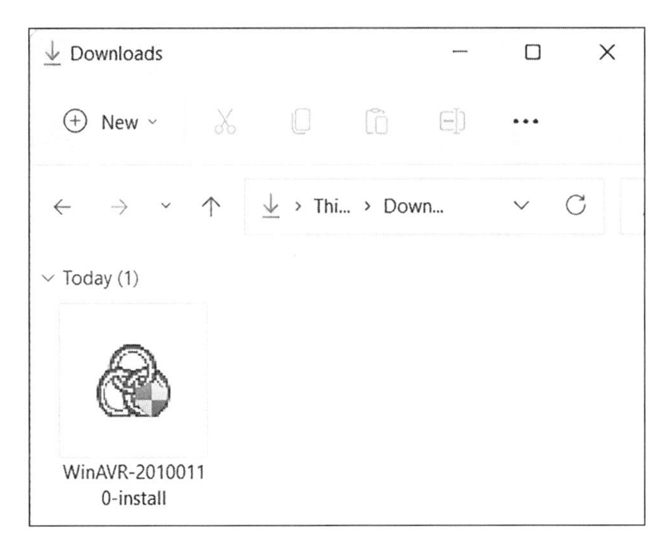


*Figure 1-10: The WinAVR package*

3. Double-click the install package, and after selecting your language, you should be presented with the WinAVR Setup Wizard, as shown in Figure 1-11.

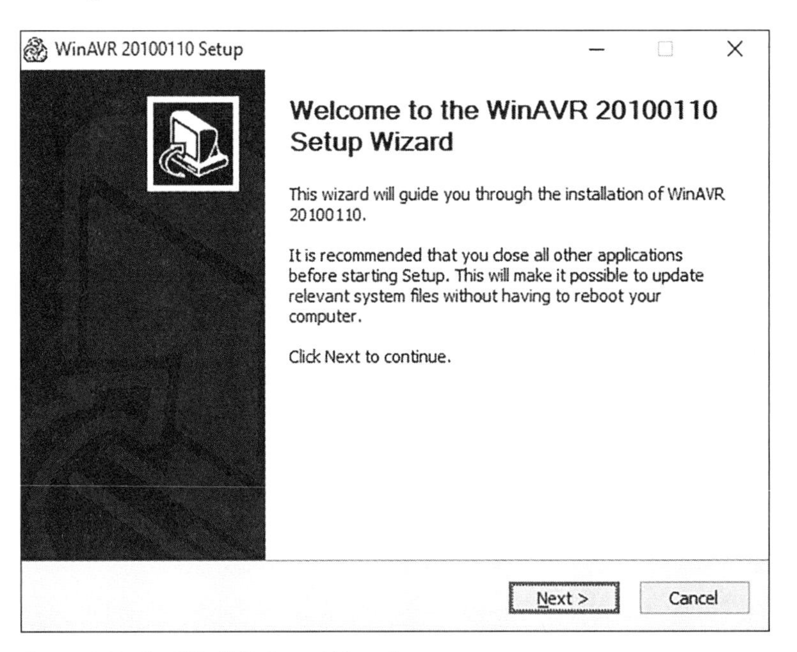


*Figure 1-11: The WinAVR Setup Wizard*

4. Click **Next** in the Setup Wizard, and click **I Agree** when the License Agreement is displayed.

5. The next window should prompt for the folder in which to install the toolchain. You can choose the default by clicking **Next**, as shown in Figure 1-12.

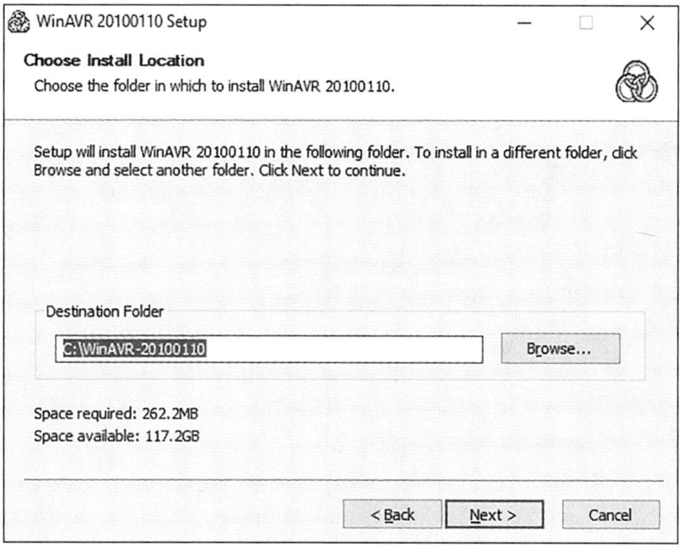

WinAVR 20100110 Setup

**Choose Install Location**

Choose the folder in which to install WinAVR 20100110.

Setup will install WinAVR 20100110 in the following folder. To install in a different folder, click Browse and select another folder. Click Next to continue.

Destination Folder

C:\WinAVR-20100110

Browse...

Space required: 262.2MB

Space available: 117.2GB

< Back | Next > | Cancel

*Figure 1-12: Choosing the file location*

6. The next window that appears, shown in Figure 1-13, should ask you which components to install. Check all three boxes and click **Install**.

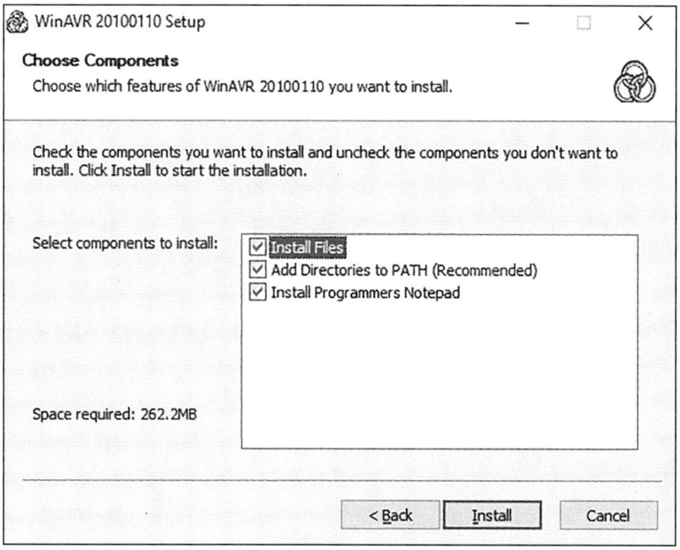

WinAVR 20100110 Setup

**Choose Components**

Choose which features of WinAVR 20100110 you want to install.

Check the components you want to install and uncheck the components you don't want to install. Click Install to start the installation.

Select components to install:

- [x] Install Files
- [x] Add Directories to PATH (Recommended)
- [x] Install Programmers Notepad

Space required: 262.2MB

< Back | Install | Cancel

*Figure 1-13: Selecting the components to install*

7. The wizard should then display a progress bar while it installs the files. Once installation has finished, close the wizard by clicking the **Finish** button.

8. You should now be presented with the WinAVR user manual page, as shown in Figure 1-14.

C:/WinAVR-20100110/WinAVR-user-manual.html

WinAVR User Manual - 20100110

Eric B. Weddington

WinAVR is a suite of executable, open source software development tools for the Atmel AVR series of RISC microprocessors and AVR32 series of microprocessors hosted on the Windows platform. It includes the GNU GCC compiler for C and C++.

1.0 What's New

- AVR32 GNU toolchain
- Splint 3.1.2 Splint is a tool for statically checking C programs for security vulnerabilities and programming mistakes. Splint does many of the traditional lint checks. More powerful checks are made possible by additional information given in source code annotations.
- New Device Support
- Component Version Upgrades

*Figure 1-14: The WinAVR user manual page*

Bookmark this page in your web browser, as it will be useful later on.

### Installing the USBasp Driver

Unlike other operating systems, Windows requires you to install a driver to enable the USBasp programmer. There are different drivers for different USBasp programmers; the driver you need will depend on the brand you buy. You can find the Windows driver and installation instructions for generic USBasps on Thomas Fischl's website: *https://www.fischl.de/usbasp/*. If you have a "branded" programmer from a company like Freetronics, however, please consult the product page for the correct driver and installation instructions.

Once you've installed the driver for your USBasp programmer, you can quickly confirm the success of the installation by plugging it in and then visiting the Windows Device Manager, as in Figure 1-15.

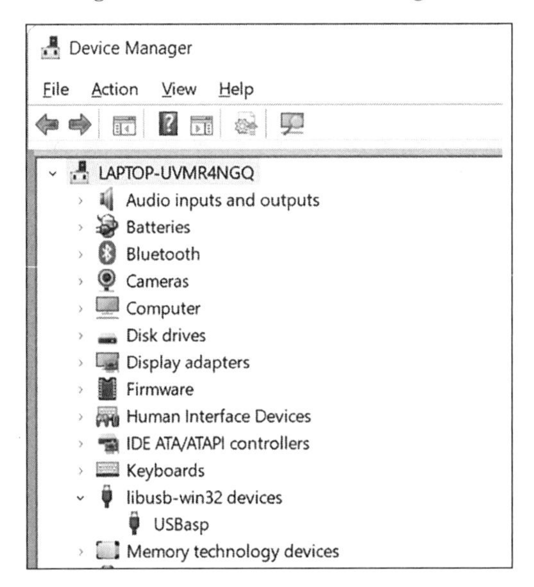


*Figure 1-15: USBasp in Device Manager*

You should see USBasp listed in the *libusb-win32 devices* category. Now that you have the software required for this book installed on your computer, it's time for you to learn about the basics of electricity!

## Current, Voltage, and Power

In order to build electronic circuits with your AVR-based projects, you'll need a basic grasp of how electricity works. In simple terms, *electricity* is a form of energy that you can harness and convert into heat, light, movement, and power. Electricity has three main properties important for our purposes: current, voltage, and power.

*Current* is the flow of electrical energy through a circuit from the positive side of a power source, such as a battery, to the negative side of the power source. In circuits that are not powered by a battery, the negative side is instead called *ground (GND)*. This kind of current is known as *direct current (DC)*. (For the purposes of this book, you won't deal with the *alternating current [AC]* that is supplied by 110 V or 230 V mains power outlets.) Current is measured in amperes (or amps), abbreviated as A. Small amounts of current are measured in milliamps (mA), where 1,000 mA equals 1 A.

*Voltage* is an indication of the difference in potential energy between a circuit's positive and negative ends, measured in *volts (V)*. The greater the voltage, the faster the current moves through a circuit.

*Power* is a measurement of the rate at which an electrical device converts energy from one form to another. Power is measured in *watts (W)*. For example, a 100 W light bulb is much brighter than a 60 W bulb because the higher-wattage bulb converts more electrical energy into light.

A simple mathematical relationship exists among voltage, current, and power:

$$\text{Power } (W) = \text{Voltage } (V) \times \text{Current } (A)$$

Later chapters of this book will explain the uses of this formula in detail.

## Electronic Components

Now that you know a little bit about the basics of electricity, let's look at how electricity interacts with electronic components and devices. Electronic *components* are the parts in a circuit that control electric current and make your designs a reality. Just as the many parts of a car work together to let us drive, electronic components work together to help us harness and control electricity to create useful devices.

Throughout this book, I'll explain specialized components as they come up. The following sections describe some of the fundamental components you'll need in any project.

## Resistors

Some components, such as LEDs, which we'll look at shortly, require only a small amount of current to function—usually around 10 mA. When a component receives excess current, it converts the excess to heat, which can damage or destroy the component. To reduce the flow of current, you can add a *resistor* between the voltage source and the component. Current flows freely along electrical wire, but when it encounters a resistor, the current flow is limited. The resistor converts some of that current into a small amount of heat energy, proportional to its value.

Figure 1-16 shows two examples of commonly used resistors.

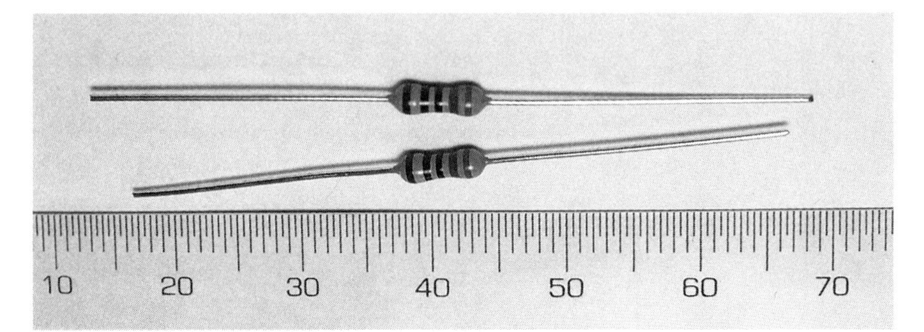


*Figure 1-16: Typical resistors*

The amount of current flow limitation is determined by the level of resistance, which can be either fixed or variable. Resistance is measured in ohms (Ω) and can range from zero to thousands of ohms (kilohms, or kΩ) or millions of ohms (megohms, or MΩ).

### Reading Resistance Values

Resistors are very small, so their resistance values usually cannot be printed on the components themselves. One common way to show a component's resistance is with a series of color-coded bands, as shown (in grayscale) in Figure 1-16, where each color represents a numerical value. The *multiplier band* dictates the number of zeros to add to the end of the previous digits to complete the value. Resistors with five bands have a higher accuracy than four-band resistors.

Here's how to read these bands, from left to right:

**First band** The first digit of the resistance

**Second band** The second digit of the resistance

**Third band** The multiplier (for four-band resistors) or the third digit of the resistance (for five-band resistors)

**Fourth band** The multiplier (for five-band resistors) or the resistor's *tolerance*, a measure of its accuracy in terms of percentage (for four-band resistors)

**Fifth band** The resistor's tolerance, for five-band resistors

To determine which band is the first and leftmost, check which band is closest to the edge of the resistor. The first band is usually closer to the left-hand edge than the last band is to the right-hand edge.

Table 1-2 lists the colors of the different bands that can appear on resistors and their corresponding values.

**Table 1-2:** Values of Bands Printed on a Resistor, in Ohms

| Color | Ohms |
|---|---|
| Black | 0 |
| Brown | 1 |
| Red | 2 |
| Orange | 3 |
| Yellow | 4 |
| Green | 5 |
| Blue | 6 |
| Violet | 7 |
| Gray | 8 |
| White | 9 |

Because it is difficult to manufacture resistors with exact values, each has a margin of error, represented by the rightmost band. A brown band indicates 1 percent tolerance, gold indicates 5 percent tolerance, and silver indicates 10 percent tolerance. The smaller the tolerance, the greater the accuracy of the resistor. That is, the value of a 1 percent resistor can only vary +/–1 percent from the stated value, whereas the value of a 5 percent resistor can vary +/–5 percent from the stated value.

Figure 1-17 shows a resistor diagram.

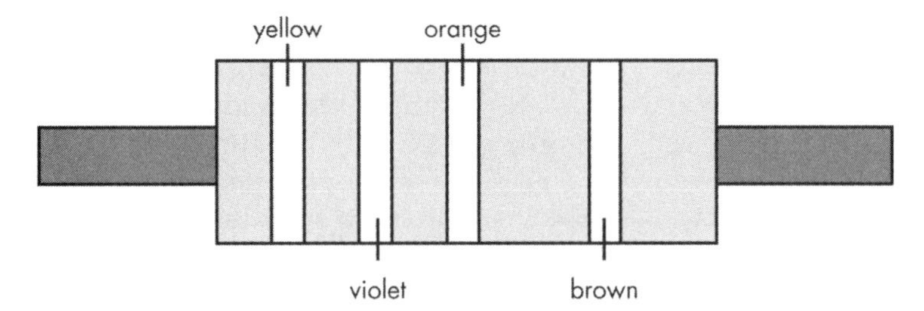

*Figure 1-17: An example resistor diagram*

The yellow, violet, and orange bands read as 4, 7, and 3, as listed in Table 1-2, with the brown band noting the tolerance (1 percent). These values translate to 47,000 Ω, more commonly written as 47 kΩ. Often, you'll see ohms written as R; for example, a 220 Ω resistor might be represented as 220 R.

Another way to read resistance values is using a *multimeter*, an incredibly useful and relatively inexpensive piece of test equipment that can measure voltage, resistance, current, and more. Figure 1-18 shows a multimeter measuring a resistor.

*Figure 1-18: A multimeter measuring a 560 Ω 1 percent tolerance resistor*

If you are colorblind, a multimeter is essential. Even if you are not, I highly recommend purchasing one: it will save you much time and reduce possible mistakes caused by misreading resistor color bands. As with other good tools, purchase your multimeter from a reputable retailer instead of fishing about on the internet for the cheapest one you can find.

### Power Rating

A resistor's *power rating* is a measurement of the power, in watts, that it will tolerate before overheating or failing. When selecting a resistor, you need to consider the relationship between power, current, and voltage. The greater the current or voltage, the greater the resistor's power rating. For example, using the formula Power (*W*) = Voltage (*V*) × Current (*A*), with a voltage of 5 V and a low current of 20 mA, the power-handling value required would be 5 × 0.02 = 0.1 W. This would work fine with the 0.25 W resistors that are most commonly used in the projects in this book (the resistors shown in Figure 1-16 are 0.25 W resistors).

If projects in this book require different power handling, we'll run through it when required and go over the best resistor to use instead. Usually, the greater a resistor's power rating, the greater it's physical size. For example, the resistor shown in Figure 1-19 is a 5 W resistor, which measures 26 mm long by 7.5 mm wide.

*Figure 1-19: A 5 W resistor*

## *Light-Emitting Diodes*

The LED is a very common yet infinitely useful component that converts electrical current into light. LEDs come in various shapes, sizes, and colors. Figure 1-20 shows a common LED.

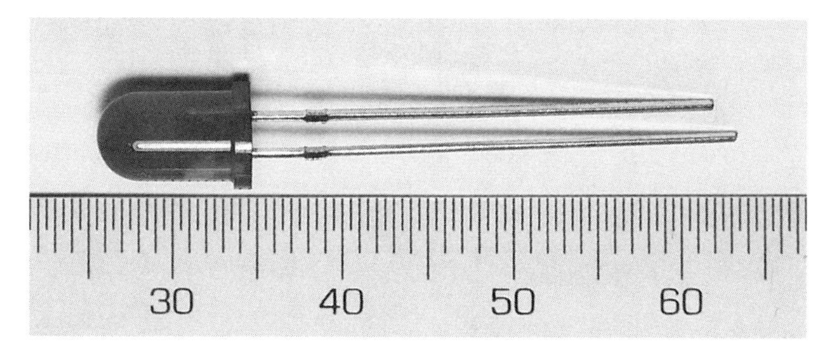

*Figure 1-20: Red LED, 5 mm diameter*

Connecting LEDs in a circuit takes some care because they are *polarized*, meaning that current can enter and leave the LED in one direction only. The current enters via the *anode* (positive) side and leaves via the *cathode* (negative) side, as shown in Figure 1-21. Any attempt to make current flow in the opposite direction will break the LED.

Thankfully, LEDs are designed so that you can tell which end is which. The leg on the anode side is longer, and on the cathode side, the rim at the base of the LED is flat, as shown in Figure 1-22.

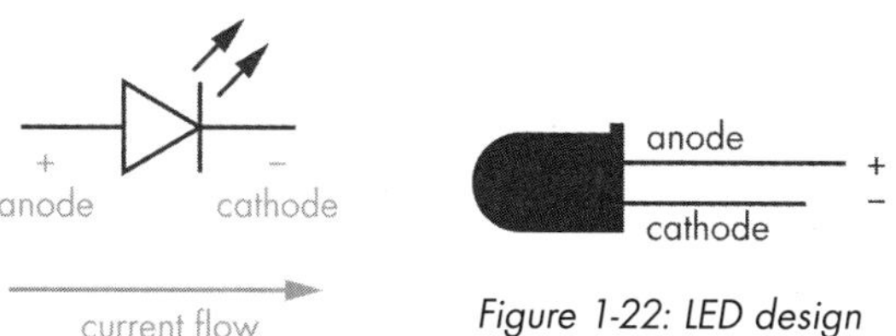


*Figure 1-21: Current flow through an LED*

*Figure 1-22: LED design indicates the anode (longer leg) and cathode (flat rim) sides.*

When adding LEDs to a project, you must consider the operating voltage and current. For example, common red LEDs require around 1.7 V and 5 to 20 mA of current. This presents a slight problem, because the power supply used in the projects in this book will output 5 V and a much higher current. Luckily, you can use a *current-limiting resistor* to reduce the current flow into an LED. But which value resistor do you use? That's where Ohm's law comes in.

*Ohm's law* states that voltage is equal to current times resistance, or $V = I \times R$. It follows that to calculate the required current-limiting resistor for an LED, you can use this formula:

$$R = (V_s - V_f) / I$$

where $V_s$ is the supply voltage (which will be 5 V when powering the circuit from our USBasp programmer), $V_f$ is the LED forward voltage drop (say, 1.7 V), and $I$ is the current required for the LED (10 mA). The value of $I$ must be in amps, so 10 mA converts to 0.01 A. You can use these values for your LEDs: substituting them into the formula gives a value for $R$ of 330 Ω. However, the LEDs will happily light when fed current less than 10 mA. It's good practice to use lower currents when possible to protect sensitive electronics, so I recommend using 560 Ω, 0.25 W resistors with your LEDs, which provide about 6 mA.

When in doubt, always choose a slightly higher value resistor, because it's better to have a dim LED than a dead one!

### THE OHM'S LAW TRIANGLE

Ohm's law states that the relationship between current, resistance, and voltage is as follows: $V = I \times R$. A popular way to remember Ohm's law is with a triangle, as shown in the figure below.

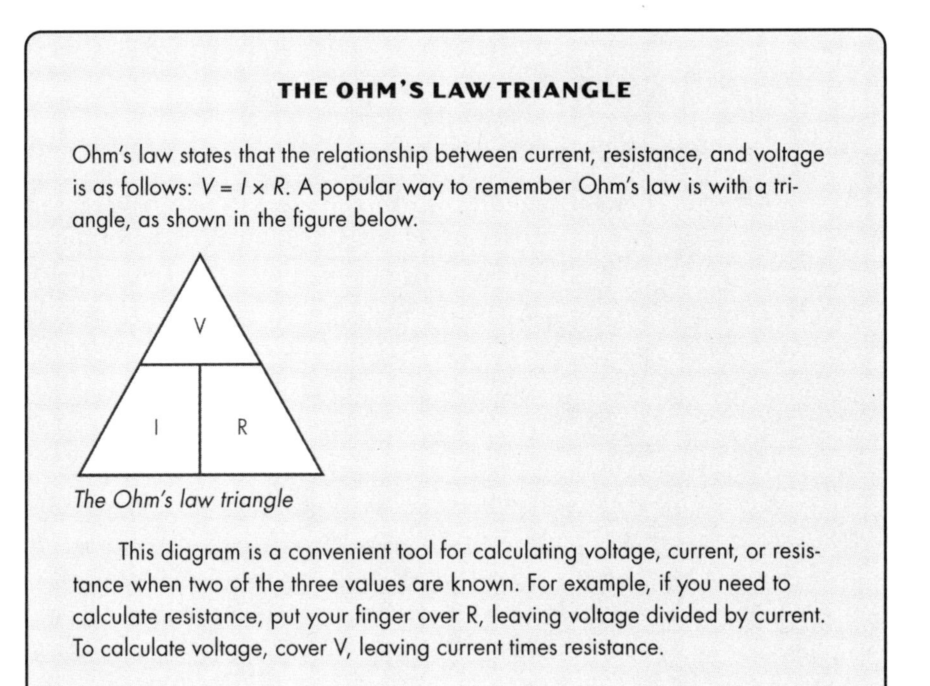


*The Ohm's law triangle*

This diagram is a convenient tool for calculating voltage, current, or resistance when two of the three values are known. For example, if you need to calculate resistance, put your finger over R, leaving voltage divided by current. To calculate voltage, cover V, leaving current times resistance.

## Power Diodes

A *power diode* is ideal for blocking current flow in one direction in a circuit. Power diodes are used in the same way as LEDs, but instead of illuminating when current flows, they protect the circuitry from current flowing in the opposite direction. There are many different types; we will use the common 1N4004 version shown in Figure 1-23.

Figure 1-24 shows the schematic symbol of the 1N4004 diode in Figure 1-23.

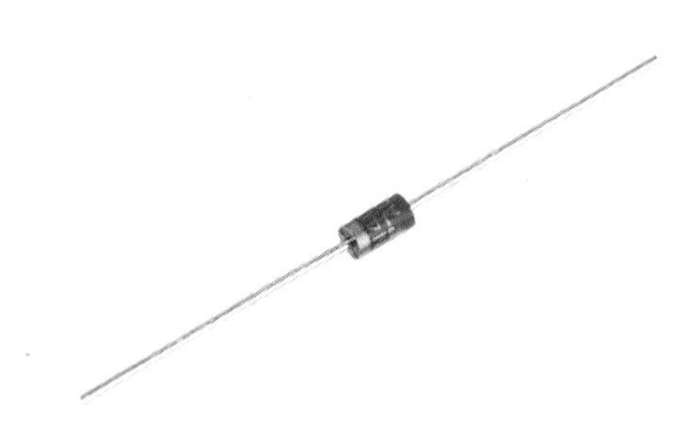

Figure 1-23: 1N4004 power diode

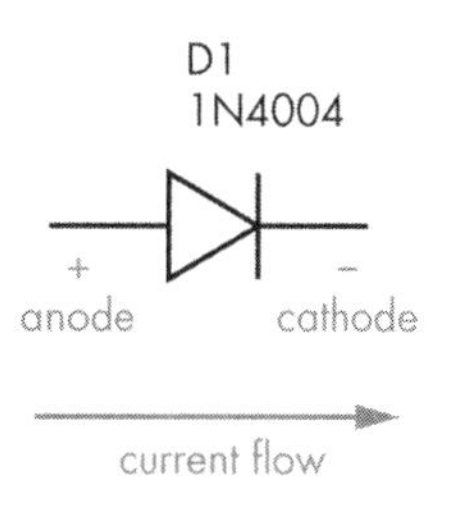


Figure 1-24: 1N4004 schematic symbol

Up to 1 A of current can flow through the 1N4004 from the anode pin to the cathode pin. The 1N4004 also causes a voltage drop of 0.7 V DC and is a convenient way to drop voltage when required.

## Capacitors

A *capacitor* is a device that can hold an electric charge. It consists of two metal plates with an insulating layer that allows an electric charge to build up between them. The electric charge builds to a maximum value as current flows to the capacitor. Once the charge hits the maximum, current stops flowing through the capacitor. However, the charge remains and flows out of the capacitor, which is called *discharging*, as soon as it is presented with a new path.

The amount of charge that a capacitor can store is measured in farads. One farad is actually a very large amount, so you'll generally find capacitors with values in picofarads or microfarads. One picofarad (pF) is 0.000000000001 of a farad, and one microfarad (μF) is 0.000001 of a farad. Capacitors are also manufactured to accept voltage maximums. For the projects in this book you'll only work with low voltages, so you'll use capacitors rated at greater than 10 or so volts. It's generally fine, however, to use capacitors with larger voltage specifications in lower-voltage circuits. Common voltage ratings for capacitors are 10, 16, 25, and 50 V.

The projects covered in this book will use two types of capacitors, ceramic and electrolytic.

### Ceramic Capacitors

Figure 1-25 shows a *ceramic capacitor*. These capacitors are very small and therefore hold a small amount of charge. They are not polarized and can be used for current flowing in either direction. Ceramic capacitors work beautifully in high-frequency circuits because they can charge and discharge very quickly due to their small capacitance.

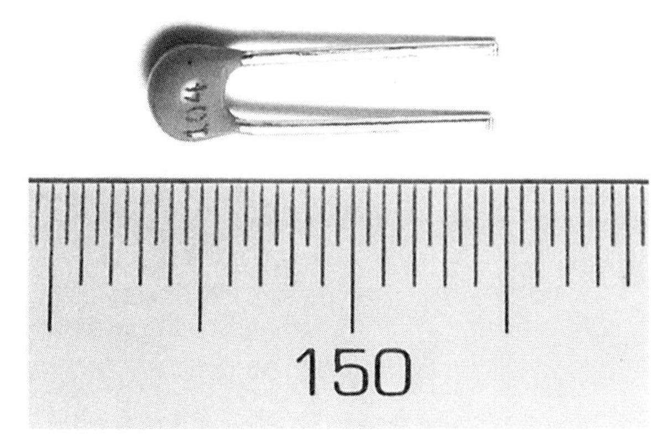


*Figure 1-25: A 0.1 μF ceramic capacitor*

Reading the value of a ceramic capacitor takes some practice, because the value is printed in a sort of code. The first two digits represent the value in picofarads, and the third digit is the multiplier in tens. For example, the capacitor shown in Figure 1-25 is labeled "104." This equates to 10, followed by four zeros, which equals 100,000 pF (equivalent to 100 nF, or 0.1 μF).

Retailers or other projects may specify capacitor values that require you to do some mental arithmetic on the fly. To simplify conversions between these units of measurement, you can print the excellent conversion chart available at *https://www.justradios.com/uFnFpF.html.*

### Electrolytic Capacitors

*Electrolytic capacitors* are physically larger than ceramic types, offer increased capacitance, and are polarized. A marking on the cover indicates the positive (+) and negative (–) sides; for example, Figure 1-26 shows the stripe and small symbol (–) marking a capacitor's negative side.

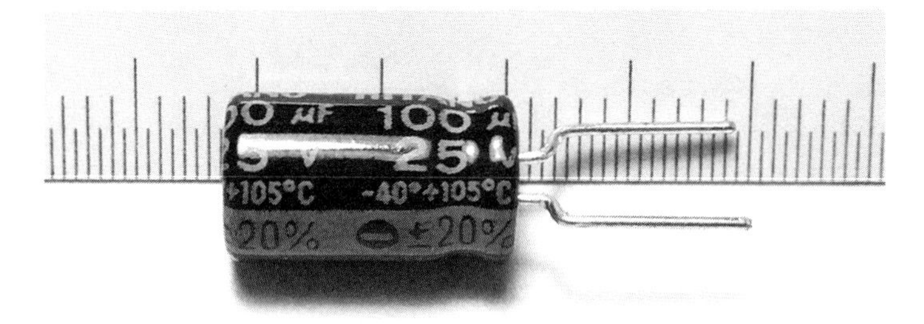


*Figure 1-26: An electrolytic capacitor*

Like resistor values, marked capacitor values are also accurate within a certain tolerance. Unlike with resistors and ceramic capacitors, the values of the electrolytic capacitor are printed on them and don't require decoding or interpretation. The capacitor in Figure 1-26 has a tolerance of 20 percent, as indicated on the stripe with the negative symbol, and a capacitance of 100 μF, shown on the darker part of the label.

Electrolytic capacitors provide power supply smoothing and stability near circuits or parts that draw high currents quickly from the supply. This prevents unwanted dropouts and noise in your circuits.

### Integrated Circuits

More commonly known by the acronym *IC*, an *integrated circuit* is a set of electronic circuits built into a piece of silicon and fitted inside a strong, usually rectangular, plastic housing. The current flows in or out of the metal legs, or pins.

IC pins are identified by numbers, and the first step in working with ICs is figuring out which pin is which. First, locate the pinout for the IC; it should be available on the supplier's website. (Check Table 1-1 to see the pinouts for the microcontrollers used in this book.) Then, look at the side of the IC that has writing on it to determine which pin is number one. This is usually the pin with a small circle closest to it. If you don't see a small circle, hold the IC in a vertical position with the notched end pointed up. Pin one is at the bottom left of the IC when it is placed horizontally in front of you, as shown in Figure 1-27.

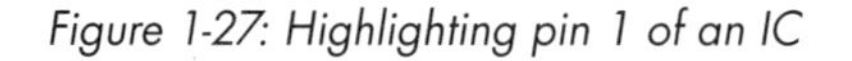

*Figure 1-27: Highlighting pin 1 of an IC*

When inserting or removing ICs from a solderless breadboard or other location, be careful not to bend the pins, as they're quite fragile and suffer from metal fatigue easily. It's a good idea to use an *IC extractor*, a simple but useful clawlike toolthat can pull both ends of an IC at the same time, as demonstrated in Figure 1-28.

*Figure 1-28: An IC extractor in use*

With this device you can slowly apply upward pressure on each end in turn, until the IC is loose enough to slowly draw up and out of the IC socket.

## Solderless Breadboards

You'll need a base to build your ever-changing circuits on, and a solderless breadboard is a great tool for this purpose. The breadboard is a plastic base with rows of electrically connected sockets. Breadboards come in many sizes, shapes, and colors, as shown in Figure 1-29. Just don't cut bread on them. The colors aren't important from an electrical perspective, they just help the end user differentiate between boards.

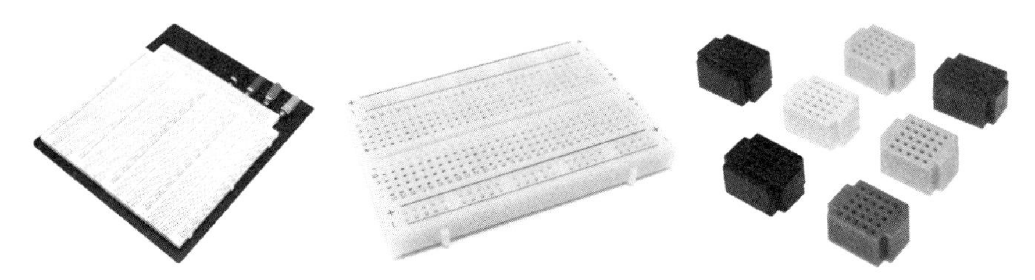

*Figure 1-29: Breadboards in various shapes and sizes*

The key to using a breadboard is understanding how the sockets are connected, whether in short columns or in long rows along the edge or in the center. The connections vary by board. For example, in the breadboard in Figure 1-30, columns of five holes are connected vertically but isolated horizontally. If you place two wires in one vertical row, they will be electrically connected. By the same token, the long rows in the center between the horizontal lines are electrically connected horizontally. You'll often need to connect a circuit to the supply voltage and ground, and these long horizontal lines of holes are ideal for those connections.

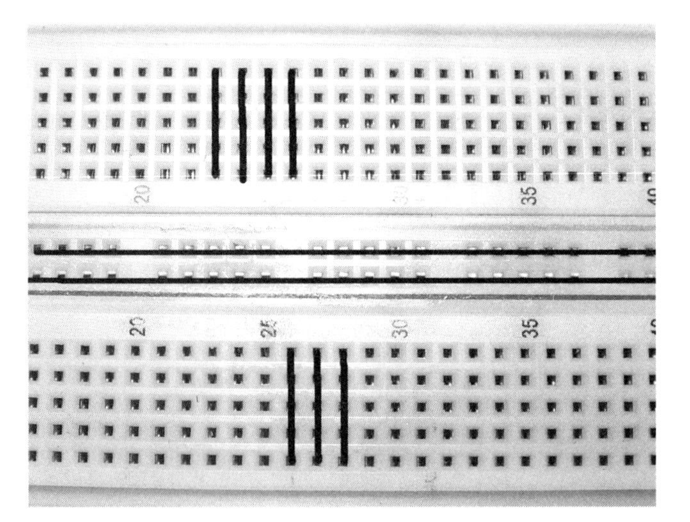


*Figure 1-30: Breadboard internal connections*

When you're building more complex circuits, a breadboard will get crowded and you won't always be able to place components exactly where you want. You can solve this problem by using short connecting wires. Retailers that sell breadboards usually also sell small boxes of wires of various lengths, such as the assortment shown in Figure 1-31.

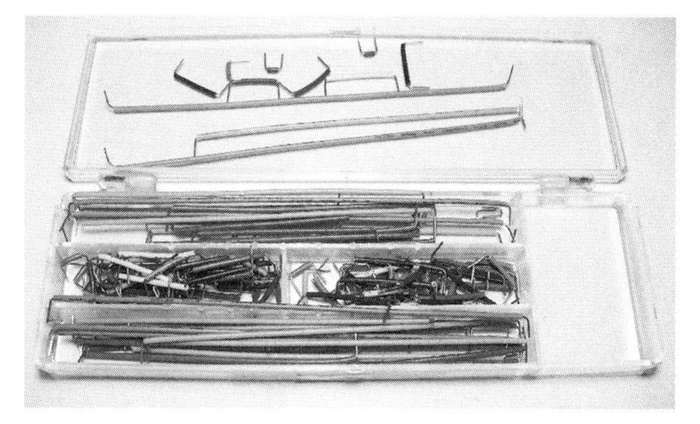

*Figure 1-31: Assorted breadboard wires*

## Powering Your Projects

You can use a multitude of methods to power your projects. In the spirit of keeping things simple and safe, I'll just show you a few easy and inexpensive options.

First, the USBasp itself can provide you with the right voltage for your AVR circuit. It provides only a limited amount of current, around 450 mA, but that's more than sufficient for small projects with a limited number of components. However, this method is convenient only when your project is close to a USB socket.

Another option is to use a *mains plugpack* with a regulated 5 V output and a *DC jack adaptor*. The adaptor, shown in Figure 1-32, allows you to connect wires to a jack that the plugpack can connect to. This lets you avoid modifying or damaging the end of the plugpack lead.

*Figure 1-32: A DC jack adaptor*

Alternatively, a *breadboard power supply* is a fantastic option, as it's both cheap and extremely easy to use. This is a small circuit board that plugs into one end of your solderless breadboard, as shown in Figure 1-33. The supply then converts current from a mains plugpack adaptor to a safe 5 V or 3.3 V.

*Figure 1-33: A solderless breadboard power supply*

Your final option is to use four rechargeable AA cells in a battery holder. When charged, these cells each have a voltage of 1.2 V. Four of them gives you 4.8 V, which is close enough to 5 V for almost any project. Don't use disposable AA cells, as the initial voltage from new will be over 6 V, which will damage your project. You can hold rechargeable AA cells together easily in a neat enclosure, such as the one shown in Figure 1-34.

*Figure 1-34: A four AA cell battery holder*

The projects in this book use all four power methods discussed in this section, so you'll need to obtain parts for each. For more detail on this, check the parts list at *https://nostarch.com/avr-workshop/*.

In this first chapter, you've built the foundations for your AVR learning by seeing some examples of what's possible, installed the required software, and learned about the basics of electricity and some of the parts you'll be using. Now you're ready for the next chapter, where you'll build your first electronic circuit based around an AVR microcontroller, upload your first code, and start to dig into the microcontroller's different operations.

# FIRST STEPS

Now that you've installed the requisite software and are prepared to enter the world of AVR microcontrollers, this chapter will ease you into your first AVR projects with some basic circuits and code.

In this chapter, you'll learn how to:

- Test your setup for making AVR-based projects.
- Read basic circuit schematics.
- Control LEDs with the digital output pins on both AVR microcontrollers, the ATtiny85 and the ATmega328P-PU.

I'll also show you how to start coding with `#define` macros and `for` loops, and how to use bitwise arithmetic and bit shifting to generate outputs efficiently.

## Testing the Hardware and Toolchain

At this point, it's wise to check that the USBasp programmer and the previously installed toolchain are working as expected. We'll do this in three stages: by building a simple circuit, testing the USBasp, and uploading code to the microcontroller.

### *Build the Circuit*

In this section you'll build a simple device that flashes an LED on and off, which is a fun and simple way to test your hardware and toolchain. To get started, you'll need the following hardware:

- USBasp programmer
- Solderless breadboard
- ATtiny85–20PU microcontroller
- One 5 mm red LED
- One 560 Ω resistor
- Seven jumper wires

Now let's focus on the practical steps of putting the circuit together to make sure your toolchain works correctly and will hold up when we dig into meatier projects. To connect the components of your test circuit, first place the solderless breadboard on a flat surface, as shown in Figure 2-1.

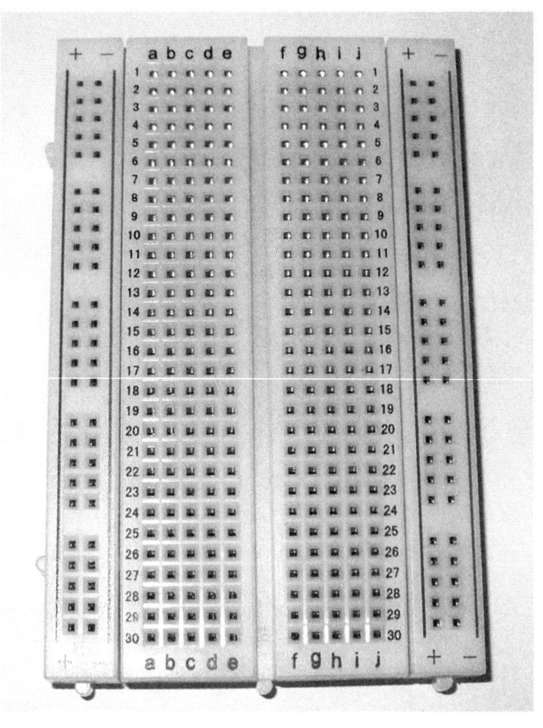


*Figure 2-1: A solderless breadboard*

Insert your ATtiny85 into the breadboard so that it sits across the vertical gap in the board's top four rows, as shown in Figure 2-2, making sure you insert the microcontroller's pin 1—indicated by the small circle beside the leg, as described in Chapter 1—into column e, row 1 of the breadboard. The pins are numbered in a counterclockwise direction from 1, so pin 4 is at the bottom left in this figure, and pin 8 is at the top right.

*Figure 2-2: The AVR in the breadboard*

Now take the 560 Ω resistor and insert one leg in the same row as pin 2 of the ATtiny85 and the other leg a few rows farther along. In Figure 2-3, I've inserted the second leg into row 8.

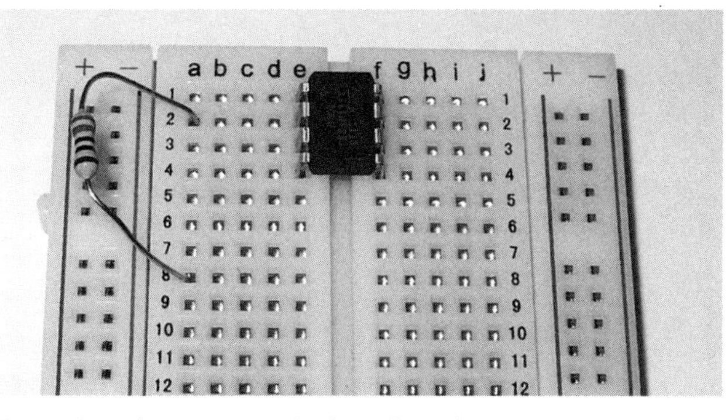


*Figure 2-3: The resistor in the breadboard*

Next, take a look at your LED. Note that one leg is longer than the other, as shown in Figure 2-4.

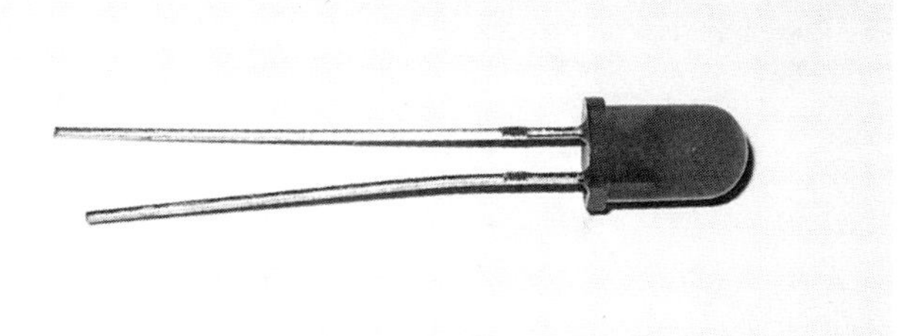

*Figure 2-4: A typical LED*

Insert the LED into the solderless breadboard, with the longer leg in the same row as the lower end of the resistor (row 8 in our example) and the shorter leg two rows farther along, using Figure 2-5 as a guide.

*Figure 2-5: Your circuit so far*

Take a jumper wire and insert one end into the same row as the LED's shorter leg, then the other end into the same row as pin 4 of the ATtiny85, as demonstrated in Figure 2-6.

*Figure 2-6: The wiring begins!*

Now that you've placed the components, I'll show you how to get the code from your computer to the microcontroller with the AVR programmer.

## *Connect and Run the Programmer*

To connect the USBasp programmer to the breadboard, first connect six male to female jumper wires to the six connection pins on the USBasp, shown in Figure 2-7.

*Figure 2-7: USBasp connection pins*

Next, connect each of the USBasp's six pins to the ATtiny85 using the mapping in Table 2-1. You'll use these same connections every time you program an ATtiny85 microcontroller. Don't worry for now about the meaning of the pin labels; I'll explain them to you as we go along in the book.

**Table 2-1**: USBasp to ATtiny85 Connections

| USBasp pin | ATtiny85 pin |
|---|---|
| RST | 1 |
| GND | 4 |
| VCC | 8 |
| SCK | 7 |
| MISO | 6 |
| MOSI | 5 |

Figure 2-8 illustrates the connections between the USBasp and an ATtiny85 described in Table 2-1. I've removed the rest of the circuit to show just an example of the connections.

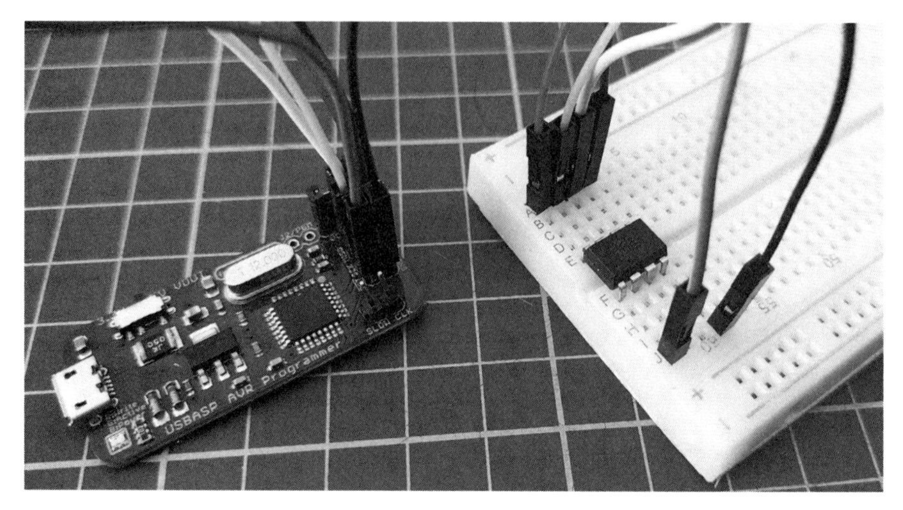


*Figure 2-8: A USBasp connected to an ATtiny85 on a solderless breadboard*

Next, check the USBasp is working correctly by connecting your USBasp programmer to your computer. To do this, you'll use a program that's part of the installed toolchain called AVRDUDE, which is a utility to upload code to AVR microcontrollers. Open a terminal window and enter the command **`avrdude -p t85 -c usbasp -B 4`**. This command includes the following options:

- `-p` selects the type of microcontroller being used in the project. You used `t85` just now for the ATtiny85, and later you'll use `m328p` for the ATmega328P-PU.
- `-c` selects the type of hardware programmer being used. In this case you've specified `usbasp`, representing your USBasp programmer.
- `-B` sets the processing speed of the microcontroller in the USBasp programmer. You set this value to `4` to bring the speed down to 187.5 kHz. This is necessary for new microcontrollers, as they are set at the factory with a clock speed of 1 MHz, which is slower than the USBasp's default speed. I'll explain more about speeds in Chapter 12.

The software toolchain should interrogate the programmer and microcontroller, and your terminal should look something like this afterward:

```
C:\> avrdude -p t85 -c usbasp -B 4
avrdude: set SCK frequency to 187500 Hz
avrdude: AVR device initialized and ready to accept instructions
Reading | ################################################## | 100% 0.03s
avrdude: Device signature = 0x1e930b
avrdude: safemode: Fuses OK
avrdude done.  Thank you.
C:\>
```

By reporting the *device signature*, a unique identifier for the type of AVR microcontroller, the software toolchain has indicated that all is well.

The AVRDUDE documentation (*http://www.nongnu.org/avrdude/user-manual/avrdude_4.html*) describes other options beyond those included in this avrdude command, if you're curious. For now, if you saw the correct output after entering that command, skip ahead to Project 0. You're ready to program your AVR!

### What If It Didn't Work?

If, after running the command in the previous step, you got an error like the following, then there was a problem with the hardware connection between the programmer and the circuit:

```
C:\> avrdude -p t85 -c usbasp -B 4
avrdude: set SCK frequency to 187500 Hz
avrdude: error: programm enable: target doesn't answer. 1
avrdude: initialization failed, rc=-1
         Double-check connections and try again, or use -F to override this check
avrdude done.  Thank you.
               C:\>
```

Don't use -F to override the check, as the error message suggests. Instead, double-check the wiring between the USBasp and the solderless breadboard to make sure no connections are loose. Then check the circuit itself to confirm the components are connected to one another according to the instructions given in the "Build the Circuit" section. In particular, check that the ATtiny85's pin 1 is aligned with the top-left corner of the breadboard. Orienting a chip incorrectly is one of the most common circuit errors!

If everything looks correct, try running the avrdude command again. It should work, but if it doesn't, walk away for a few moments, then review the process again.

**USING AVR SAFELY**

As we move on to our first project, a reminder about safety: as with any hobby or craft, it's up to you to take care of yourself and those around you. In this book, I discuss working with basic hand tools, battery-powered electrical devices, sharp knives, cutters, and sometimes hot soldering irons. *At no point in your projects should you work directly with AC mains current.* That is, don't directly wire anything you make to the wall outlets. Leave that to a licensed electrician who is trained for such work. Remember that direct contact with AC current could kill you.

Once you've got your circuit working, keep it together, including the USBasp, as you'll need all of this for the first project.

## Project 0: Blinking an LED

Now that you have a working circuit and programmer connection, it's time to create and upload your first *program* (also known as *code*), a set of instructions that tells the microcontroller how to accomplish a particular task.

Over the course of this book, our programs will increase in length as the projects get more complex. Instead of typing out each piece of code included in the book, please download the ZIP file containing the book's code from *http://www.nostarch.com/avr-workshop*. This file includes folders for each project (organized by chapter) that contain the code and anything else required to program the projects.

**NOTE** If you're reading an electronic version of this book, don't copy and paste the code from the book into your text editor. Use the downloaded code files instead.

Projects in this book use the C programming language. Since C is popular across many microcontroller and computer platforms, it should be easy to find help if you need it and to share your work with others.

### Uploading Your First AVR Code

Once you've downloaded and extracted the book's code files, use a text editor to open the *main.c* file located in the *Project 0* subfolder of the *Chapter 2* folder. This C file contains a small program that should cause your LED to blink on and off once you compile the code and upload it to the microcontroller. However, to get the desired effect, you'll need to pair the C files for each of your projects with a *Makefile*.

A Makefile contains a set of instructions for the toolchain to use when compiling and uploading your code to the microcontroller, including the microcontroller type, the CPU speed the microcontroller requires, and the type of programmer you plan to use. Every time you start a new project of your own, you should create a new folder for that project and place the *main.c* file and Makefile inside that folder. To save you time I've already done this for the Project 0 files and for all the other projects in this book.

If you're curious about the contents of the Makefile, open the one for Project 0 with a text editor and take a look. I'll introduce any necessary changes to the Makefiles in the download bundle for this book as you progress with the projects.

Now that you're familiar with these file types, it's time to bring your first project to life using the circuit you constructed earlier in this chapter. If you closed it earlier, open a terminal window just as you did when testing the toolchain. Next, navigate to the folder containing the two files for Project 0, and enter the command **`make flash`**.

After a moment, the toolchain should compile the program file and create the required data file to upload to the microcontroller. The microcontroller should then start running the program; in this case, your LED should start blinking. During this process, the status should appear in the terminal window, as shown here:

```
C:\> make flash
avrdude -c USBasp  -p attiny85 -B 4 -U flash:w:main.hex:i
avrdude: set SCK frequency to 187500 Hz
avrdude: AVR device initialized and ready to accept instructions
Reading | ################################################## | 100% 0.02s
avrdude: Device signature = 0x1e930b
avrdude: NOTE: FLASH memory has been specified, an erase cycle will be
performed
         To disable this feature, specify the -D option.
avrdude: erasing chip
avrdude: set SCK frequency to 187500 Hz
avrdude: reading input file "main.hex"
avrdude: writing flash (108 bytes):
Writing | ################################################## | 100% 0.05s
avrdude: 108 bytes of flash written
avrdude: verifying flash memory against main.hex:
avrdude: load data flash data from input file main.hex:
avrdude: input file main.hex contains 108 bytes
avrdude: reading on-chip flash data:
Reading | ################################################## | 100% 0.06s
avrdude: verifying . . .
avrdude: 108 bytes of flash verified
avrdude: safemode: Fuses OK
avrdude done.  Thank you.
C:\>
```

Let's take a look at the code to see how this program works:

```
❶ // Project 0 - Blinking an LED

❷ #include <avr/io.h>
  #include <util/delay.h>

❸ int main(void)
  {
  ❹ DDRB = 0b00001000; // Set PB3 as output
  ❺ while(1)
    {
      PORTB = 0b11111111;
      _delay_ms(1000);
      PORTB = 0b00000000;
      _delay_ms(1000);
    }
    return 0;
  }
```

The first line of *main.c* ❶ is a *comment* naming the program and describing what it's intended to accomplish. When writing programs, it's a good idea to add comments like these explaining how to use the program or highlighting other important details; they may prove useful when you revisit your code or share it with others. Comments can be any length you like, and you can use them anywhere in your program. To add a comment on a single line, enter two forward slashes and then the comment, like this:

```
// Project 0 - Blinking an LED
```

The forward slashes tell the compiler in the software toolchain to ignore the rest of the text on the line when compiling the program. In your own projects, you can include comments spanning two or more lines by entering the characters /* on a line before the comment text, then ending the comment with the characters */, as follows:

```
/*
Project 0
Blinking an LED
by Mary Smith, created 20/10/2022
*/
```

Returning to *main.c*, the `include` statements ❷ tell the compiler to look inside a library file, like *avr/io.h*, for more information required to compile the program. There are many libraries, each allowing you to use different functions in your code, and you can even create your own if necessary. You'll learn about that in Chapter 10.

All the instructions to run the program appear between the curly brackets after `int main(void)`, where the main section of the program begins ❸. Within these brackets, the program configures certain *parameters* of the microcontroller—that is, certain settings to make various operations take place. First, the program tells the microcontroller which physical pins will be inputs or outputs ❹. You hooked your LED up to pin 2, which the AVR knows as PB3, so the code activates that pin as an output. (Don't worry if this is a little confusing right now; I'll cover inputs and outputs in detail in the next few chapters.)

Finally, the code between the curly brackets after `while(1)` ❺ should execute repeatedly, blinking the LED by continuously toggling whether pin 2 outputs a 1 or a 0, until the microcontroller loses power or you reset it. A 1 supplies power to the LED, so that should turn the LED on, while a 0 should turn the LED off.

To experiment with the speed of the LED's blinking, go back to *main.c* and replace `1000` in the two lines that read `_delay_ms(1000);` with any non-negative number you like. Then save the *main.c* file and rerun the **`make flash`** command. The LED should blink faster or slower, depending on the value you use.

## What If It Didn't Work?

If there is an error in your code, the compiler will indicate which line in the code contains the error, or a line very close to it. For example, here's what happened when I ran the `make flash` command with a spelling mistake in line 10 of the Project 0 code. The compiler found the error and gave the resulting output:

```
avr-gcc -Wall -Os -Iusbdrv -DF_CPU=1000000 -mmcu=attiny85 -c main.c -o main.o
main.c: In function 'main':
main.c:17: error: 'return' undeclared (first use in this function)
main.c:17: error: (Each undeclared identifier is reported only once
main.c:17: error: for each function it appears in.)
main.c:17: error: expected ';' before numeric constant
make: *** [main.o] Error 1
C:>
```

If this happens to you, open the *main.c* file in your text editor and locate and fix the mistake before trying another upload. To compile your program for this purpose without uploading it, just run the **make** command by itself in the terminal window.

Running make is a good way to check for errors like typos in the code, but it may not help you catch errors in logic—that is, whether you've correctly told the microcontroller what you want it to do. As a general rule, remember that even if your program compiles, it may not behave as you expect if you don't carefully plan your instructions to the microcontroller before you write the code.

## Controlling Digital Outputs

Now that you've seen the electronic components you'll use in this book, let's talk a bit more about the digital output pins on the ATtiny85 and the ATmega328P-PU.

To recap, a digital output pin is a source of electrical current that you can control; it can be either on or off. On the ATtiny85 up to six pins can operate as outputs, and on the ATmega328P-PU up to eight pins can operate as outputs. I say "up to" because some pins can have more than one function, depending on how you set them up. I'll explain how to select pin functions later in this chapter.

Each output pin offers a maximum amount of current. On the ATtiny85, that maximum is 40 mA. However, the total maximum current you can run through the IC is 200 mA. Drawing too much current can cause problems, so to avoid any issues, assume you can have a maximum of 20 mA per output pin on the ATtiny85 and the ATmega328P-PU. Keep these ratings in the back of your mind when creating your own projects; all the projects in this book are designed to avoid this problem.

### *Hardware Registers*

The key to understanding digital outputs is to learn about the *hardware registers*. Our AVR microcontrollers all have multiple registers that store information related to all the possible settings for the microcontroller's operations. The numerical values placed inside these registers control digital outputs.

The first AVR register to consider is called the *DDR*, for *data-direction register*. This is used to tell the microcontroller which pins are outputs and which are inputs. Some microcontrollers, such as the ATtiny85, will have only one DDR register, and some, like the ATmega328P-PU, will have more. The second register to consider is called *PORT*. You'll use this to set which pins are on or off.

Each register is 8 bits in size, where a bit can be either a 0 or a 1, just like a binary number. Each bit relates to a physical pin on the microcontroller. In the DDR*x* registers, 1s indicate that a pin is an output and 0s indicate that a pin is an input. In the PORT*x* registers, 1s indicate that a pin is on and 0s indicate that a pin is off.

You can find out how many pins and registers are available by looking at the data sheet for your microcontroller (you can download the ATtiny85's data sheet from the Microchip website at *https://www.microchip.com/wwwproducts/en/ATtiny85*). For example, the diagram in Figure 2-9 shows that there is one PORT register on the ATtiny85: the PORTB register, which spans pins 5, 6, 7, 3, 2, and 1.

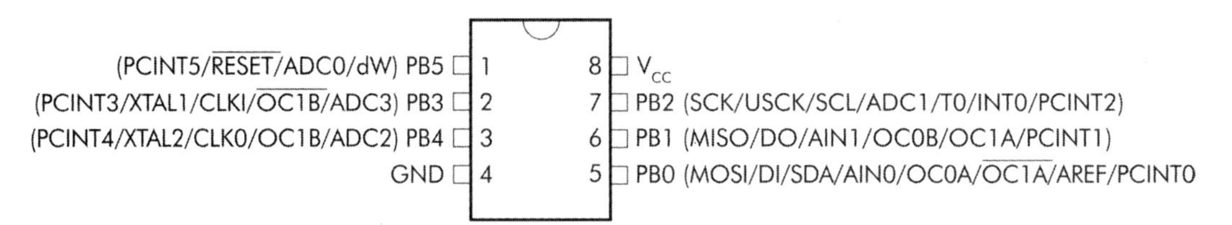


*Figure 2-9: Pinouts and PORT registers for the ATtiny85*

PORT register names are usually abbreviated on diagrams and data sheets; in Figure 2-9, the PORTB register is referred to as PB, with each pin labeled as PB*x*, where *x* is the number of the pin. You'll refer to this register as `PORTB` in your code. The matching DDR register is called DDRB, as it controls the data direction for PORTB. Please note that you can only use pins PB0 through PB4, as PB5 has other functions, which we'll examine later.

To set some of the pins in the DDRB register to outputs, just set the respective bits to 1s. As an example, let's revisit the code from Project 0:

```
// Project 0 - Blinking an LED

#include <avr/io.h>
#include <util/delay.h>

int main(void)
{
❶ DDRB = 0b00001000;
  for(;;)
  {
    _delay_ms(1000);
    PORTB = 0b11111111;
    _delay_ms(1000);
    PORTB = 0b00000000;
  }
  return 0;
}
```

This code sets the physical pin 2 to an output ❶. To set all the PORTB pins to outputs, you could use:

```
DDRB = 0b11111111;
```

Let's examine how this works. The physical pins of the DDRB register of your ATtiny85 are 1, 3, 2, 7, 6, and 5. Each pin corresponds to a single bit in the register; from left to right, those pins are bits 5, 3, 4, 2, 1, and 0, respectively. So, for example, to set physical pin 6 (PB1) as an output and the rest as inputs you would use:

```
// Bits: 76543210
// Pins: 44132765
DDRB = 0b00000010;
```

And, as we saw in the Project 0 code, to set physical pin 2 (PB3) as an output and the rest as inputs you would use:

```
// Bits: 76543210
// Pins: 44132765
DDRB = 0b00001000;
```

We call the bit on the right-hand side of the register the *first bit* or *least significant bit (LSB)* and the bit on the left-hand side the *last bit* or *most significant bit (MSB)*. This may seem backward at first, but bits are the equivalent of binary numbers, whose contents are referenced using the same method.

Because the ATtiny85 has only six outputs, you can leave the last two bits (6 and 7) as 0s or 1s in the DDRB statement. Once you've set a pin to an output, use the `PORTx` function to switch the output on or off. To turn all the outputs on, use:

```
PORTB = 0b11111111;
```

To turn them all off, use:

```
PORTB = 0b00000000;
```

Or to turn, say, pins 3 (PB4) and 5 (PB0) on and the rest off, use:

```
// Bits:   76543210
// Pins:   44132765
PORTB = 0b00010001;
```

To experiment with these functions, in the next project you'll build a new circuit like the one in Project 0, but this time with four LEDs.

## Project 1: Experimenting with ATtiny85 Digital Outputs

To practice and increase your understanding of using the ATtiny85's DDRB and PORTB registers, in this project you'll control four output devices

(LEDs). Although blinking LEDs may seem a somewhat trivial example, the ability to control digital outputs is the foundation for controlling a wide range of objects.

### The Hardware

You will need the following hardware:

- USBasp programmer
- Solderless breadboard
- ATtiny85–20PU microcontroller
- Four LEDs
- Four 560 Ω resistors
- Jumper wires

The circuit for this project is similar to the one in Project 0, but with three more LEDs. Assemble your circuit as shown in Figure 2-10.

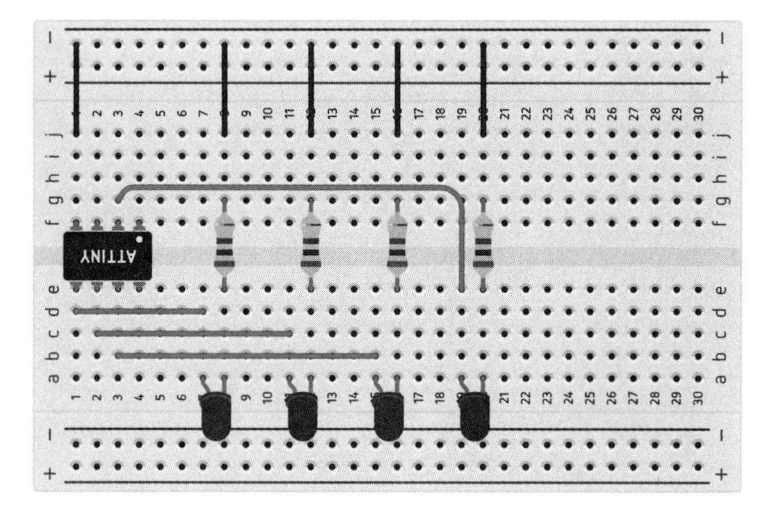


*Figure 2-10: A physical layout diagram of the main circuit for Project 1*

Once you've assembled your circuit, connect the USBasp programmer. Repeat the connections as shown in Table 2-1.

**NOTE** *If you'd like to try creating diagrams like the one in Figure 2-10, use Autodesk Tinkercad, available at* https://www.tinkercad.com/.

### The Code

Open a terminal window, navigate to the folder containing the two files for Project 1, and enter the command `make flash`. After a moment, the toolchain should compile the program file and create the required data file to upload to the microcontroller. Then the microcontroller should run the program that turns the digital outputs on and off, causing all four of the LEDs to blink.

To see how this works, open the *main.c* file located in the *Project 1* subfolder of the *Chapter 2* folder:

```
// Project 1 - Experimenting with ATtiny85 Digital Outputs

#include <avr/io.h>
#include <util/delay.h>

int main(void)
{
❶ DDRB = 0b11111111; // Set pins as outputs
  for(;;)
  {
  ❷ _delay_ms(250);
  ❸ PORTB = 0b00011111;
    _delay_ms(250);
  ❹ PORTB = 0b00000000;
  }
  return 0;
}
```

First we use `DDRB` to set all the pins to outputs ❶. The `_delay_ms()` function ❷ tells the microcontroller to stop everything for a set period of time. To use a delay, just enter the number of milliseconds you want the program to pause inside the parentheses: `_delay_ms(250)` sets the delay to 250 milliseconds. Next, we turn on the digital outputs so current will flow from the pins, through the resistors and LEDs, and then to GND, completing the electrical circuit ❸. After another delay we turn off the digital outputs, causing the LEDs to turn off ❹.

By now you should understand how to set the pins on the ATtiny85 to outputs, turn them on and off, and create a delay. Experiment with the delay and pins by changing which LEDs turn on and off, the length of the delay, and so on.

## Using Schematic Diagrams

Projects 0 and 1 showed you how to build circuits using a picture and a physical layout diagram, respectively. Physical layout diagrams like the one in Figure 2-10 may seem like the easiest way to diagram a circuit, but as you add more components, direct representations make physical diagrams a real mess. Because your circuits are about to get more complicated, from now on I'll use *schematic diagrams* (also known as *circuit diagrams*) to illustrate them, like the one shown in Figure 2-11.

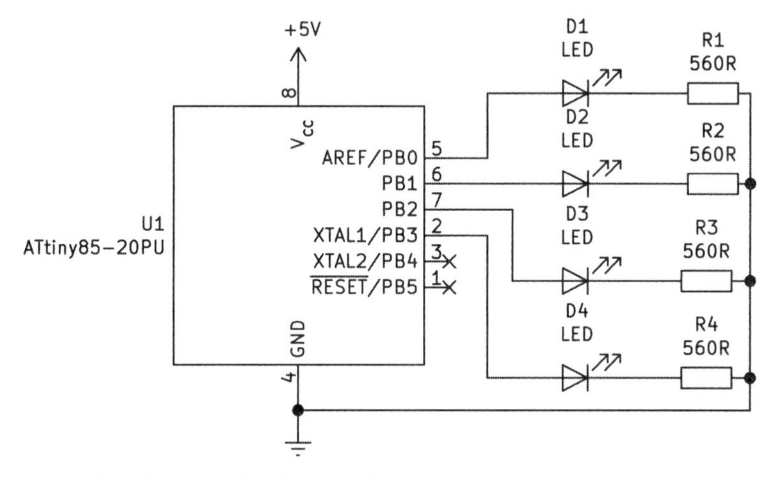


*Figure 2-11: An example schematic diagram*

Schematics are road maps that show where electrical current flows through various components, with the lines between components indicating those paths. Instead of showing components and wires, a schematic uses symbols and lines.

## Components in Schematics

Once you know what the symbols mean, reading a schematic is easy. To begin, let's examine the symbols for the components you've already used.

Figure 2-12 shows the ATtiny85 microcontroller symbol. The pin numbers are labeled clearly; don't forget that pin 1 is at the top left on the physical chip. Other microcontrollers and ICs use similar symbols, but their size depends on the number of pins.

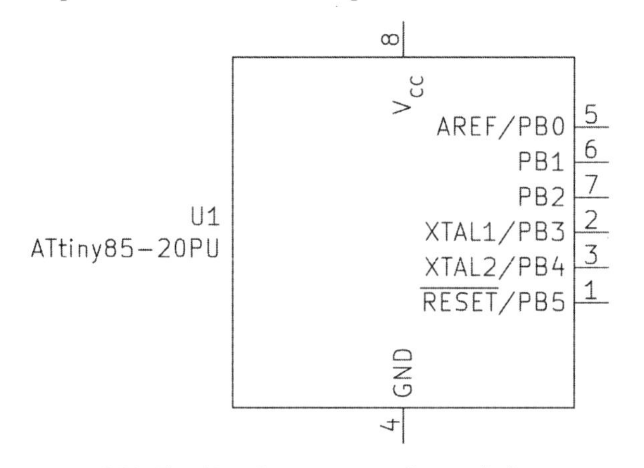


*Figure 2-12: The ATtiny85 microcontroller symbol*

Figure 2-13 shows the resistor symbol. It's good practice to display the resistor value and part designator along with the resistor symbol (220 Ω and R1, in this case). This makes life a lot easier for everyone trying to make sense of the schematic, including you!

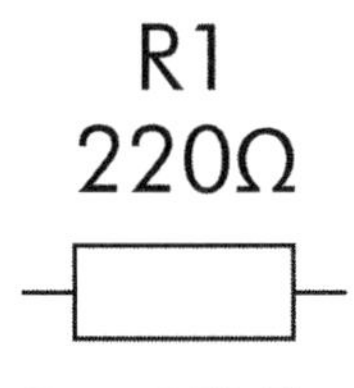


*Figure 2-13: The resistor symbol*

Figure 2-14 shows the LED symbol. All members of the diode family share a common symbol, the triangle and vertical line, but LED symbols show two parallel arrows pointing away from the triangle to indicate that light is being emitted.

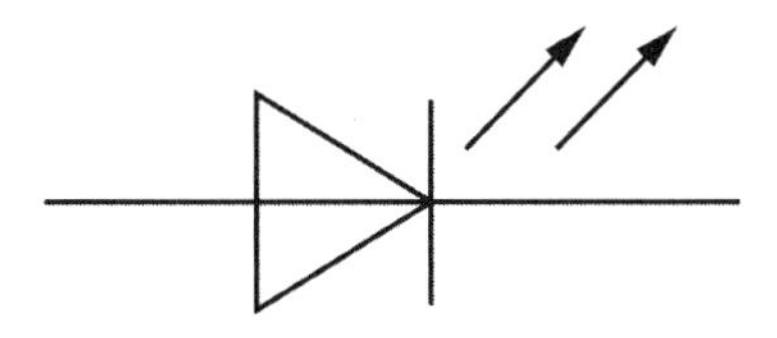

*Figure 2-14: The LED symbol*

Now that you have an understanding of the various component schematic symbols, I'll show you how the wired connections between the components are shown in circuit schematics.

## Wires in Schematics

When wires cross or connect in schematics, they are drawn in the following ways:

**Crossing but not connected wires**

When two wires cross but are not connected, the crossing can be represented in one of two ways, as shown in Figure 2-15. Either way is correct; it's just a matter of preference.

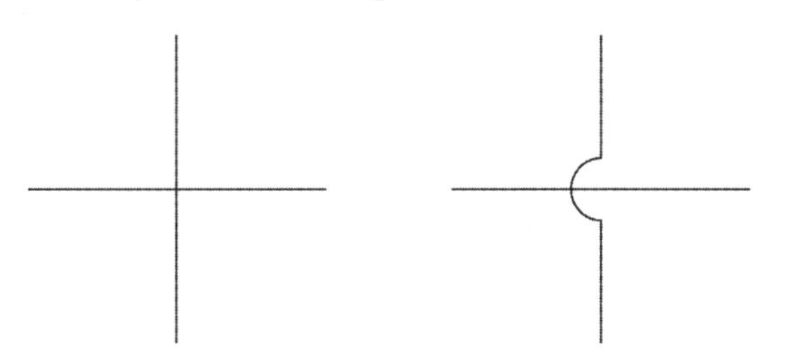

*Figure 2-15: Non-connecting crossed wires*

**Connected wires**

When wires are physically connected, a *junction dot* is drawn at the point of connection, as shown in Figure 2-16.

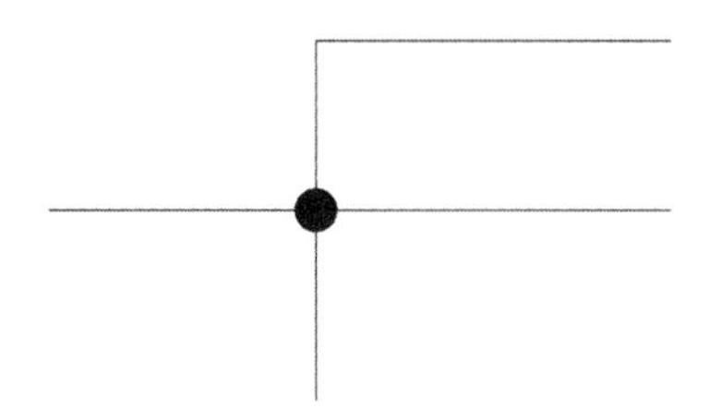

Figure 2-16: Two wires that are connected

**Wire connected to ground**

The symbol shown in Figure 2-17 indicates when a wire is connected to ground (GND).

Figure 2-17: The GND symbol

The GND symbol at the end of a line in a schematic tells you that the wire is physically connected to the microcontroller's GND pin. For your circuits, this is also known as the *negative*.

## Dissecting a Schematic

Now that you know the symbols for the parts you've used so far, let's dissect the schematic you'd draw for Project 1. Compare the schematic shown in Figure 2-18 with physical image of the circuit in Figure 2-10.

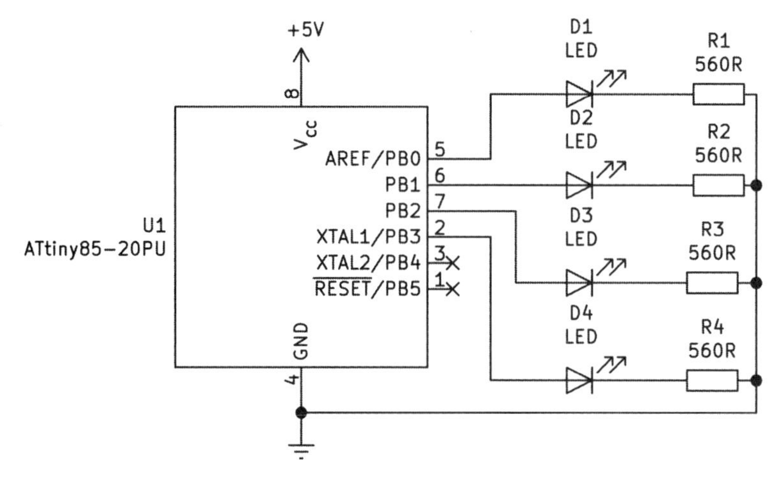


Figure 2-18: Schematic for Project 1

The arrow with +5V written above it represents the 5 V power supply on the breadboard, and the ATtiny85 is labeled with its chip name. LEDs 1 through 4 are connected to the ATtiny85 at the same pins you saw in Figure 2-10, and just like in the original circuit, all four resistors attached to the LEDs go to GND, pin 4 on the microcontroller. Note the dots at the junctions of R1, R2, R3, and R4, which indicate that those resistors are all connected to the same place (GND). The Xs on pins 1 and 3 in Figure 2-18 signal that those pins are not connected to anything.

In this schematic, you can trace the flow of current from the power supply through the circuit to ground. Current is sourced from the 5 V power supply and enters the microcontroller. Our code then allows the current to flow from digital output pins when required. This current goes through an LED (which causes the LED to glow) and the resistor (which regulates the current) before reaching ground and completing the circuit.

**NOTE** *If you'd like to create your own computer-drawn schematics, try the open source KiCad package, available for free or with a donation from* https://www.kicad.org/.

Along with using the ATmega328P-PU for the first time, the following project will also put your new knowledge of reading circuit schematics to use.

## Project 2: Experimenting with ATmega328P-PU Digital Outputs

You can tell at a glance that the ATmega328P-PU has more digital outputs than the ATtiny85. The pinout diagram from the data sheet in Figure 2-19 gives the details.

| Left pin function | Pin | Pin | Right pin function |
|---|---|---|---|
| (PCINT14/RESET) PC6 | 1 | 28 | PC5 (ADC5/SCL/PCINT13) |
| (PCINT16/RXD) PD0 | 2 | 27 | PC4 (ADC4/SDA/PCINT12) |
| (PCINT17/TXD) PD1 | 3 | 26 | PC3 (ADC3/PCINT11) |
| (PCINT18/INT0) PD2 | 4 | 25 | PC2 (ADC2/PCINT10) |
| (PCINT19/OC2B/INT1) PD3 | 5 | 24 | PC1 (ADC1/PCINT9) |
| (PCINT20/XCK/T0) PD4 | 6 | 23 | PC0 (ADC0/PCINT8) |
| VCC | 7 | 22 | GND |
| GND | 8 | 21 | AREF |
| (PCINT6/XTAL1/TOSC1) PB6 | 9 | 20 | AVCC |
| (PCINT7/XTAL2/TOSC2) PB7 | 10 | 19 | PB5 (SCK/PCINT5) |
| (PCINT21/OC0B/T1) PD5 | 11 | 18 | PB4 (MISO/PCINT4) |
| (PCINT22/OC0A/AIN0) PD6 | 12 | 17 | PB3 (MOSI/OC2A/PCINT3) |
| (PCINT23/AIN1) PD7 | 13 | 16 | PB2 (SS/OC1B/PCINT2) |
| (PCINT0/CLKO/ICP1) PB0 | 14 | 15 | PB1 (OC1A/PCINT1) |

*Figure 2-19: The ATmega328P-PU pinout and port register diagram*

There are two registers you can use for digital outputs: PORTB (PB) and PORTD (PD). For this and the following few projects, you'll use the PORTB register. In this project, you'll put the ATmega328P-PU's digital outputs to work using more LEDs.

### The Hardware

You will need the following hardware:

- USBasp programmer
- Solderless breadboard
- ATmega328P-PU microcontroller

- Eight LEDs
- Eight 560 Ω resistors
- Jumper wires

I've provided both a physical layout diagram and the schematic for the required circuit, which you can use to assemble your circuit. The diagram in Figure 2-20 indicates the physical connections you see with your own eyes, and the schematic in Figure 2-21 indicates the same electrical connections between the various components in a more compact and easier-to-follow form.

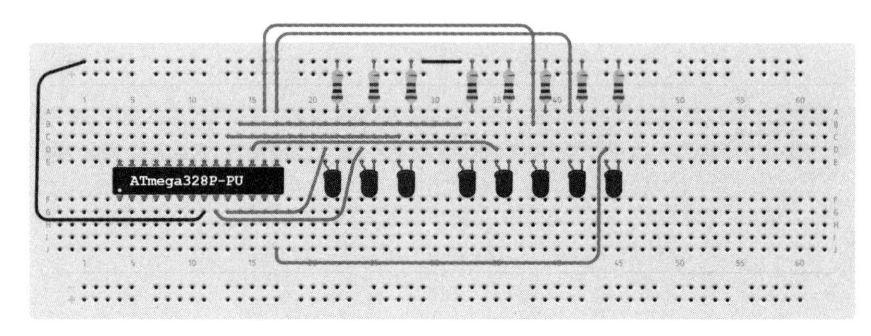


*Figure 2-20: Diagram for Project 2*

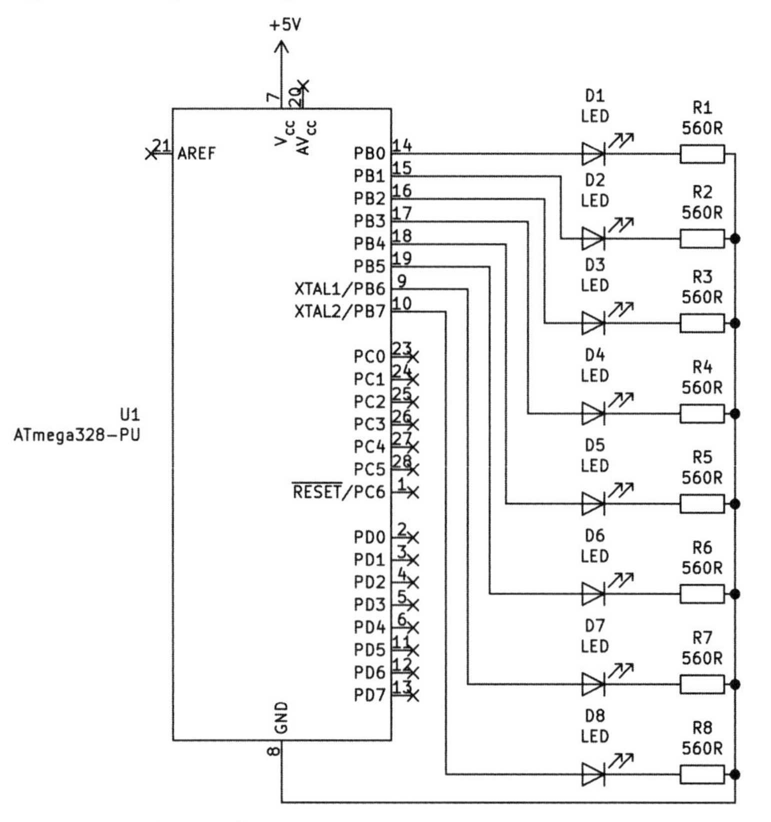


*Figure 2-21: Schematic for Project 2*

Connect your USBasp programmer to the circuit, then connect jumper wires between the programmer and the microcontroller using the information in Table 2-2. Take note of the connections, as they will be the same every time you program an ATmega328P-PU microcontroller.

**Table 2-2**: USBasp to ATmega328P-PU Connections

| USBasp pin | ATmega328P-PU pin |
|---|---|
| RST | 1 |
| GND | 8 |
| VCC | 7 |
| SCK | 19 |
| MISO | 18 |
| MOSI | 17 |

Before uploading code to the microcontroller, test that the programmer is communicating with the ATmega328P-PU. Connect your USBasp programmer to your computer and open a terminal window. Enter the command **avrdude -p m328p -c usbasp -B 4** and press ENTER. The software toolchain should interrogate the programmer and microcontroller, and you should see the following output in your terminal:

```
C:\> avrdude -p m328p -c usbasp -B 4
avrdude: set SCK frequency to 187500 Hz
avrdude: AVR device initialized and ready to accept instructions
Reading | ################################################## | 100% 0.03s
avrdude: Device signature = 0x1e950f
avrdude: safemode: Fuses OK
avrdude done.  Thank you.
C:\>
```

The only difference between this `avrdude` command and the command used earlier to test your setup with the ATtiny85 is that we changed the microcontroller parameter `-p` to `m328p`. If you saw the correct output after entering this command, you can move forward; otherwise, review the "What If It Didn't Work?" section on page 35.

### *The Code*

Open a terminal window, navigate to the folder containing the two files for Project 2, and enter the command **make flash**. After a moment, the toolchain should compile the program file and create the required data file to upload to the microcontroller. Then the microcontroller should run the program, causing all eight of the LEDs to blink together.

To see how this works, open the *main.c* file located in the *Project 2* subfolder of the *Chapter 2* folder.

```
// Project 2 - Experimenting with ATmega328P-PU Digital Outputs

#include <avr/io.h>
#include <util/delay.h>

int main(void)
{
  DDRB = 0b11111111; // Set PORTB register as outputs
  for(;;)
  {
    _delay_ms(250);
❶   PORTB = 0b11111111;
    _delay_ms(250);
❷   PORTB = 0b00000000;
  }
  return 0;
}
```

The code in this project is identical to that of Project 1. However, because the ATmega328P-PU has a full PORTB register (that is, eight outputs), you can control all of them with the PORTB function. For some practice, change the PORTB lines to experiment with blinking LEDs. Try turning half of them on and half of them off by changing the line at ❶ to:

```
PORTB = 0b11110000;
```

and the line at ❷ to:

```
PORTB = 0b00001111;
```

You can also increase the effects by adding more PORTB lines with different on/off states. Go crazy! Keep this circuit together when you're done experimenting, since you'll use it for the rest of this chapter.

You'll soon see how each bit of the PORTB register relates to a digital output on the microcontroller. However, there's a better way to control the outputs to create complicated patterns. In the next project, you'll use the same circuit to control outputs more efficiently with variables, functions, bit shifting, and bitwise arithmetic.

## Project 3: Bit-Shifting Digital Outputs

In this project, you'll learn more efficient ways to control the digital outputs. These techniques will give you more control over the outputs without excessive code, so that you don't waste program memory in the microcontroller.

This project uses the same hardware as Project 2, so you should have that set up already. Open a terminal window, navigate to the folder containing the two files for Project 3, and enter the command **make flash**. Once again, the toolchain should process the code, and in a moment the LEDs should start blinking on and off in a repeating pattern from left to right, then from right to left, and so on.

Now open the *main.c* file located in the *Project 3* subfolder of the *Chapter 2* folder for a closer look at how this works:

```
// Project 3 - Bit-Shifting Digital Outputs

#include <avr/io.h>
#include <util/delay.h>

❶ #define TIME 100 // Delay in milliseconds

int main(void)
{
  ❷ uint8_t i;          // 8-bit integer variable "i"
    DDRB = 0b11111111; // Set PORTB register as outputs
    while (1)
    {
      ❸ for (i = 0; i < 8; i++)
        {
          ❹ _delay_ms(TIME);
            PORTB = 0b00000001 << i;
        }

      ❺ for (i = 1; i < 7; i++)
        {
            _delay_ms(TIME);
            PORTB = 0b10000000 >> i;
        }
    }
    return 0;
}
```

This code introduces some new concepts. First, the `#define` macro lets you assign values to words, which are called *constant values* or just *constants* for short. Constants make it possible to reference values later in your code and make your code easier to read. For example, `#define TIME 100` ❶ tells the compiler to replace the word `TIME` with the value 100 anywhere you use `TIME` in your code, as in the `_delay_ms_` lines ❹. To change the blink delay you only need to change the value in the original `#define` macro, and the compiler takes care of the rest. Whenever you use `#define`, you must place it before the main `int main(void)` loop in the code.

Inside the main loop we define a *variable*, which that represents data. A variable's value can change while the code is being executed, whereas the value of a constant defined by the `#define` macro cannot. You can think of a variable as a part of the microcontroller's memory that stores a number you can change as needed during program execution. The first type of variable you'll use in this book is an *integer*. In programming terms, this type can hold a whole number; that is, a number that can be positive, negative, or zero, without a fractional or decimal part.

To define a variable, first enter the type and then the label. The line `uint8_t i;` ❷ defines a variable called `i` of type `uint8_t`. This type of variable can store a whole number between 0 and 255 (the `u` stands for *unsigned*; an unsigned integer cannot store negative numbers). The letter `i` now represents an integer whose value you can change.

There are six types of integer variables you can make use of:

- `uint8_t` is an 8-bit unsigned integer (0 to 255).
- `int8_t` is an 8-bit signed integer (–128 to 127).
- `uint16_t` is a 16-bit unsigned integer (0 to 65,535).
- `int16_t` is a 16-bit signed integer (–32,768 to 32,767).
- `uint32_t` is a 32-bit unsigned integer (0 to 4,294,967,295).
- `int32_t` is a 32-bit signed integer (–2,147,483,648 to 2,147,483,647).

At first glance, you may think that the smaller integer types are redundant and that you should just use `int32_t` for all your integer needs. However, the larger the integer type, the more time it takes your microcontroller to process those numbers. To maximize efficiency, bear in mind the needs of your project when selecting an integer type, and use the smallest type that will accommodate the largest possible value.

This code also introduces *for loops*, which allow you to repeat a section of code without retyping it. Retyping is inefficient and wastes memory; `for` loops simply let you set how many times the code inside the loop will repeat. There are two `for` loops in the Project 3 code. Let's look at the first one ❸:

```
for (i = 0; i < 8; i++)
{
    _delay_ms(TIME);
    PORTB = 0b00000001 << i;
}
```

A `for` loop repeats the code between the curly brackets as long as a certain condition is true. In this case, this loop will repeat until the value of the variable `i` is 8. In the first section of the loop, we set the initial value of `i` to 0 using `i = 0`. The second section of the loop checks to see if the condition is true: in this case, if `i < 8`. The third section of the loop keeps track of how many times the code loops; `i++` means "add one to the value of `i` after each loop." Every time the code between the curly brackets is executed, the value of `i` increases by one and the code checks to see if `i` is less than 8. When `i` equals 8, the looping stops and the microcontroller moves on to the code after the `for` loop.

The final concept this project introduces is *bit shifting*, a technique that moves the bits in a binary number to the left or right. This helps you to efficiently use binary numbers in your `PORTx` functions to turn the output pins on and off. The `for` loop at ❸ shifts the first bit one to the left every time the loop completes. This is faster than using the equivalent eight `PORTB` functions:

```
PORTB = 0b00000001;
PORTB = 0b00000010;
PORTB = 0b00000100;
PORTB = 0b00001000;
PORTB = 0b00010000;
PORTB = 0b00100000;
PORTB = 0b01000000;
PORTB = 0b10000000;
```

Instead of wasting code on such functions, you can shift the first bit to the left with << or to the right with >>, followed by the number of bits to move. For example, to turn the first three outputs in PORTB on and off in sequence, you could enter:

```
PORTB = 0b00000001;
PORTB = 0b00000001 << 1; // equivalent to 0b00000010
PORTB = 0b00000001 << 2; // equivalent to 0b00000100
```

The `for` loop at ❸ demonstrates shifting bits to the left with `<< i`. Here, `i` starts with a value of 0 in the first loop, meaning the first output is on. When the code loops again, `i` has a value of 1, meaning the second output is on, and so forth. In the same manner, the `for` loop at ❺ turns on the LEDs from left to right with >>.

You can manipulate outputs even more effectively with some deeper manipulation of the PORT*x* register using *bitwise arithmetic*. This is a way of manipulating numbers in the form of their individual bits that is directly supported by the microcontroller. Don't let past experiences in math class scare you; it's quite simple. The four operators you can use with numbers or variables to change the bits in a register are NOT, AND, OR, and XOR. You'll see how they work in Projects 4 through 7.

## Project 4: Experimenting with NOT

The NOT (~) operator inverts all the bits in a number or register. If you place a tilde (~) in front of a number, it will be interpreted as the binary opposite. For example:

```
~0b00001111 = 0b11110000
```

Try this yourself by replacing the main loop from Project 2 with the following:

```
for(;;)
  {
    PORTB = 0b00001111;    // PORTB pins 3, 2, 1, and 0 turned on
    _delay_ms(250);
    PORTB = ~0b00001111;   // PORTB pins 7, 6, 5, and 4 turned on
    _delay_ms(250);
  }
```

After the ~ line, all the pins on PORTB should be turned on. NOT (and all the other bitwise operators) is a useful tool to keep in your arsenal when planning out projects.

## Project 5: Experimenting with AND

The AND (&) operator compares two binary numbers and returns a new binary number. If both numbers have 1s at the same position, the new number will have a 1 at that position, and the other bits will be 0s. For example:

```
  0b00100010 &
  0b10101011
= 0b00100010
```

AND is useful when you only want to turn outputs on depending on a certain value. Project 5 demonstrates this by displaying binary numbers from 0 to 255. Use the same circuit from Project 3, and open the *main.c* file located in the *Project 5* subfolder of the *Chapter 2* folder:

```
// Project 5 - AND & Demonstration

#include <avr/io.h>
#include <util/delay.h>

#define TIME 5                 // Delay in milliseconds
int main(void)
{
  uint8_t i;                   // 8-bit integer variable "i"
  DDRB = 0b11111111;           // Set PORTB register as outputs
  for(;;)
  {
    for (i = 0; i < 256; i++)
    {
    _delay_ms(TIME);
    PORTB = 0b11111111 & i; // Displays value of i in binary using LEDs
    }
  }
  return 0;
}
```

In this code, the `for` loop counts from 0 to 255. Every time the code loops, it performs an AND on `0b11111111` and the variable `i` and sets `PORTB` to the result of the operation. For example, let's say `i` has a value of 9, which is `0b00001001`. The result of `0b00001001 & 0b11111111` will be `0b00001001` because the bits in the ones and eights columns match. Thus, the `PORTB` setting will be `0b00001001` and all the LEDs for 1 and 4 will turn on.

## Project 6: Experimenting with OR

The OR (|) operator compares two binary numbers and returns another binary number with 1s in any position where either operand had a bit set to 1. For example:

```
  0b00100110 |
  0b10101011
= 0b10101111
```

This operator is useful when you want to turn outputs on when bits in either of two numbers have a certain value. To try it out, use the same circuit from Project 3 and open the *main.c* file in the *Project 6* subfolder of the *Chapter 2* folder:

```
// Project 6 - Experimenting with OR

#include <avr/io.h>
#include <util/delay.h>

#define TIME 50       // Delay in milliseconds
int main(void)
{
  uint8_t i;          // 8-bit integer variable "i"
  DDRB = 0b11111111; // Set PORTB register as outputs

  for(;;)
  {
    for (i = 0; i < 255; i++)
    {
    _delay_ms(TIME);
    PORTB = 0b00001111 | i;
    }
  }
  return 0;
}
```

The main loop counts from 0 to 255, as in the previous example. Every time the code loops, it performs an OR on `0b00001111` and the variable `i` and sets `PORTB` to the result of the operation. If `i` has a value of, for example, 0, which is `0b00000000`, the result of `0b00001111 | 0b00000000` will be `0b00001111`. Thus, the `PORTB` setting will be `0b0001111`, and the four LEDs on the right side will stay on.

As the value of `i` increases, the number of bits in `i` increases and more LEDs will turn on. For example, when `i` has a value of 128, or `0b10000000`, the resulting `PORTB` is `0b10001111`. Load this code onto your AVR to see it in action, then experiment with the code and create your own OR situations to practice.

## Project 7: Experimenting with XOR

The final operator, XOR (^), compares the same bits in two numbers and returns a new binary number. Anywhere the two numbers had opposite bits in the same position, the new number will have a 1; anywhere the two numbers had identical bits, the new number will have a 0. For example:

```
  0b00100110 ^
  0b10101011
= 0b10001101
```

XOR operators are useful when you want to turn outputs on when bits in two numbers have different values. To see how this works, use the same circuit from Project 3 and open the *main.c* file located in the *Project 7* subfolder of the *Chapter 2* folder:

```
// Project 7 - Experimenting with XOR

#include <avr/io.h>
#include <util/delay.h>
#define TIME 250     // Delay in milliseconds

int main(void)
{
  uint8_t i;         // 8-bit integer variable "i"
  DDRB = 0b11111111; // Set PORTB register as outputs

  for(;;)
  {
    for (i = 0; i < 255; i++)
    {
    _delay_ms(TIME);
    PORTB = 0b11111111 ^ i;
    }
  }
  return 0;
}
```

Once again, the code's main loop counts from 0 to 255. Every time the code loops, it performs an XOR on `0b11111111` and the variable `i` and sets `PORTB` to the result of the operation. For example, if `i` has a value of 15, or `0b00001111`, the result of `0b11111111 ^ 0b00001111` is `0b11110000`.

When you run this code on your AVR, it should demonstrate counting in binary from 0 to 255. However, in this case the LEDs are lit in an inverse fashion—that is, numbers are shown with LEDs that are off, not on.

As this chapter concludes, I encourage you to experiment with the code samples. Enjoy creating patterns, learning about bitwise arithmetic, and bit shifting using your newfound knowledge. In the next chapter, you'll learn to use the inputs of your microcontrollers to create interactive devices.

# GETTING AND DISPLAYING INPUT

AVR microcontrollers can process input from the outside world and react with output, which offers huge potential for interactive projects—for example, those that react to the surrounding temperature. In this chapter, you'll program ATtiny85s and ATmega328P-PUs to detect input signals from external devices and put them to work.

In particular, you will:

- Learn about digital inputs and buttons.
- Make decisions with `if...else` and `switch...case` statements.
- Use seven-segment LED displays to make a digital counter.
- Make a single-cell battery tester.
- Learn about the TMP36 temperature sensor.
- Make a digital thermometer.

Along the way, you'll learn how to counter switch bounce with resistors, as well as gaining more experience with programming in C by creating your own functions and using floating-point variable and analog inputs.

## Digital Inputs

In Chapter 2, you learned to use digital I/O pins as outputs. You can use the same pins to accept input from users and other components. Just like digital outputs, digital inputs have two states: instead of *on* or *off*, they're *high* and *low*. Using digital I/O pins for input is similar to controlling the outputs, too. In this section, you'll set up the DDR*x*, then monitor the value of another register called the *PINx* that stores the status of the digital input pins. Let's get started!

### Introducing the Pushbutton

One of the simplest forms of digital input is the *pushbutton*, shown in Figure 3-1. Pushbuttons are easy to insert into a solderless breadboard. When pressed, they allow an electrical current to pass, which your microcontroller can detect through a digital input.

Figure 3-1: Simple pushbuttons on a breadboard

You'll use pushbuttons in the next project, so note how the pushbutton at the bottom of Figure 3-1 is inserted into the breadboard. The legs bridge rows 23 and 25 so that when you press the button, an electrical connection is made between those two rows. Figure 3-2 shows the schematic symbol for this type of pushbutton.

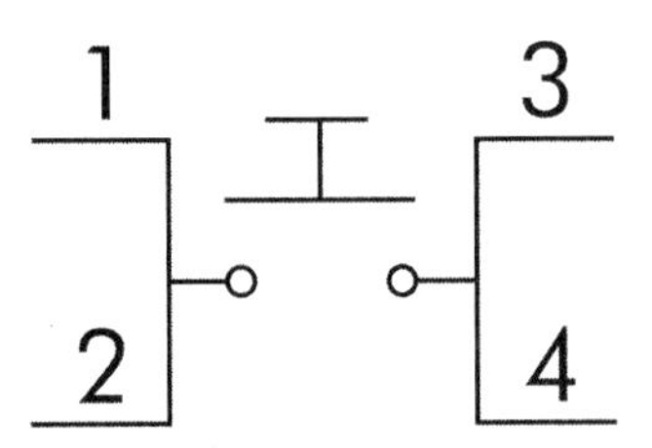


Figure 3-2: Pushbutton schematic symbol

The lines labeled 1 and 2 represent the pushbutton's legs on one side, with 3 and 4 representing the legs on the other side. When you compare the schematic symbol against the real pushbutton in Figure 3-1, the legs labeled 1 and 2 are in row 23, and the legs labeled 3 and 4 are in row 25. The broken line represents an open switch inside the pushbutton. Pins 1 and 2 are electrically connected, as are 3 and 4, so you don't need to run a wire to both pins on each side of the button. When you press the button, the switch closes, allowing current to flow.

### *Reading the Status of Digital Input Pins*

Once you set an I/O pin to an input by assigning a value to `DDRB`, the status of each input pin is stored in the PIN*x* register. Just like the other registers, the PIN*x* register is 8 bits wide, and each bit corresponds to a physical I/O pin.

Think of the PIN*x* register as a binary number, where each bit represents the state of the matching physical pin. If a bit is 1, there is current at the pin and the pin is *high*; if the bit is 0, there's no current and the pin is *low*. For example, on an ATtiny85 with all I/O pins set to inputs, if pins 5 (PB0) and 6 (PB1) were high, then the PINB register would have a value of `0b00000011`.

The numerical value of the PIN*x* register is assigned to an integer variable, which is then compared against another number. For example, to check whether input pins PB0 and PB1 are high, you would compare the value of PINB to `0b00000011`.

But enough theory—let's build some simple circuits that demonstrate inputs and outputs with microcontrollers!

## Project 8: Blinking an LED on Command

In this project, you'll experiment with ATtiny85 digital inputs by blinking an LED, as you did in Chapter 2. This time, however, that LED will blink only when you press a pushbutton.

### *The Hardware*

You will need the following hardware:

- USBasp programmer
- Solderless breadboard
- ATtiny85–20PU microcontroller
- One LED
- One 560 Ω resistor
- One pushbutton
- One 10 kΩ resistor
- Jumper wires

Assemble the circuit shown in Figure 3-3 on your breadboard. When you're done experimenting with this project, keep the circuit together; you'll use it for Projects 9, 10, and 11 as well.

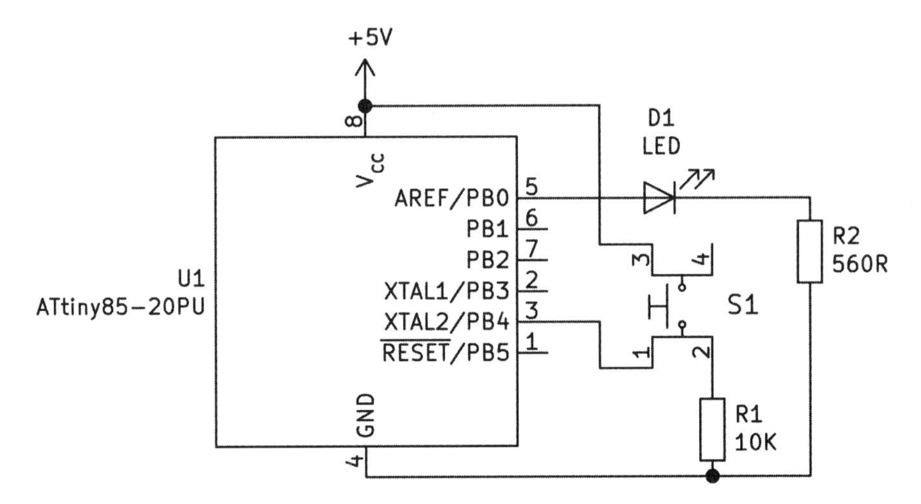


*Figure 3-3: The main circuit for Project 8*

With your circuit assembled, connect the USBasp programmer to the ATtiny85. Repeat the connections as shown in Table 2-1 in Chapter 2.

## The Code

Open a terminal window, navigate to the *Project 8* subfolder of this book's *Chapter 3* folder, and enter the command **make flash**. The toolchain will compile the program file and then upload the data to the microcontroller. At this point, the circuit won't do anything until you press the button. When you do so, the LED should stay on for about one second.

Let's see how this works. Open the *main.c* file for Project 8:

```
// Project 8 - Blinking an LED on Command

#include <avr/io.h>
#include <util/delay.h>

int main(void)
{
❶ DDRB =  0b00001111;
❷ PORTB = 0b00000000;
  for (;;)
  {
  ❸ if (PINB == 0b00010000) // If PB4 is HIGH . . .
  ❹ {
        PORTB = 0b00000001;  // then turn on PB0 output
        _delay_ms(1000);     // Wait a moment
        PORTB = 0b00000000;  // Turn off PB0 output
     }
  }
  return 0;
}
```

This code tells the microcontroller to continually check whether pin 3 (PB4) is high. If it is, we turn on an LED connected to pin 5 (PB0) for about one second.

First we set the DDRB register so that pins PB0 to PB3 are outputs and pins PB4 to PB7 are inputs ❶. Although the ATtiny85 only has eight pins in total, and up to six for inputs and outputs, we still include all eight bits in the `DDRB` statement. Next, we turn off the internal pullup resistors ❷. (I'll return to pullup resistors later in this chapter.)

After that, the program compares the value of the PINB register to the value `0b00010000` ❸. If an electrical current is present at PB4 (pin 3), the fourth bit will be 1 and thus the PINB register will match `0b00010000`. If the comparison results in two equal numbers, the code between the curly brackets will run ❹, turning the LED on for a second and then off again.

In the next section we'll take a closer look at `if` statements like the one in this project, which are used to make comparisons and decisions.

## Making Decisions in Code

Sometimes you'll want certain code to run only if a condition is true or false, such as whether or not a button is pressed. You can use `if` statements, `if...else` statements, and `switch...case` statements to test those conditions and decide what code to execute next.

### *if Statements*

The first line of an `if` statement like the one in Project 8 tests for a condition. If the condition is true (in Project 8, if the value of `PINB` matches the number `0b00010000`), then the code in the curly brackets executes. If the condition is false, then the code in the curly brackets is ignored.

To test conditions, you'll use one or more of the following *comparison operators* inside the `if` statement:

- Equal to: `==`
- Not equal to: `!=`
- Greater than: `>`
- Less than: `<`
- Greater than or equal to: `>=`
- Less than or equal to: `<=`

Over time, you'll use the comparison operators more often and they'll become second nature.

**WARNING** *One common mistake is to use a single equal sign (`=`), which means "make equal to," in a test statement instead of a double equal sign (`==`), which says, "test if it is equal to." You may not get an error message from the toolchain, but your code probably won't work properly!*

You can also make various types of comparisons where there are two or more options to choose from, which can save code space. These are explained in the following sections.

### if . . . else Statements

You can add another action to an if statement using else. For example, you can rewrite the code in Project 8 as follows:

```
if (PINB == 0b00010000) // If PB4 is high . . .
{
   PORTB = 0b00000001;  // turn on PB0 output
} else
{
   PORTB = 0b00000000;  // turn off PB0 output
}
```

With this modification, the LED turns on if you press the button and off if you don't press the button.

### Making Two or More Comparisons

You can also use two or more comparisons in the same if using *comparison operators*. For example, to compare the value of the integer variable counter against a range between 23 and 42, use two comparisons joined with the AND operator, &&:

```
if (counter>=23 && counter <42) // If counter is between 23 and 42 . . .
{
   PORTB = 0b00000001;          // turn on PB0 output
} else
{
   PORTB = 0b00000000;          // turn off PB0 output
}
```

Note that the AND operator used for comparisons (&&) is different from the bitwise arithmetic AND operator (&) introduced in Chapter 2.

You can also use an OR comparison. For example, if you need to test for cases in which the value of the counter variable is less than 100 or greater than 115, use two comparisons joined with the operator ||:

```
if (counter<100 || counter >115) // If counter is under 100 or over 115 . . .
{
   PORTB = 0b00000001;           // turn on PB0 output
} else
{
   PORTB = 0b00000000;           // turn off PB0 output
}
```

You'll expand your knowledge of these useful operators in later projects.

### *switch . . . case Statements*

To compare two or more variables, it's easier to use a `switch...case` statement instead of several `if...else` statements. The `switch...case` statement runs code when one of the defined comparisons is true.

For example, say you want to run different code for each possible value of the integer variable `counter`, 1, 2, or 3. You could use one `switch...case` statement instead of multiple `if...else` statements:

```
switch(counter)
{
   case 1:   // Do something if the value of counter is 1
      break; // Finish and exit the switch statement
   case 2:   // Do something if the value of counter is 2
      break; // Finish and exit the switch statement
   case 3:   // Do something if the value of counter is 3
      break; // Finish and exit the switch statement
   default:  // Do something if counter isn't 1, 2, or 3
             // (the "default" section is optional)
}
```

The optional `default` section at the end of this code lets you run code when there aren't any true comparisons in the `switch...case` statement.

## Creating Your Own Functions

Sooner or later, you'll want to repeat sections of code more than once or define your own set of instructions. You can achieve both goals by creating your own functions that either take care of a task, accept variables and act upon them, or return a value as their result, like a mathematical function. We'll discuss these three types of functions in the next three projects.

The first type of function simply repeats some code:

```
void name()
{
   // Insert your code to run here.
}
```

In this example, *name()* is a placeholder. You can name your function almost anything you'd like, but the name must always be preceded with `void`. Additionally, you can't use *reserved keywords* in your own creations, since the language already uses them. For example, you can't call a function `void void()` because `void` is a reserved keyword in C, C++, and other languages. You can find a complete list of reserved keywords in C at *https://en.cppreference.com/w/c/keyword.*

The function's code goes inside the curly brackets. Always put your custom functions before the `int main(void)` section of your code.

## Project 9: A Simple Custom Function

This project demonstrates the creation of a simple custom function that performs a task. Using the hardware from Project 8, open a terminal window, navigate to the *Project 9* subfolder of this book's *Chapter 3* folder, and enter the command **`make flash`** to upload the code for Project 9 as usual. You should see the LED blink twice every five seconds.

Let's take a look at the code:

```
// Project 9 - A Simple Custom Function

#include <avr/io.h>
#include <util/delay.h>

❶ void blinkTwice()
{
   PORTB = 0b11111111;
   _delay_ms(100);
   PORTB = 0b00000000;
   _delay_ms(100);
   PORTB = 0b11111111;
   _delay_ms(100);
   PORTB = 0b00000000;
   _delay_ms(100);
}

int main(void)
{
   // Set PB3 (and all other pins on PORTB) as output
   DDRB = 0b11111111;
   for(;;)
   {
    ❷ blinkTwice();
      _delay_ms(5000);
   }
return 0;
}
```

The custom function `blinkTwice()` ❶ makes the LED blink twice, as it sets the entire PORTB register on and off twice with a short delay. Once you've created a function like this, you can call it anywhere in the code ❷.

What if you want to be able to easily change how many times the LED blinks? That's where the second type of custom function—functions to which you can pass values—comes in handy:

```
void name(type variable, type variable2, . . .)
{
   // Insert your code to run here.
}
```

Again, you'll give your function a name, but this time the parentheses will contain two parameters that will be used in the code inside the parentheses after the function name: *type* and *variable*, which specify the type and name of the variable being passed to the function, respectively.

## Project 10: Custom Functions with Internal Variables

This project demonstrates the creation of custom functions that accept variables as parameters, then act upon those variables. Using the same hardware from Project 8, navigate to the *Project 10* subfolder of this book's *Chapter 3* folder in your terminal window and enter the command **make flash** to upload the code for Project 10. You should see the LED blink on and off 11 times.

Let's take a look at the code in this project's *main.c* file:

```
// Project 10 - Custom Functions with Internal Variables

#include <avr/io.h>
#include <util/delay.h>

void delay_ms(int ms)
{
   uint8_t i;
   for (i = 0; i < ms; i++)
   {
      _delay_ms(1);
   }
}

❶ void blinkLED(uint8_t blinks)
{
   uint8_t i;
   for (i = 0; i < blinks; i++)
   {
      PORTB = 0b11111111;
      delay_ms(100);
      PORTB = 0b00000000;
      delay_ms(100);
   }
}

int main(void)
{
   DDRB = 0b11111111; // Set PB3 as output
   for(;;)
   {
   ❷ blinkLED(10);
      _delay_ms(3000);
   }
   return 0;
}
```

The function `blinkLED(uint8_t blinks)` ❶ accepts an unsigned integer and uses it in a `for` loop to blink the LED that number of times. Now you can call the `blinkLED()` function with different values anywhere in your code. For example, in this project's code we call `blinkLED(10)` ❷ to make the LED blink 11 times.

To pass more than one variable into a custom function, just add a comma between each variable within the parentheses after the function name. For example, here I've added a second parameter, `blinkDelay`, to the `blinkLED()` function, which allows you to set the value of the delay between the LED turning on and off. This parameter is then passed into the `delay_ms()` function:

```
void blinkLED(uint8_t blinks, uint8_t blinkDelay)
{
   uint8_t i;
   for (i = 0; i < blinks; i++)
   {
      PORTB = 0b11111111;
      delay_ms(blinkDelay);
      PORTB = 0b00000000;
      delay_ms(blinkDelay);
   }
```

The code for Project 10 actually defines two custom functions: the first one is `delay_ms(int ms)`. Sometimes you might want a delay function in your code, and you'll want to specify the length of that delay using a variable. That isn't possible with the standard `_delay_ms()` function, so just create your own!

## Project 11: Custom Functions That Return Values

In this project I'll demonstrate the third type of custom function you can create: one that accepts one or more variables, uses them in a mathematical operation, and then returns the result.

Functions that return the result of an operation can be incredibly useful. Think of them as mathematical formula black boxes: values go in on one end, operations are performed on them, and the result pops out the other end. You can create such a function as follows:

```
  type variable (type variable, type variable 2, . . .)
{
   // Declare a variable to hold the results of the calculations.
   // Insert your code to run here.
   // Return the declared variable.
}
```

Note that the variable you declare must be of the same type as the function.

Let's put this into action. Using the same hardware from Project 8, navigate to the *Project 11* subfolder of this book's *Chapter 3* folder in your terminal window and enter the command **make flash** to upload the code for Project 11. You should see the LED blink 12 times every 2 seconds.

To see how this works, open the *main.c* file for this project:

```
// Project 11 - Custom Functions That Return Values

#include <avr/io.h>
#include <util/delay.h>

void blinkLED(uint8_t blinks)
{
   uint8_t i;
   for (i = 0; i < blinks; i++)
   {
      PORTB = 0b11111111;
      _delay_ms(100);
      PORTB = 0b00000000;
      _delay_ms(100);
   }
}

❶ uint8_t timesThree(uint8_t subject)
{
   uint8_t product;
 ❷ product = subject * 3;
 ❸ return product;
}

int main(void)
{
   DDRB = 0b11111111; // Set PB3 as output
   uint8_t j;
   uint8_t k;
   for(;;)
   {
      j = 4;
      k = timesThree(j);
      blinkLED(k);
      _delay_ms(2000);
   }
   return 0;
}
```

We're still blinking the LED (this is the last time, I promise!), but now we call `blinkLED` with `k`, which is set to the result of the `timesThree()` function. The `timesThree()` function multiplies an integer by three. First we declare what type of variable the function will return—in this case, an integer (`uint8_t`) ❶. This is followed by the function name, `timesThree()`, and the variable that will contain the number we pass into the function.

Inside the function, we declare another variable to hold the result of the multiplication operation ❷. Next, we call `return` to pass the result back to the code ❸.

To actually use `timesThree()` in other parts of your program, just call it with an argument and set a variable equal to it:

```
k = timesThree(j);
```

As you gain more experience with microcontrollers, you'll find that creating your own functions can save you quite a bit of time. For now, though, let's turn to discussing possible problems that can arise when using buttons, and how to solve them in order to create more reliable projects.

## Switch Bounce

Pushbuttons, which you first encountered in Project 8, are prone to *switch bounce* (also known as *switch bouncing*), a button's tendency to turn on and off several times after being pressed only once by the user. Switch bounce occurs because the metal contacts inside a pushbutton are so small that they can vibrate after the button has been released, causing the switch to quickly and repeatedly open and close until the vibration ends.

Switch bounce can be observed using a *digital storage oscilloscope (DSO)*, a device that plots the change in a voltage over a period of time. Figure 3-4 shows a DSO displaying the voltage measured across a pushbutton during a switch bounce.

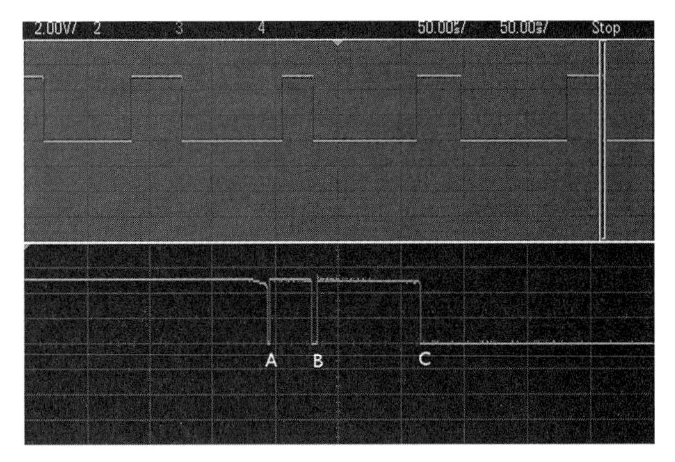


*Figure 3-4: Measuring switch bounce*

The top half of the display in Figure 3-4 shows the result of the button being pressed several times. When the voltage line in the lower half of the image is at the higher horizontal position (5 V), the button is on.

Underneath the word Stop at the top right, two vertical lines highlight a slice of time. The pushbutton voltage during this time is magnified in the bottom half of the screen. At A, the user releases the button and the line drops down to 0 V. However, due to physical vibration, it immediately jumps back up to 5 V again until B, when it vibrates off and then on again until C, where it settles at the low (off) state. In effect, instead of relaying one button press to our microcontroller, we have unwittingly sent three.

You can't prevent switch bounce, but you can prevent your program from reacting to it: just use the `_delay_ms()` function to force the program to wait before executing any more code after detecting a button press. About 50 ms should be long enough, but test this with your own hardware to find the precise length of time that meets your needs.

## Protecting Your AVR from Fluctuating Voltages

In a perfect world, a digital input pin would see either a 5 V electrical signal (high) or no electrical signal (low). In reality, switch bounce and other imperfections can cause the voltage at an input pin to vary wildly between 5 V and 0 V.

Adding delays in your program helps prevent software malfunctions due to switch bounce, but these fluctuations can happen even when you're not using the pin, which can confuse or even damage your AVR. Fortunately, you can protect the AVR with pullup or pulldown resistors.

### Pullup Resistors

A *pullup resistor*, illustrated in Figure 3-5, keeps the voltage at a digital input pin as close as possible to a high state.

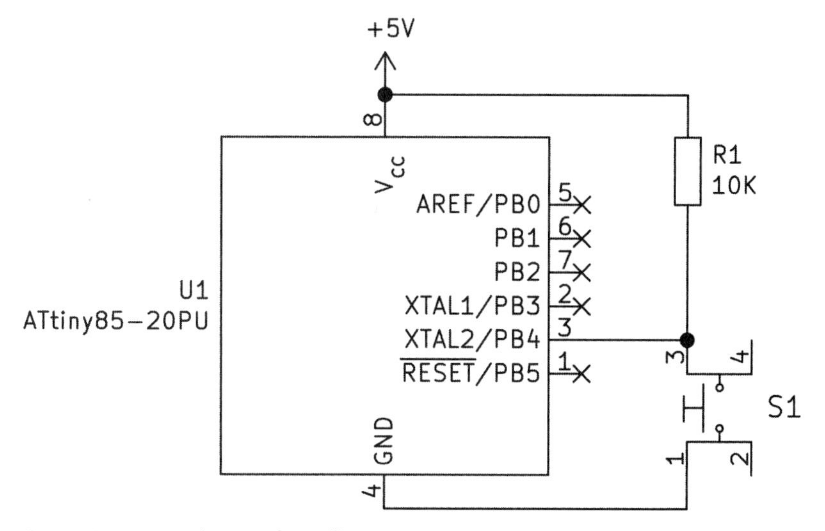

*Figure 3-5: Example use of a pullup resistor*

The voltage at input pin 3 on the microcontroller will always be connected to a high signal, until the button is pressed, at which point pin 3 will be directly connected to GND and become low. The resistor prevents you from shorting 5 V to GND (which can damage your hardware) when you press the button.

AVR microcontrollers have internal pullup resistors. This is a neat way to reduce the size of a circuit, but the catch is that you'll have to invert the logic in your programs. For example, if you have a button connected between a digital input pin and GND and the internal pullups are enabled, the input goes low (instead of high) when the button is pressed. This is a small price to pay for convenience. To turn on the internal pullup resistors for pins that are set as inputs, write 1s to the corresponding bits in PORT*x*; to turn internal pullups off, write 0s to those bits.

### Pulldown Resistors

A *pulldown* resistor, illustrated in Figure 3-6, keeps the voltage at a digital input pin as close as possible to a low state.

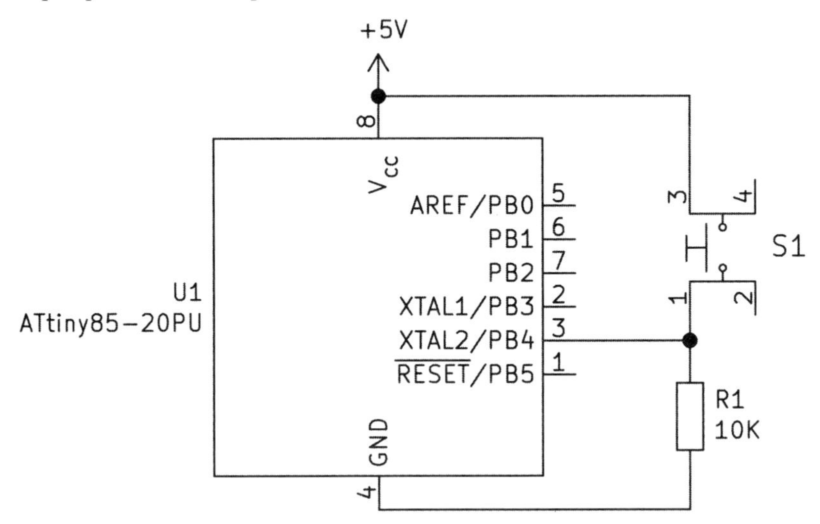


*Figure 3-6: Example use of a pulldown resistor*

The voltage at pin 3 on the microcontroller connects to a low signal until the button is pressed. Then pin 3 directly connects to 5 V (high). Again, we use the resistor to avoid a dead short between the 5 V and GND when the button is pressed.

## Introducing Seven-Segment LED Displays

One way to make your microcontroller react to input is by displaying a number. To do that, you can add a new component to your toolbox: the *seven-segment LED display*, shown in Figure 3-7.

*Figure 3-7: Seven-segment LED display modules*

These small plastic bricks contain eight LEDs, arranged into a familiar digital-number display with a decimal point. You'll find them in various household appliances, and they're great for displaying numbers, letters, or even symbols.

Seven-segment displays are available in a myriad of sizes and colors, and electrically they are the same as eight separate LEDs, with one catch: to reduce the number of pins used by the display, all of the anodes or cathodes of the LEDs are connected together. These are called *common-anode* and *common-cathode* configurations, respectively. All seven-segment displays in this book use common-cathode modules. The schematic symbol for this example display is shown in Figure 3-8.

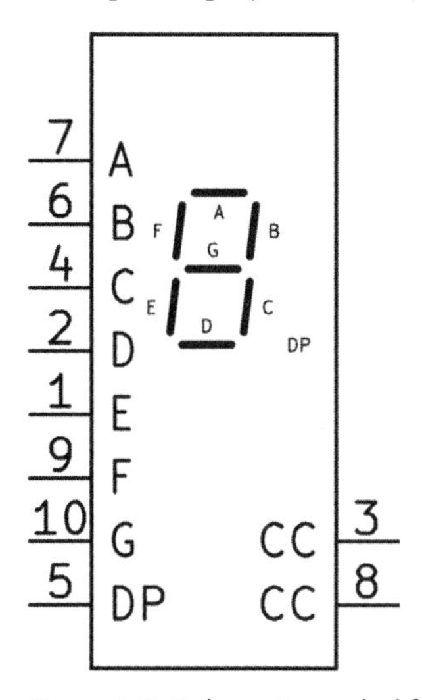


*Figure 3-8: Schematic symbol for a seven-segment common-cathode LED display module*

Next to each LED's cathode is its matching segment. The display's LEDs are labeled A to G and DP (for the decimal point) with an anode pin for each LED segment, and the cathodes are connected to one or two common cathode pins.

Seven-segment display pin layouts vary by manufacturer, so when you purchase them, make sure the supplier gives you the data sheet showing which pins are the anodes for each segment, and which pin is the cathode. If in doubt, note that most common models have pin 1 at the bottom-left corner of the display with the rest numbered counterclockwise. Remember that they're still individual LEDs, and you'll still need a current-limiting resistor for each one.

## Project 12: Building a Single-Digit Numerical Counter

Let's consolidate what you've learned so far by making an interactive device that I hope will spur your imagination: a single-digit numerical counter. Your counter will have two buttons (one to increase the count, and another to reset the counter to zero) and a seven-segment display.

### The Hardware

You will need the following hardware:

- USBasp programmer
- Solderless breadboard
- ATmega328P-PU microcontroller
- One common-cathode seven-segment LED display
- Seven 560 Ω resistors (R1–R7)
- Two pushbuttons
- Two 10 kΩ resistors (R8, R9)
- Jumper wires

Assemble your circuit as shown in Figure 3-9.

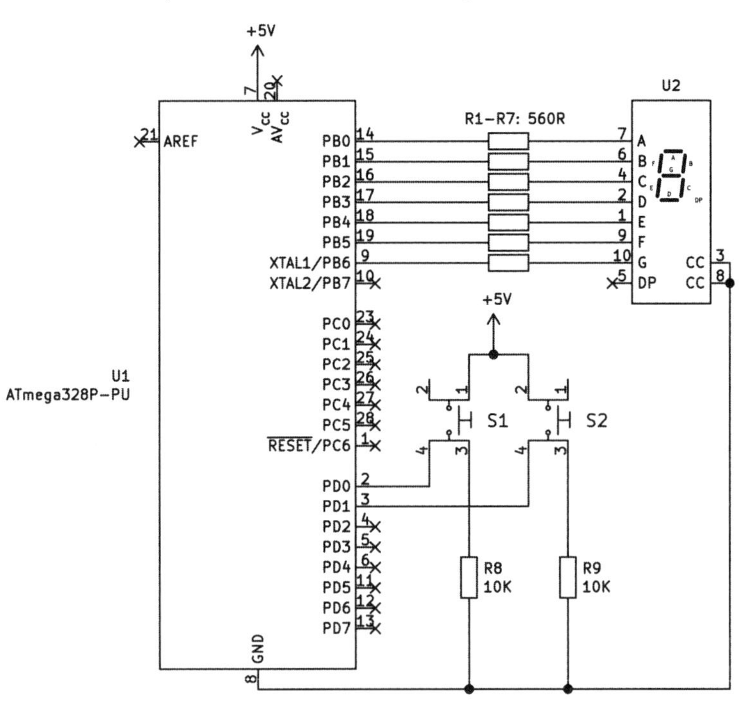


Figure 3-9: The main circuit for Project 12

By now you should remember on your own to connect the USBasp and wire it to the circuit before getting started, so after this project I'll stop reminding you.

### The Code

Open a terminal window, navigate to the *Project 12* subfolder of this book's *Chapter 3* folder, and enter the command `make flash`. Once you've uploaded the code to your microcontroller, press the button connected to PD0 to increase the counter, and press the other button to reset the count to zero.

To see how this works, open the *main.c* file for Project 12:

```
// Project 12 - Building a Single-Digit Numerical Counter

#include <avr/io.h>
#include <util/delay.h>
#define TIME 150

❶ void displayNumber(uint8_t value)
// Displays numbers 0-9 on a seven-segment LED display
{
   switch(value)
   {
      case 0 : PORTB = 0b00111111; break; // 0
      case 1 : PORTB = 0b00000110; break; // 1
      case 2 : PORTB = 0b01011011; break; // 2
      case 3 : PORTB = 0b01001111; break; // 3
      case 4 : PORTB = 0b01100110; break; // 4
      case 5 : PORTB = 0b01101101; break; // 5
      case 6 : PORTB = 0b01111101; break; // 6
      case 7 : PORTB = 0b00000111; break; // 7
      case 8 : PORTB = 0b01111111; break; // 8
      case 9 : PORTB = 0b01101111; break; // 9
   }
}

int main(void)
{
 ❷ uint8_t i = 0;              // Counter value
 ❸ DDRB = 0b11111111;          // Set PORTB to outputs
   DDRD = 0b00000000;          // Set PORTD to inputs
 ❹ PORTD = 0b11111100;         // Turn off internal pullups for PD0 and PD1
   for(;;)
   {
    ❺ displayNumber(i);        // Display count
      _delay_ms(TIME);
    ❻ if (PIND == 0b11111110) // If reset button pressed . . .
      {
         i = 0;                // set counter to zero
      }
    ❼ if (PIND == 0b11111101) // If count button pressed . . .
      {
         i++;                  // increase counter
         if (i > 9)            // If counter is greater than 9 . . .
         {
            i = 0;             // set counter to zero
         }
      }
   }
   return 0;
}
```

This code defines a new function, `displayNumber()`, that receives an integer and sets the PORTB outputs to turn individual segments of the LED display on or off, thus displaying a digit from 0 to 9 ❶. By using a `switch... case` statement, the code can neatly decide which `PORTB` command to run based on what digit you want to display.

The variable `i` keeps track of what number you want to display ❷. This variable is initialized with a value of zero so that the counter starts at zero when the power is turned on.

Next, we set up the I/O pins, using PORTB as outputs for the LEDs and PORTD as inputs that detect when the buttons are pressed ❸. The internal pullup resistors are turned off for pins PD0 and PD1 ❹ because the buttons are connected to those pins, and the pullups are turned on for the rest of PORTD. This ensures that the unused pins in PORTD will always be high.

Every time the code loops, it first displays the value of the count ❺, then checks if the reset or count buttons have been pressed. Note that when the code compares `PIND` against the binary values representing button presses on PD0 ❻ or PD1 ❼, the unused bits (or inputs) in the comparison are 1s, not 0s. This is because the internal pullup resistors have been activated for the unused inputs, keeping them at 1.

When you press the count button, the count variable `i` should increase by one. If the count is greater than 9, it should reset back to 0 because you're working with a single-digit display.

Once you've completed this project, you'll have a nifty display—but remember, you learn by doing. Experiment with this program! For example, try changing it to a countdown timer or creating changing patterns with the LEDs instead of numbers.

If possible, keep this project assembled, as you'll reuse most of it for Project 14.

## Analog Inputs

Up to this point, your projects have used digital electrical signals with just two levels: high and low. For your microcontrollers, high is close to 5 V and low is close to 0 V (or GND). We used the PORT*x* register to blink LEDs and the PIN*x* register to measure whether a digital input was high or low. Figure 3-10 illustrates a digital signal measured with a DSO.

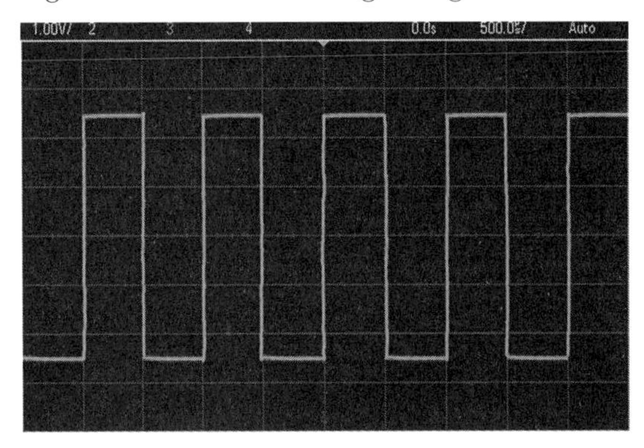


*Figure 3-10: A digital signal, with highs appearing as horizontal lines at the top and lows appearing at the bottom*

Unlike digital signals, *analog signals* can vary with an indefinite number of steps between high and low. For example, Figure 3-11 shows an analog voltage signal that looks like a sine wave.

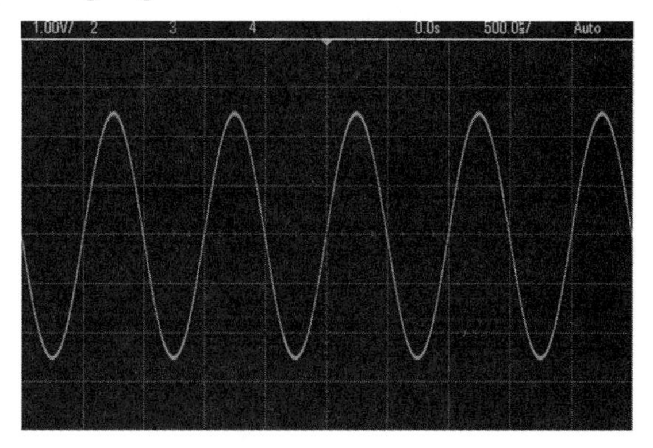

*Figure 3-11: A sinusoidal analog signal*

Notice that as time progresses, the voltage moves fluidly between high and low levels. Analog signals can represent all sorts of information from various devices, such as temperature or distance sensors. To use analog signals in your projects, you measure the voltage values of the signal using certain I/O pins in your microcontrollers, which are connected to *analog-to-digital converters (ADCs)*. An ADC can convert a voltage to a number, which you can then use in your code. Refer to the pinout and port register diagrams in Chapter 2; the analog inputs are marked as ADC*x* for *analog-to-digital converter x*.

On the ATtiny85 you have PB0, PB1, PB2, and PB5 (physical pins 5–7 and 1). On the ATmega328P-PU you have a whole new register (PORTC), which has six ADCs from PC0 to PC5.

## Using ATtiny85 ADCs

To set up the ADC pins on the ATtiny85, you'll need to set two new registers. (Yes, more registers! You'll get used to them the more you use them.) The first, ADMUX, selects which pin you'll connect to the ADC. You'll leave the first six bits as `001000` and use the last two bits to select the physical pin for the ADC. They're `00` for ADC0 (pin 1), `01` for ADC1 (pin 7), `10` for ADC2 (pin 3), and `11` for ADC3 (pin 2). For example, to use physical pin 3 as the analog input, set ADMUX as follows:

```
ADMUX = 0b00100010;
```

The second register to set, called ADCSRA, is responsible for several settings, including the speed of the ADC. All ATtiny85 projects in this book set the speed to 1 MHz. As the ADC section of the microcontroller operates at a different speed, you'll use ADCSRA to set a prescale value that determines the ADC speed. You'll typically use a prescaler of 8, which brings the ADC speed down to 125 kHz (which we calculate by dividing the speed of 1 MHz by 8). The matching ADCSRA line is:

```
ADCSRA = 0b10000011;
```

It's a good idea to put the two register settings in their own custom function, which you could name `startADC()`, for example. This keeps your code neat and makes it easy to call the function wherever you need it.

When using ADCs, it can be helpful to use a different way of addressing bits in a register—using bitwise operations. For now I'll keep it simple and just show you how this works; we'll revisit this topic (and the use of prescalers) in more detail in Chapter 6.

To measure a value from your preset ADC pin, first start the ADC with the line:

```
ADCSRA |= (1 << ADSC);
```

This sets the ADSC bit in the ADCSRA register to 1, which tells your ATtiny85 to read the analog input and convert it to a value. When the ATtiny85 has finished, the ADSC bit returns to 0. You'll need to tell your code to wait a moment until this happens, like so:

```
while (ADCSRA & (1 << ADSC));
```

This compares the ADSC bit in the ADCSRA register to 1 and does nothing if they're both 1. When the ADC process is complete, the ADSC bit returns to 0, the `while()` function finishes, and the microcontroller progresses.

Finally, the value from the ADC is stored in the variable `ADCH`. The value of `ADCH` should be between 0 and 255. You can then work with `ADCH` as required.

## Project 13: Making a Single-Cell Battery Tester

In this project, you'll put your ATtiny85's ADC to work by making a simple battery tester. You can use this to check the voltage of AA, AAA, C, or D cell batteries.

**WARNING** *Do not connect any battery (or other power source) with a voltage greater than 5 V to your tester, and don't connect the battery the wrong way around (check the schematic in Figure 3-12). Doing either will damage your ATtiny85.*

This tester uses two LEDs to indicate whether the battery is good (with a voltage greater than or equal to 1.4 V) or bad (with a voltage less than 1.4 V).

### The Hardware

You'll need the following hardware:

- USBasp programmer
- Solderless breadboard

- ATtiny85–20PU microcontroller
- Two LEDs
- One 560 Ω resistor
- Jumper wires

Assemble your circuit as shown in Figure 3-12. Note that the two wires labeled positive (+) and negative (–) are jumper wires used to contact the battery you want to test. Connect the + and – wires to the matching points on the battery.

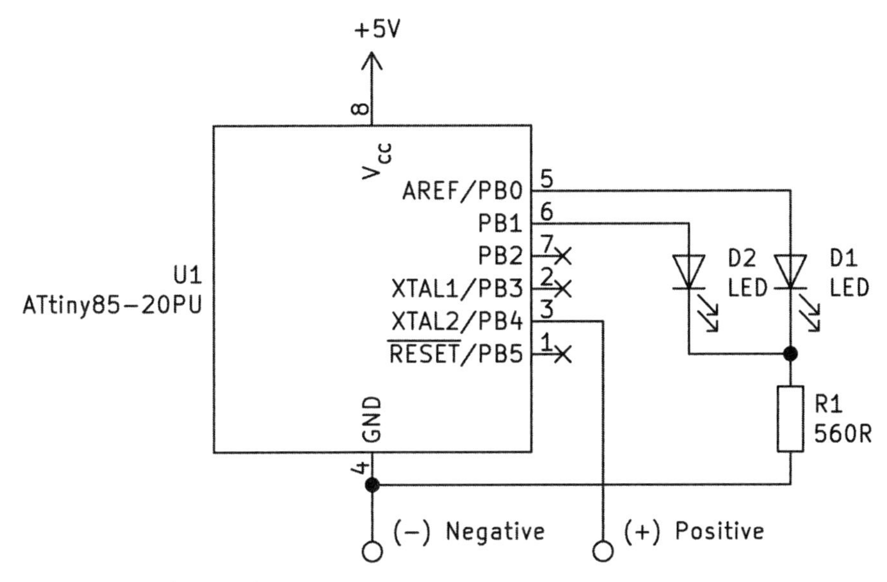

*Figure 3-12: Schematic for Project 13*

You may find that it helps to use some insulation tape to keep the wires connected to the battery under test. You might also use red and green LEDs to indicate whether the battery under test is "bad" or "good."

## The Code

Open a terminal window, navigate to the *Project 13* subfolder of the *Chapter 3* folder for this book, and enter the command **`make flash`**. Once you've uploaded the code, find a AA, AAA, C, or D cell battery, and connect the positive and negative leads to the circuit as shown in the schematic. If the voltage is greater than or equal to 1.4 V, LED2 should turn on; if it's less than 1.4 V, LED1 should turn on.

To see how this is implemented, open the *main.c* file for Project 13:

```
// Project 13 - Making a Single-Cell Battery Tester

#include <avr/io.h>
#include <util/delay.h>

❶ void startADC()
```

```
// Set up the ADC
{
   ADMUX = 0b00100010;              // Set ADC pin to 3
   ADCSRA = 0b10000011;             // Set prescaler speed for 1 MHz
}

int main(void)
{
   DDRB = 0b00000011;               // Set pins 5 and 6 as outputs
❷  startADC();
   for(;;)
   {
    ❸ ADCSRA |= (1 << ADSC);        // Start ADC measurement
      while (ADCSRA & (1 << ADSC) ); // Wait until conversion completes
      _delay_ms(5);

    ❹ if (ADCH >= 71)
      {
         // If ADC input voltage is more than or equal to ~1.4 V . . .
         PORTB = 0b00000010;         // Turn on "battery OK" LED2

    ❺ } else if (ADCH < 71)
      {
         // Else, if ADC input voltage is less than ~1.4 V . . .
         PORTB = 0b00000001;         // Turn on "battery not OK" LED1
      }
   }
   return 0;
```

}The function startADC() sets physical pin 3 to use its ADC function and sets the prescaler for 1 MHz operation ❶. We need to call this function before using the ADC ❷. We then activate the ADC for a reading and wait for it to complete ❸.

The value from the ADC—a number between 0 and 255—is stored in the variable ADCH. This value maps to the voltage range of the ADC, which is 0 to 5 V. You can find the ADC value with some basic math:

(Map Voltage × 256) / Supply Voltage = ADC Value

For our example, we calculate the matching ADC value for 1.4 V as follows:

(1.4 V × 256 / 5 V) = 71.68

Based on this calculation, 1.4 V maps to an ADC value of 71.4, which we round to 71 in our code because we're using whole numbers in this project. This is the value used in the if statements at and to determine whether the battery is good to use ❹ or not ❺.

At this point, you should understand how to read analog signals in the form of varying voltages from external devices. This is incredibly useful, as there are many types of sensors whose values are returned as a varying voltage and thus are easy to read with the ADC pin of your microcontroller.

Next, let's look at the ADCs on the ATmega328P-PU, along with some more information about variable types.

## Using the ATmega328P-PU ADCs

Setting up the ADC pins on the ATmega328P-PU is similar to setting them up on the ATtiny85. You'll start by setting up some registers. The first is ADMUX, which has two functions: indicating which ADC pin you want to use and selecting the source of the reference voltage the ADC compares against the analog signal being measured.

First you'll set the REFS0 bit in the ADMUX register to 1, which tells the microcontroller to use the voltage connected to $AV_{CC}$ (pin 20) for comparisons with analog signals. Again, you can use a bitwise operation for this:

```
ADMUX |= (1 << REFS0);
```

This is a simpler and less error-prone method of setting individual bits in a register, as it lets you avoid having to deal with all eight bits at once—you only set the bit you want to change. Also remember that you only need to set bits to 1, as by default they're all 0.

Next, you'll set the MUX2 and MUX0 bits to 1, which tells the ADC to read signals coming in on pin 28:

```
ADMUX |= (1 << MUX2) | (1 << MUX0);
```

The second register you'll set, ADCSRA, activates the ADC and sets the speed of the ADC in the microcontroller. All your ATmega328P-PU projects from here until Chapter 13 will use the speed 1 MHz, and the matching ADCSRA line is:

```
ADCSRA |= (1 << ADPS1) | (1 << ADPS0);
```

Finally, you need to activate the ADC by setting the ADEN bit to 1:

```
ADCSRA |= (1 << ADEN);
```

As in Project 12, it's a good idea to put the register settings in their own custom function, which I've named `startADC()`.

When you want to measure a value from the preset ADC pin, you first need to start the ADC as follows:

```
ADCSRA |= (1 << ADSC);
```

This sets the ADSC bit in the ADCSRA register to 1, which tells your ATmega328P-PU to read the analog input and convert it to a value. This is not instantaneous; your code needs to wait until the ATmega328P-PU has finished the ADC reading, at which point the ADSC bit returns to 0. The following is a convenient space-saving function that can be used to monitor changes of bits in registers:

```
loop_until_bit_is_clear(ADCSRA, ADSC);
```

In this case, it forces the code to wait until the ADSC bit returns to 0; when this happens, the ADC process is complete, and the code can then continue.

The value from the ADC is a 10-bit number, which the toolchain makes available in a virtual register variable called, you guessed it, `ADC`. However, for purposes where accuracy isn't entirely necessary, you can just use an 8-bit value, dropping the last 2 bits off the ADC register like so:

```
ADCvalue = ADC;
```

where `ADCvalue` is an integer variable used to hold the value from the ADC.

Finally, when using the same power to run the microcontroller and its internal ADC, it's wise to use a small smoothing capacitor over the positive and negative power supply lines, as shown in the following project.

## Introducing the Variable Resistor

Variable resistors, also known as *potentiometers*, can generally be adjusted from 0 Ω up to their rated value. Figure 3-13 shows their schematic symbol.

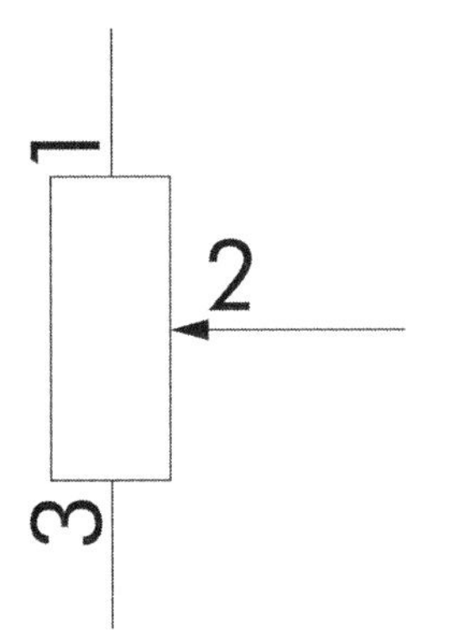


*Figure 3-13: Variable resistor (potentiometer) symbol*

Variable resistors have three electrical connections, one in the center and one on each side. As the shaft of the variable resistor is turned, it increases the resistance between one side and the center and decreases the resistance between the opposite side and the center.

Variable resistors can be either *linear* or *logarithmic*. The resistance of linear models changes at a constant rate when turning, while the resistance of logarithmic models changes slowly at first and then increases rapidly. Logarithmic potentiometers are used more often in audio amplifier circuits because they model the human hearing response. You can generally identify whether a potentiometer is logarithmic or linear via the marking on the rear. Most will have either an A or a B next to the resistance value: A for logarithmic, B for linear. Most projects use linear variable resistors such as the one shown in Figure 3-14.

*Figure 3-14: A typical linear variable resistor*

Miniature variable resistors are known as *trimpots* or *trimmers* (see Figure 3-15). Because of their size, trimpots are more useful for making adjustments in circuits, but they're also very useful for breadboard work because they can be slotted in.

*Figure 3-15: Various trimpots*

When shopping for trimpots, take note of the type. If possible, you'll want one that is easy to adjust with the screwdriver you have on hand. The enclosed types pictured in Figure 3-15 are also preferable, as they last longer than the cheaper, open-contact types.

## Project 14: Experimenting with an ATmega328P-PU ADC

In this project, you'll experiment with an ADC on the larger ATmega328P-PU microcontroller, along with practicing with more involved decision-making code. This project measures the signal from a trimpot, which varies between 0 V and 5 V. The value falls into one of four ranges and is indicated by one of the four LEDs.

### The Hardware

You'll need the following hardware:

- USBasp programmer
- Solderless breadboard
- ATtiny328P-PU microcontroller
- Four LEDs
- Four 560 Ω resistors
- 0.1 μF ceramic capacitor
- 10 kΩ breadboard-compatible linear trimpot
- Jumper wires

Assemble your circuit as shown in Figure 3-16.

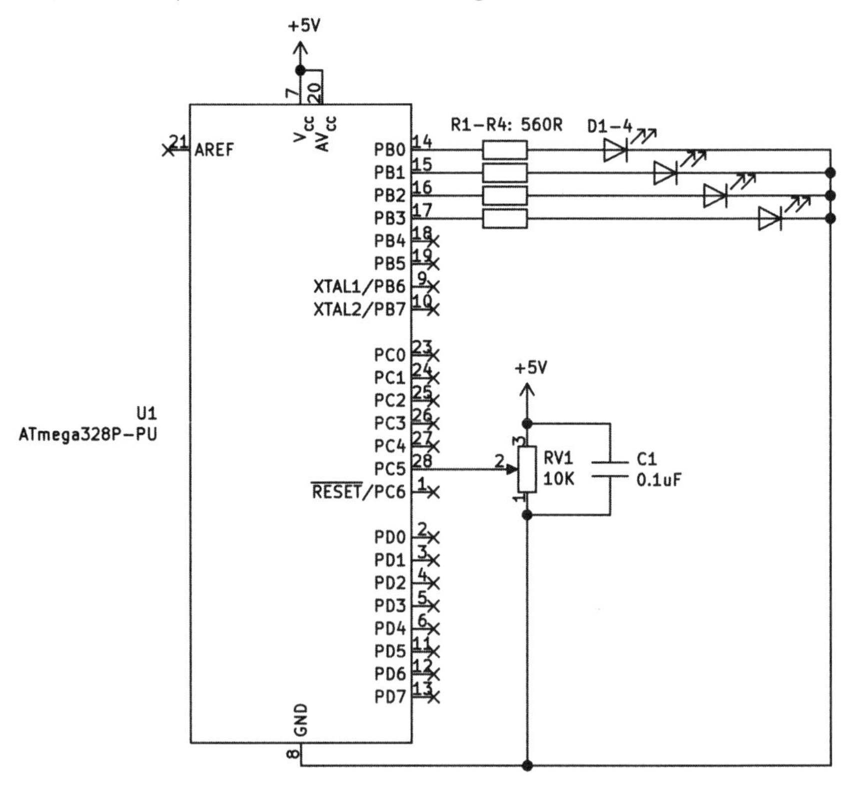


*Figure 3-16: Schematic for Project 14*

If you can't get a breadboard-compatible trimpot, you can use a full-size potentiometer, although you'll need to solder jumper wires to the potentiometer's three pins in order to make contact with the solderless breadboard.

## The Code

Open a terminal window, navigate to the *Project 14* subfolder of this book's *Chapter 3* folder, and enter the command **make flash** to upload the code for Project 14 as usual. Once you've uploaded the code, start slowly moving the trimpot toward one limit, then turn it through the other direction to the other limit. The LEDs should indicate which quartile of the trimpot range you are currently turning through.

Let's examine how this works. Open the *main.c* file for Project 14 and take a look at the code:

```
// Project 14 - Experimenting with an ATmega328P-PU ADC

#include <avr/io.h>
#include <math.h>
#include <util/delay.h>

❶ void startADC()
// Set up the ADC
{
   ADMUX |= (1 << REFS0);                     // Use AVcc pin with ADC
   ADMUX |= (1 << MUX2) | (1 << MUX0);        // Use ADC5 (pin 28)
   ADCSRA |= (1 << ADPS1) | (1 << ADPS0);     // Prescaler for 1MHz (/8)
   ADCSRA |= (1 << ADEN);                     // Enable ADC
}

int main(void)
{
  uint16_t ADCvalue;
❷ DDRB = 0b11111111;                          // Set PORTB to outputs
  DDRC = 0b00000000;                          // Set PORTC to inputs
❸ startADC();
  for(;;)
  {
❹ // Take reading from potentiometer via ADC
    ADCSRA |= (1 << ADSC);                    // Start ADC measurement
    loop_until_bit_is_clear(ADCSRA, ADSC);
    // Wait for conversion to finish
  ❺ _delay_ms(10);

 ❻ // Assign ADC value to "ADCvalue"
   ADCvalue = ADC;
 ❼ if (ADCvalue>=0 && ADCvalue <256)
   {
      PORTB = 0b00000001;
   }
   else if (ADCvalue>=256 && ADCvalue<512)
   {
```

```
      PORTB = 0b00000010;
    }
    else if (ADCvalue>=512 && ADCvalue<768)
    {
      PORTB = 0b00000100;
    }
    else if (ADCvalue>=768 && ADCvalue<1023)
    {
      PORTB = 0b00001000;
    }
     // Turn off the LEDs in preparation for the next reading
     _delay_ms(100);
    PORTB = 0b00000000;
  }
  return 0;
}
```

First we specify the pins used for outputs (LEDs—PORTB) and inputs for ADC ❷, then we call the function `startADC()` ❶ to set up the ADC ❸. We measure the value fed from the trimpot to the ADC via pin PC5 ❹ and store it into the integer variable `ADCvalue` ❻ after a short delay ❺ to give the ADC time to complete its conversion of data.

Next, the code evaluates the value of the ADC using a series of `if...else` functions ❼. Each of these checks if the ADC value falls within a certain range using the `AND` (`&&`) conditional operator, then activates an LED to provide a visual indication if the result of the test is true.

Finally, the LED is turned off at the end of the main loop after a short delay to allow time for indication, and the process starts again.

## Doing Arithmetic with an AVR

Like a pocket calculator, the AVR can perform basic calculations for you. This is really handy when you're dealing with analog-to-digital conversions. Here are some of the mathematical operations available for your AVR:

```
432 + 956; // Addition
100 / 20;  // Division
5 * 200;   // Multiplication
25 - 25;   // Subtraction
10 % 4;    // Modulo
```

The C language handles some kinds of calculations a little differently than a pocket calculator, though. For example, when dividing two integers, the AVR simply discards the remainder rather than rounding the quotient up or down: 16 divided by 2 equals 8, 10 divided by 3 equals 3, and 18 divided by 8 equals 2. I'll explain a few other oddities as they come up.

When working with numbers that have or will result in a decimal point (for example, dividing 1 by 3), you will need to use a new type of variable called a *float*. The values that can be stored in a float can fall between $-3.39 \times 10^{38}$ and $3.39 \times 10^{38}$. To use floating-point math in your code, you'll need to include a new library:

```
#include <math.h>
```

You'll get used to this math by using it in Project 15, along with your first analog sensor, which I'll introduce next.

## Using External Power

Up to this point you've been powering your projects directly from the AVR programmer, which is a neat solution for small projects and experiments. However, this method leaves less than 5 V of output voltage available to your circuit, as the internal circuitry of the programmer reduces the voltage.

If you use a multimeter to measure the voltage across pins 7 and 8 of your ATmega328P-PU in Project 14 or others created earlier, you'll find it's less than 5 V. When working with parts that expect 5 V (such as the TMP36 used in the next project), you'll need an external power supply for accuracy and reliability.

One easy way to add external 5 V power is to use a breadboard power supply module, such as the one from PMD Way (part number 20250303) shown in Figure 3-17.

*Figure 3-17: A breadboard power supply module*

The PMD Way module is inserted into the end of your breadboard and is powered by a common AC to DC wall wart power supply. The unit supplies 5 V or 3.3 V to both sides of the breadboard and has a neat power switch for control. It is a small and very convenient outlay.

## The TMP36 Temperature Sensor

With the TMP36 temperature sensor (shown in Figure 3-18) and a little math, you can turn your AVR into a thermometer. This inexpensive and easy-to-use analog sensor outputs a voltage that changes with the temperature around it.

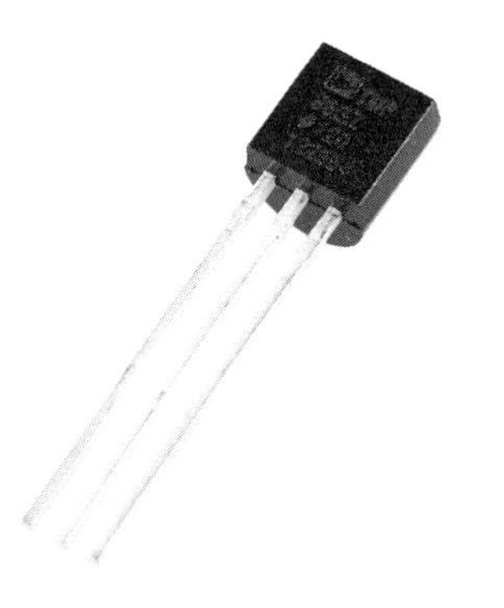

*Figure 3-18: The TMP36 temperature sensor*

The TMP36 has three legs. When you're looking at the flat side of the sensor with the writing on it, the legs are (from left to right) voltage in, voltage out, and GND. You'll connect pin 1 to the 5 V power in your projects, pin 2 to an analog input on the microcontroller, and pin 3 to GND. Figure 3-19 shows the schematic symbol for the TMP36.

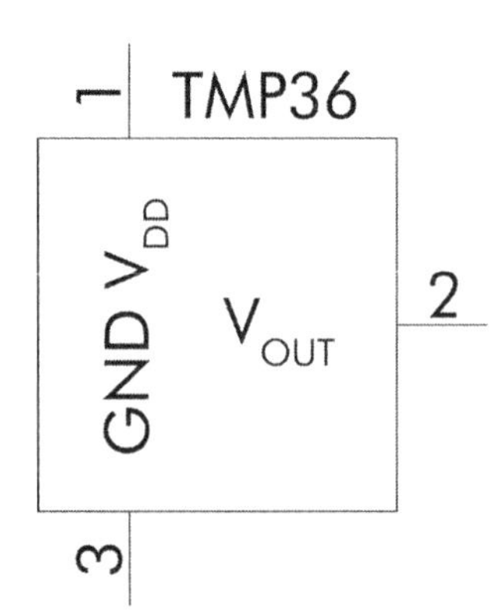


*Figure 3-19: Schematic symbol for the TMP36*

The voltage output from the TMP36 is a representation of the temperature around the sensor; for example, at 25 degrees Celsius, the output is 750 mV (or 0.75 V), and for every change of one degree, the voltage output changes by 10 mV. The TMP36 can measure temperatures from −40 to 125 degrees Celsius, but for the next project you'll just measure room temperatures.

To determine the temperature from the voltage, multiply the value in ADC by 5, then divide by 1,024, which gives you the actual voltage returned by the sensor. Next, subtract 0.5 (0.5 V is the offset used by the TMP36 to allow for temperatures below 0) then multiply by 100, which gives you the temperature in degrees Celsius.

**NOTE** *If you want to work in Fahrenheit, multiply the Celsius value by 1.8 and add 32 to the result.*

As this is an analog device, the output voltage is determined by the input voltage. If you don't have 5 V or very close to 5 V on the input, your output and thus the temperature reading will not be correct.

## Project 15: Creating a Digital Thermometer

In this project, you'll use what you learned from Project 12 to create a numeric display with the ADC on the ATmega328P-PU that acts as a digital thermometer. To keep things simple, this project will display temperatures starting from 0 degrees Celsius and going up; negative readings are not included.

### The Hardware

To build your thermometer, you'll read the TMP36 analog temperature sensor with the microcontroller, which will then display the temperature one digit at a time using the seven-segment LED display from Project 12.

You'll need the following hardware:

- USBasp programmer
- Solderless breadboard
- 5 V breadboard power supply
- ATmega328P-PU microcontroller
- One TMP36 temperature sensor
- One common-cathode seven-segment LED display
- Seven 560 Ω resistors (R1–R7)
- 0.1 µF ceramic capacitor
- Jumper wires

Assemble your circuit as shown in Figure 3-20.

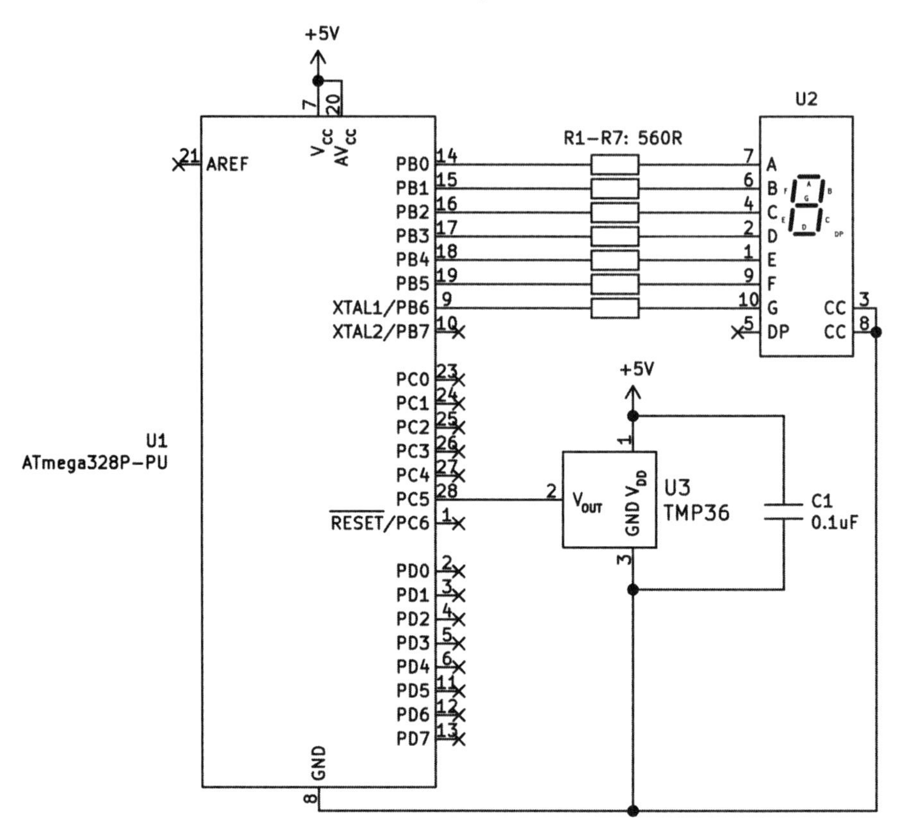


*Figure 3-20: Schematic for Project 15*

You'll use the 0.1 µF capacitor to help maintain a smooth power supply to the TMP36 temperature sensor; it should be mounted as close as possible to the TMP36's 5 V and GND pins.

## The Code

Open a terminal window, navigate to the *Project 15* subfolder of this book's *Chapter 3* folder, and enter the command `make flash` to upload the code for Project 15 as usual. Once you've uploaded the code, the LED module should display the approximate temperature one digit at a time. For example, if the temperature were 8 degrees Celsius, the display would show a 0, followed by a short delay, and then an 8.

To see how this works, open the *main.c* file for Project 15 and take a look at the code:

```
// Project 15 - Creating a Digital Thermometer

❶ #include <avr/io.h>
#include <math.h>
#include <util/delay.h>
```

```
❷ void startADC()
  // Set up the ADC
  {
     ADMUX |= (1 << REFS0);                // Use AVcc pin with  ADC
     ADMUX |= (1 << MUX2) | (1 << MUX0);    // Use ADC5 (pin 28)
     ADCSRA |= (1 << ADPS1) | (1 << ADPS0); // Prescaler for 1MHz (/8)
     ADCSRA |= (1 << ADEN);                // enable ADC}
  }

❸ void displayNumber(uint8_t value)
  // Displays a number from 0-9 on the seven-segment LED display
  {
     switch(value)
     {
        case 0 : PORTB = 0b00111111; break; // 0
        case 1 : PORTB = 0b00110000; break; // 1
        case 2 : PORTB = 0b01011011; break; // 2
        case 3 : PORTB = 0b01111001; break; // 3
        case 4 : PORTB = 0b01110100; break; // 4
        case 5 : PORTB = 0b01101101; break; // 5
        case 6 : PORTB = 0b01101111; break; // 6
        case 7 : PORTB = 0b00111000; break; // 7
        case 8 : PORTB = 0b01111111; break; // 8
        case 9 : PORTB = 0b01111101; break; // 9
     }
  }

  int main(void)
  {
   ❹ uint8_t tens = 0;  // Holds tens digit for temperature
     uint8_t ones = 0;  // Holds ones digit for temperature
     float temperature;
     float voltage;
     uint16_t ADCvalue;
     uint8_t finalTemp;

     DDRB = 0b11111111; // Set PORTB to outputs
     DDRC = 0b00000000; // Set PORTC to inputs
     startADC();
     for(;;)
     {
      ❺ // Take reading from TMP36 via ADC
        ADCSRA |= (1 << ADSC);          // Start ADC measurement
        while (ADCSRA & (1 << ADSC) ); // Wait for conversion to finish
        _delay_ms(10);

        // Get value from ADC register, store in ADCvalue
        ADCvalue = ADC;

      ❻ // Convert reading to temperature value (Celsius)
        voltage = (ADCvalue * 5);
        voltage = voltage / 1024;
        temperature = ((voltage - 0.5) * 100);

      ❼ // Display temperature on LED module
```

```
        finalTemp = (uint8_t) round(temperature);

        tens = finalTemp / 10;
        ones = finalTemp % 10;

    ❽ displayNumber(tens);  // Display tens digit
        _delay_ms(250);
        displayNumber(ones);  // Display ones digit
        _delay_ms(250);

    ❾ // Turn off the LED display in preparation for the next reading
        PORTB = 0b00000000;
        _delay_ms(1000);
    }

    return 0;
}
```

In this code, we first include the necessary libraries (among them, *math.h*, for the floating-point math) ❶. We add the function `startADC()` to start the ADC ❷ (this function is called at the start of the main part of the code), and we reuse the `displayNumber()` function from Project 12 ❸.

In the main section of the code, we declare the required variables, define the input and output pins, and initialize the ADC ❹. The main loop of the code is broken into five steps:

1. The voltage from the TMP36 is measured by the ADC and stored in the variable `ADCvalue` ❺.
2. Using the formula described in "Introducing the TMP36 Temperature Sensor," the code converts the value of the ADC to a voltage. This voltage is then converted to the temperature in degrees Celsius ❻.
3. The digits used to represent the temperature are extracted from `finalTemp` and then rounded up or down to the nearest whole number with `round()`. The code determines the left digit (for tens) by dividing the temperature by 10. If the temperature is less than 10 degrees, this will be 0. It determines the right digit (for ones) by taking the remainder from dividing the temperature by 10 with modulo ❼.
4. The `displayNumber()` function is used to display the tens and ones digits of the temperature, respectively, with a quarter-second delay between the digits ❽.
5. Finally, the display is turned off for a second ❾, giving a visual break between the displayed value and the new value to come.

This seemingly complex project just combines your existing knowledge in new ways, which I hope is starting to fire up your imagination. In Chapter 4, we'll turn to a new topic: enabling bidirectional communication between your microcontroller and a PC for the purposes of data capture and control.

# 4

# COMMUNICATING WITH THE OUTSIDE WORLD USING THE USART

This chapter will teach you to use the *universal synchronous and asynchronous receiver-transmitter (USART)*, a dedicated two-way port that transfers information from the AVR to a computer and allows the two to communicate. The USART lets you control your AVR projects from a computer. It can also help you debug your projects, since you can send status reports from the AVR back to the computer to keep track of the code's progress.

In this chapter you will:

- Use terminal emulation software on your computer to act as an input and output device for your AVR-based projects.
- Send serial data between your AVR and your computer.

- Transfer data, including numbers and letters, from an AVR to a computer.
- Familiarize yourself with ASCII codes.

Along the way, you'll learn to log temperature readings from a thermometer to a PC for later analysis and build a simple calculator.

## Introducing the USART

There are many ways for computers to communicate with one another. One method is to use *serial data*, the process of sequentially sending data one bit at a time. AVR-based projects do this using the AVR's USART. The USART on the ATmega328P-PU microcontroller uses pin 2 to receive data and pin 3 to send data. Data is sent and received in *bytes*, where each byte represents 8 bits of data.

Instead of sending 1s and 0s, which is how computers represent bytes, their values are represented by changing voltage levels over a certain time period. A high voltage represents a 1, while a low voltage represents a 0. Each byte begins with a start bit, which is always a 0, and an end bit, which is always a 1. The byte of data is sent and received with the rightmost or least significant bit first.

I'll demonstrate what these bytes of data look like with a digital storage oscilloscope, which, as you saw in Chapter 3, is a device that can display the change in voltage over a period of time. For example, consider Figure 4-1, which shows a byte of data representing the number 10 sent from the USART.

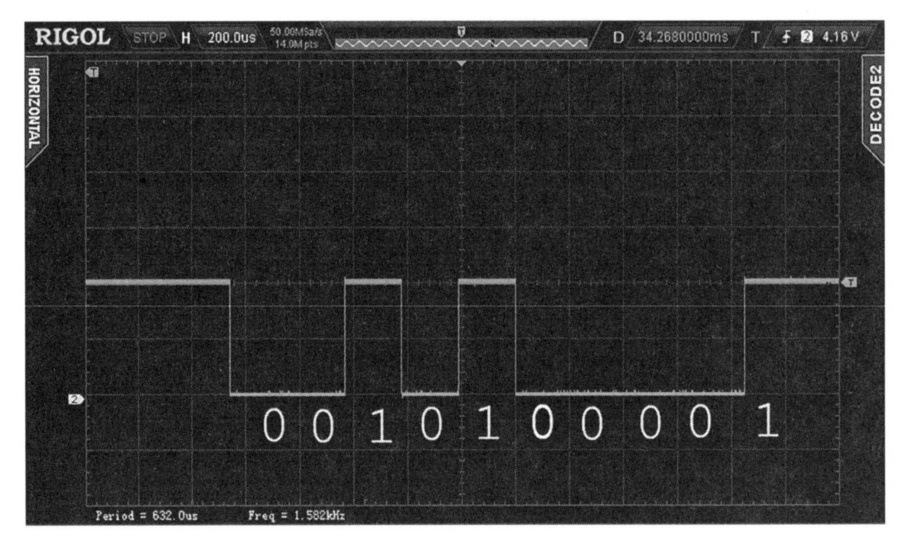


*Figure 4-1: A byte of data represented on the DSO*

Let's see how to translate this into the decimal number 10. The start bit is always 0, so the voltage is first low, then low again (0), then high (1), then low, then high, then low for four periods, followed by high (the end bit, which is always 1). This gives us the binary number 01010000, but because bytes are sent and received with the LSB first, we have to flip it around. This leaves us with 00001010, the binary representation of the decimal number 10.

Data is sent and received at various speeds. In Figure 4-1, and all our projects in this chapter, the data speed is 4,800bps (bits per second).

## Hardware and Software for USART Communication

To prepare your computer to send and receive data to and from your AVR microcontroller, you'll need two things: a USB-to-serial converter and suitable terminal software on your computer.

Let's begin with the *USB-to-serial converter*, the easiest way to connect your AVR projects to a computer. There are many different types of these converters, but for this book, I recommend using one built into a cable for convenience. I use a PL2303TA-type USB-to-serial cable, like the one shown in Figure 4-2. CP2102 and CP2104 cables are also popular. Install the drivers for your converter cable as instructed by the supplier.

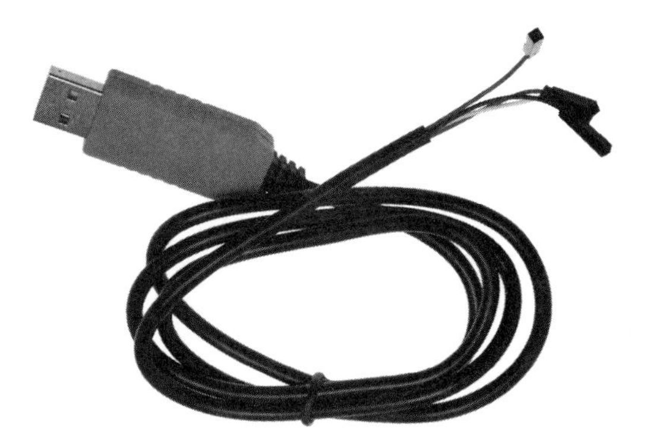

*Figure 4-2: A PL2303TA-type USB-to-serial converter cable*

To interact with your computer via your AVR-based project, you'll also need a *terminal emulator*, a simple program that captures and displays incoming data from your AVR and lets you send data from your computer to the AVR. CoolTerm is an excellent terminal emulator by Roger Meier available for various platforms; you can download it from his website at *http://freeware.the-meiers.org/*. The software is free, but please consider donating via the website to support Meier's efforts.

Once you have downloaded and installed CoolTerm, plug your USB-to-serial converter into your computer, open CoolTerm, and click the **Options** button located at the top of the window. The screen in Figure 4-3 should appear.

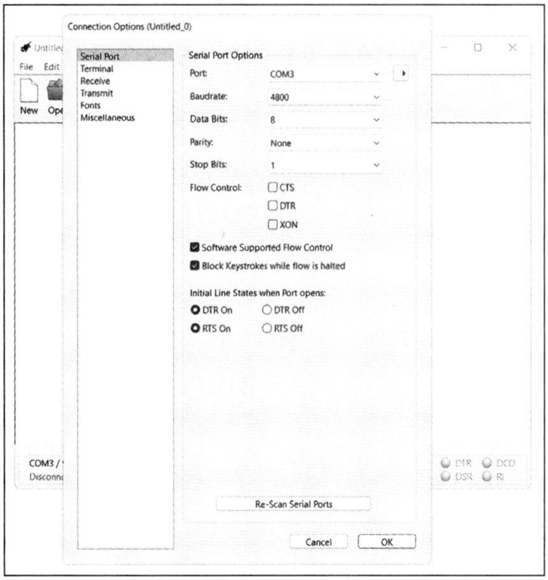

Connection Options (Untitled_0)

Serial Port
Terminal
Receive
Transmit
Fonts
Miscellaneous

Serial Port Options
Port: COM3
Baudrate: 4800
Data Bits: 8
Parity: None
Stop Bits: 1
Flow Control: CTS
DTR
XON
Software Supported Flow Control
Block Keystrokes while flow is halted
Initial Line States when Port opens:
DTR On DTR Off
RTS On RTS Off
Re-Scan Serial Ports
Cancel OK

Figure 4-3: CoolTerm's Serial Port Options configuration screen

I use Windows in my examples throughout this book, but CoolTerm should look similar when running on other platforms. Change the serial port options settings to match those shown in Figure 4-3, except for Port, which will vary depending on your computer—change it to match the name of your USB converter. For example, for Windows PCs, use the drop-down menu beside Port to select the COM port your USB-to-serial converter is using.

Next, select the **Terminal** option from the list on the left, change the settings to match those shown in Figure 4-4, and click **OK**.

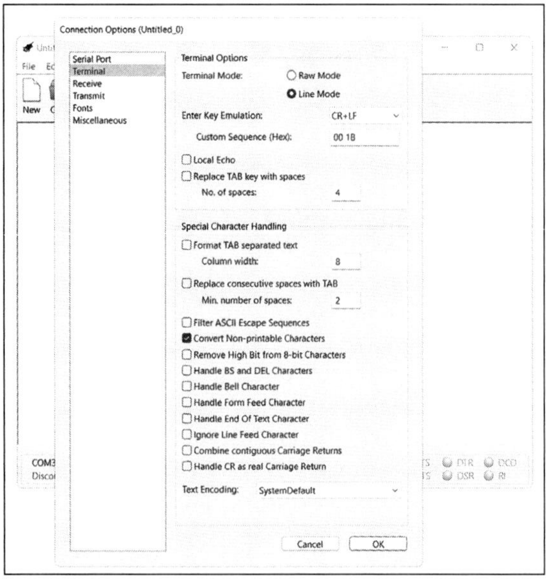

Connection Options (Untitled_0)

Serial Port
Terminal
Receive
Transmit
Fonts
Miscellaneous

Terminal Options
Terminal Mode: Raw Mode
Line Mode
Enter Key Emulation: CR+LF
Custom Sequence (Hex): 00 1B
Local Echo
Replace TAB key with spaces
No. of spaces: 4
Special Character Handling
Format TAB separated text
Column width: 8
Replace consecutive spaces with TAB
Min. number of spaces: 2
Filter ASCII Escape Sequences
Convert Non-printable Characters
Remove High Bit from 8-bit Characters
Handle BS and DEL Characters
Handle Bell Character
Handle Form Feed Character
Handle End Of Text Character
Ignore Line Feed Character
Combine contiguous Carriage Returns
Handle CR as real Carriage Return
Text Encoding: SystemDefault
Cancel OK

Figure 4-4: CoolTerm's Terminal Options configuration screen

Once CoolTerm is configured correctly, the window shown in Figure 4-5 should appear, indicating CoolTerm is ready for use.

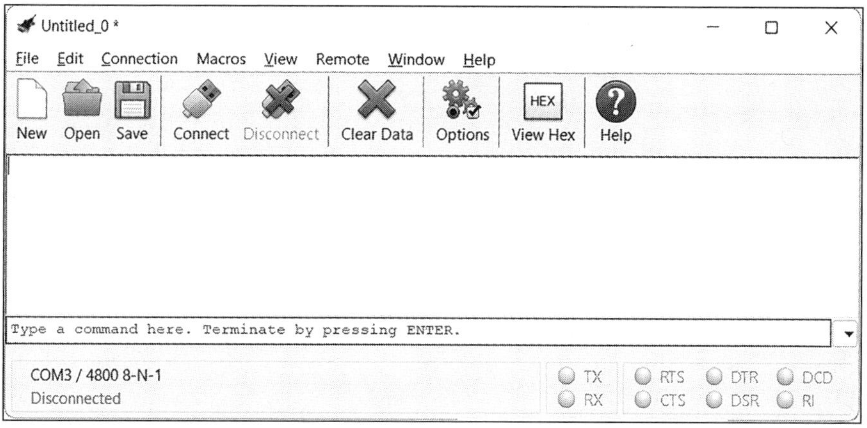

Figure 4-5: CoolTerm ready for use

You're now ready to put your USB-to-serial converter and terminal software to the test.

## Project 16: Testing the USART

In this project you'll use the USART for the first time, testing your USB-to-serial connection and hardware. This will prepare you for more advanced projects in which you send data from your AVR back to the computer to debug your code and check its progress.

### *The Hardware*

For this project, you'll need the following hardware:

- USBasp programmer
- Solderless breadboard
- ATmega328P-PU microcontroller
- USB-to-serial converter
- Jumper wires

Connect the USBasp to your microcontroller via the solderless breadboard, as you have for the previous projects. Next, find the four connections on your USB-to-serial converter: GND, TX, RX, and $V_{CC}$ (or 5 V). Connect the first three pins to the microcontroller as directed in Table 4-1. If you are using the PL2303TA cable shown in Figure 4-2, the white wire is RX and the green wire is TX. If you are using a different model, consult the supplier's instructions to determine the correct wires.

**Table 4-1**: USB-to-Serial Converter to Microcontroller Connections

| USB-to-serial converter pin | ATmega328P-PU pin |
|---|---|
| GND | 8 GND |
| TX | 2 RX |
| RX | 3 TX |

You'll be using this hardware for the next three projects in this chapter, so keep it intact once you've assembled it.

## *The Code*

Open a terminal window, navigate to the *Project 16* subfolder of this book's *Chapter 4* folder, and enter the **`make flash`** command. The toolchain should compile the program files in the subfolder and then upload the data to the microcontroller.

Next, switch over to the terminal software and click the **Connect** button. After a moment, the CoolTerm window should fill with the timeless message `Hello world`, as shown in Figure 4-6.

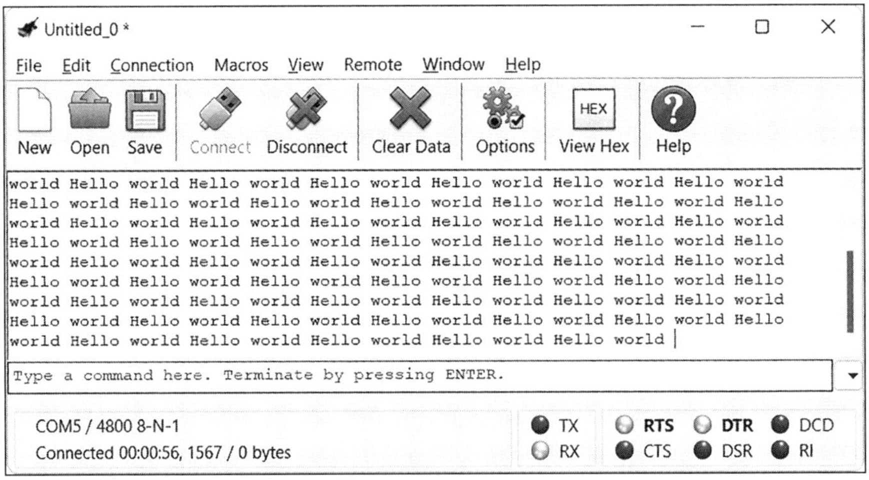

*Figure 4-6: Success! The Project 16 code prints "Hello world" across the terminal.*

To see how this is accomplished, open the *main.c* file located in the *Project 16* subfolder, which contains the following code:

```
// Project 16 - Testing the USART

#include <avr/io.h>
❶ #define USART_BAUDRATE 4800
#define UBRR_VALUE 12

❷ void USARTInit(void)
{
   // Set baud rate registers
   UBRR0H = (uint8_t)(UBRR_VALUE>>8);
```

```
    UBRR0L = (uint8_t)UBRR_VALUE;
    // Set data type to 8 data bits, no parity, 1 stop bit
    UCSR0C |= (1<<UCSZ01)|(1<<UCSZ00);
    // Enable transmission and reception
    UCSR0B |= (1<<RXEN0)|(1<<TXEN0);
}

❸ void USARTSendByte(uint8_t u8Data)
{
    // Wait while previous byte sent
    while(!(UCSR0A&(1<<UDRE0))){};
    // Transmit data
    UDR0 = u8Data;
}
❹ void HelloWorld(void)
{
    USARTSendByte(72);  // H
    USARTSendByte(101); // e
    USARTSendByte(108); // l
    USARTSendByte(108); // l
    USARTSendByte(111); // o
    USARTSendByte(32);  // space
    USARTSendByte(119); // w
    USARTSendByte(111); // o
    USARTSendByte(114); // r
    USARTSendByte(108); // l
    USARTSendByte(100); // d
    USARTSendByte(32);  // space
}
int main(void)
{
    // Initialize USART0
  ❺ USARTInit();
    while(1)
    {
        HelloWorld();
    }
}
```

Before using the USART, you must initialize it and set the data speed (4800bps in this example) ❶. All the initialization code is inside the `USARTInit()` function ❷, which needs to be called once during the main loop in the code ❺.

The `USARTSendByte()` function ❸ sends a byte of data from the USART to your computer. This function waits for the USART to clear old data before sending the new byte of data in the form of an 8-bit integer (the data type `uint8_t`).

Finally, the text "Hello world" is sent using the `HelloWorld()` function ❹. Notice that instead of directly sending letters, we send numbers that each represent a letter. For reference, I've commented in the code which letters each number corresponds to. These numbers are part of *ASCII code*, originally devised for sending messages between telegraph and older communications systems. You can find a copy of the ASCII control code chart at *https://en.wikipedia.org/wiki/ASCII.*

You can experiment with this code by changing the text sent to the computer; just substitute your own ASCII codes in the `USARTSendByte()` function calls. Don't spend too long on this, though, because the next project shows you a better way to transmit text.

Finally, always click **Disconnect** in CoolTerm when you've finished monitoring the USART.

## Project 17: Sending Text with the USART

This project uses the same hardware as Project 16. Open a terminal window, navigate to the *Project 17* subfolder of this book's *Chapter 4* folder, and enter the **`make flash`** command to upload the code from Project 17 as usual.

Next, switch over to the terminal software and click the **Connect** button. After a moment, the screen should once again fill with `Hello, world`—this time in a single column, as shown in Figure 4-7.

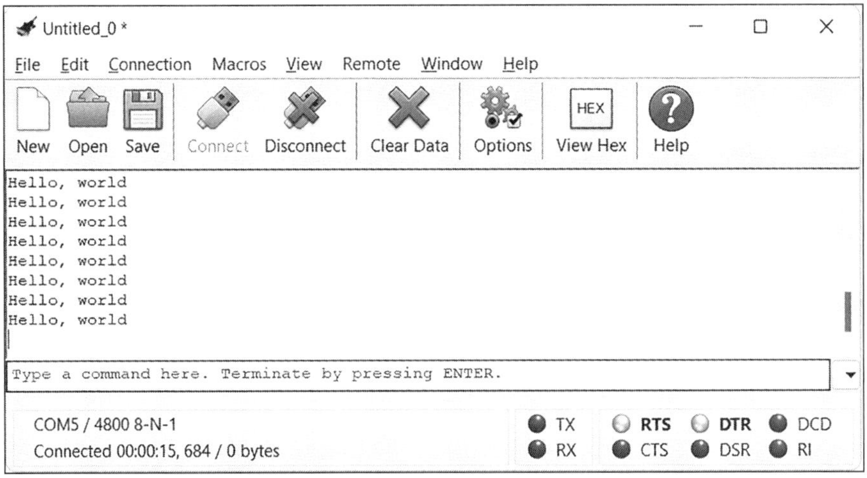

*Figure 4-7: Example result from Project 17*

The code for Project 17 is the same as that for Project 16, except that it except it uses a *character array* to simplify the process of sending text. These arrays store one or more characters, which can be letters, numbers, symbols, and anything else you can generate with your keyboard. They are defined as follows:

```
char i[x] = ""
```

where *x* is the maximum number of characters that can appear in the array (it's always a good idea to include this).

To see how to transmit text in this way, open the *main.c* file located in the *Project 17* subfolder, which contains the following code:

```
// Project 17 - Sending Text with the USART

#include <avr/io.h>
#include <util/delay.h>
#define USART_BAUDRATE 4800
#define UBRR_VALUE 12

void USARTInit(void)
{
   // Set baud rate registers
   UBRR0H = (uint8_t)(UBRR_VALUE>>8);
   UBRR0L = (uint8_t)UBRR_VALUE;
   // Set data frame format to 8 data bits, no parity, 1 stop bit
   UCSR0C |= (1<<UCSZ01)|(1<<UCSZ00);
   // Enable transmission and reception
   UCSR0B |= (1<<RXEN0)|(1<<TXEN0);
}
void USARTSendByte(unsigned char u8Data)
{
   // Wait while previous byte is sent
   while(!(UCSR0A&(1<<UDRE0))){};
   // Transmit data
   UDR0 = u8Data;
}
❶ void sendString(char myString[])
{
   uint8_t a = 0;
   while (myString[a])
   {
      USARTSendByte(myString[a]);
      a++;
   }
}
int main(void)
{
❷ char z[15] = "Hello, world\r\n"; // Make sure you use " instead of ”
   // Initialize USART
   USARTInit();
   while(1)
   {
      sendString(z);
      _delay_ms(1000);
   }
}
```

In the main loop, we define a character array with our `Hello, world` message ❷. The `\r` and `\n` beside that message are *silent control codes*, also known as *escape sequences*, which send information to the terminal software but aren't themselves displayed. `\r` instructs the software to move the cursor to the start of the line and `\n` instructs it to move the cursor down to the next vertical position; the combination `\r\n` thus moves the cursor to the start of the next line in the terminal display, so that the output is printed in an organized column.

We use a new function called `sendString()` ❶ to read each character of the array we defined and send them to the USART one by one, by looping from zero (the first position of an array is always zero) until there are no more characters. During each iteration of the loop, the AVR sends the current byte in the array to the USART.

If you receive an error message after entering your code, such as:

```
main.c:42:2: error: stray '\342' in program
```

that means you're using the wrong kind of quotes to define your character array. Make sure you are using straight quotes (") and not curly quotes (”). You may need to change the auto-correct settings in your text editor to prevent the incorrect quotes from appearing.

In the next project, you'll learn how to send data from the AVR to the terminal software on your computer.

## Project 18: Sending Numbers with the USART

You'll often need to send numbers between your AVR and your computer. For example, you may want to log data generated by your hardware, send output from an interface you created, or just send simple status reports from your AVR when debugging a project. In this project, you'll learn how to send both integers and floating-point numbers.

This project once again uses the same hardware as Project 16. Open a terminal window, navigate to the *Project 18* subfolder of this book's *Chapter 4* folder, and enter the **`make flash`** command to upload the code as usual.

Next, switch over to the terminal software and click the **Connect** button. After a moment the terminal software should display an integer and a floating-point number in turn, as shown in Figure 4-8.

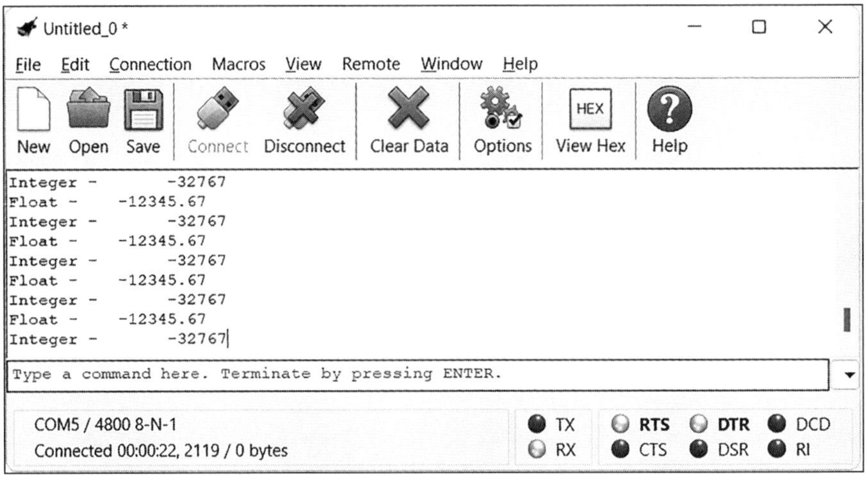

*Figure 4-8: Results of Project 18*

Now open the *main.c* file located in the *Project 18* subfolder, which contains the following code:

```
// Project 18 - Sending Numbers with the USART

#include <avr/io.h>
#include <stdlib.h>
#include <stdio.h>
#include <util/delay.h>
#include <string.h>
#define USART_BAUDRATE 4800
#define UBRR_VALUE 12

void USARTInit(void)
{
   // Set baud rate registers
   UBRR0H = (uint8_t)(UBRR_VALUE>>8);
   UBRR0L = (uint8_t)UBRR_VALUE;

   // Set data frame format to 8 data bits, no parity, 1 stop bit
   UCSR0C |= (1<<UCSZ01)|(1<<UCSZ00);

   // Enable transmission and reception
   UCSR0B |= (1<<RXEN0)|(1<<TXEN0);
}

void USARTSendByte(unsigned char u8Data) // Send a byte to the USART
{
   // Wait while previous byte is sent
   while(!(UCSR0A&(1<<UDRE0))){};
   // Transmit data
   UDR0 = u8Data;
}

void sendString(char myString[])
{
   uint8_t a = 0;
   while (myString[a])
   {
      USARTSendByte(myString[a]);
      a++;
   }
}

int main(void)
{
   char a[10] = "Float - ";
   char b[10] = "Integer - ";
   char t[10] = "";                         // For our dtostrf test
   char newline[4] = "\r\n";
   int16_t  i = -32767;
   float j = -12345.67;

   // Initialize USART
   USARTInit();
```

```
    while(1)
    {
    ❶ dtostrf(j,12,2,t);
       sendString(a);                      // "Float - "
       sendString(t);                      // Send float
       sendString(newline);
       _delay_ms(1000);
    ❷ dtostrf((float)i,12,0,t);
       sendString(b);                      // "Integer - "
       printf(t);                          // Send integer
       sendString(newline);
       _delay_ms(1000);
    }
}
```

This code sends text stored in character arrays to the computer, just like in Project 17. However, it includes some incredibly useful new functions. Before we can send floating-point and integer variables to the USART, we need to convert them into character arrays themselves. We do so with the `dtostrf()` function, included by default in AVR C compilers:

```
dtostrf(float j, w, d, char t[]);
```

This function takes the floating-point number `j` and places it into character array `t[]`. The variable `d` sets the number of decimal places for the fraction, and the variable `w` sets the total number of characters that display the array. We use the `dtostrf()` function in the project code to convert our floating-point variable to a character array before sending it to the USART ❶. To convert an integer to a character array, we use the same function but place the prefix `(float)` before the integer variable ❷.

In the next project, you'll put your newfound knowledge of converting integers and floats into character arrays to good use by sending thermometer data to your PC.

## Project 19: Creating a Temperature Data Logger

In this project you'll send readings from the TMP36 temperature sensor you used in Project 15 in Chapter 3 to your computer. The terminal software will capture the data in a text file, which you can then open in a spreadsheet for further analysis.

### The Hardware

You will need the following hardware:

- USBasp programmer
- Solderless breadboard
- 5 V breadboard power supply

- ATmega328P-PU microcontroller
- One TMP36 temperature sensor
- 0.1 µF ceramic capacitor
- Jumper wires
- USB-to-serial converter

Assemble your circuit as shown in Figure 4-9, then connect your USB-to-serial converter as usual.

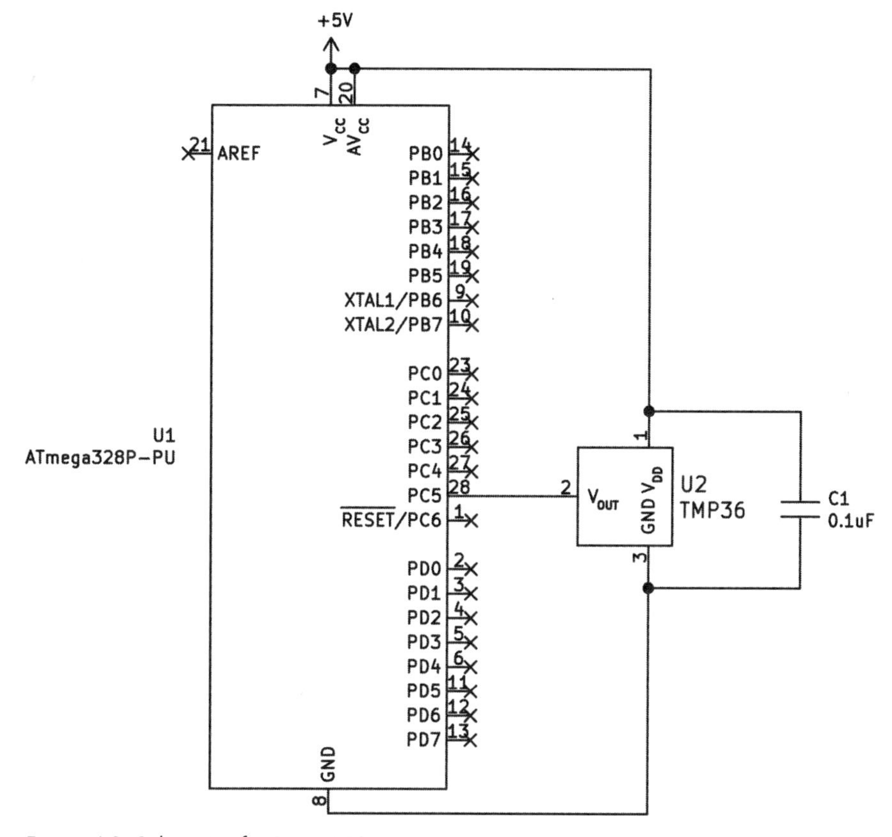


*Figure 4-9: Schematic for Project 19*

Open a terminal window, navigate to the *Project 19* subfolder of this book's *Chapter 4* folder, and enter the `make flash` command to upload the code for Project 19.

Next, switch over to the terminal software and click the **Connect** button. After a moment, the terminal software should begin displaying the ambient temperature as measured by the sensor, as shown in Figure 4-10.

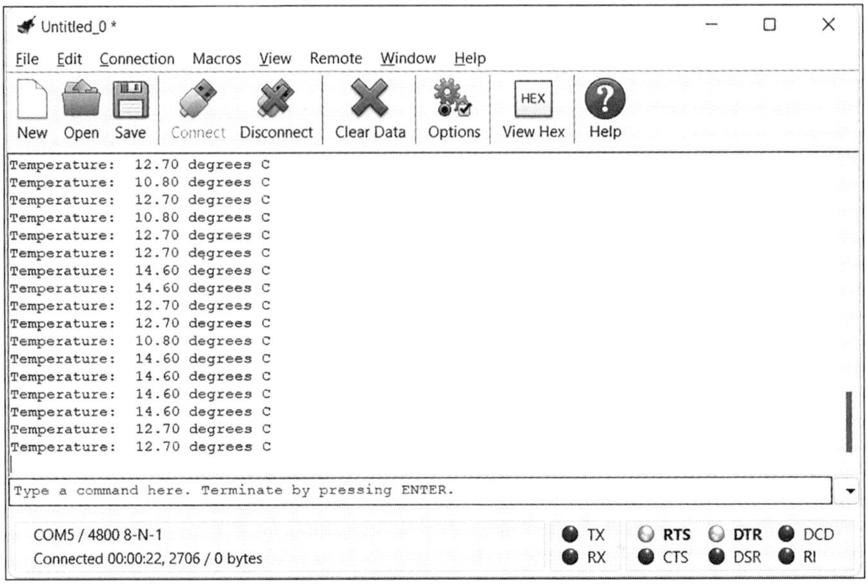

Figure 4-10: The temperature sensor at work

As you can see, this project combines your existing knowledge of the temperature sensor and sending text and numbers to the terminal software via the USART. To show some variance in the example output, I used a small fan to change the airflow around the temperature sensor. This helped change the measured temperature, which is captured once per second.

## The Code

To see how the functions used in this project convert the temperature into an array and send it out, open the *main.c* file located in the *Project 19* subfolder and take a look at the code:

```
// Project 19 - Creating a Temperature Data Logger

#include <avr/io.h>
#include <stdlib.h>
#include <stdio.h>
#include <math.h>
#include <util/delay.h>

#define USART_BAUDRATE 4800
#define UBRR_VALUE 12

❶ void startADC()          // Set up the ADC
{
   ADMUX =  0b01000101; // Set ADC pin to 28
   ADCSRA = 0b10000011; // Set prescaler speed for 1 MHz
}
void USARTInit(void)
{
   // Set baud rate registers
```

```
   UBRR0H = (uint8_t)(UBRR_VALUE>>8);
   UBRR0L = (uint8_t)UBRR_VALUE;

   // Set data frame format to 8 data bits, no parity, 1 stop bit
   UCSR0C |= (1<<UCSZ01)|(1<<UCSZ00);

   // Enable transmission and reception
   UCSR0B |= (1<<RXEN0)|(1<<TXEN0);
}

void USARTSendByte(unsigned char u8Data)
{
   // Wait while previous byte is sent
   while(!(UCSR0A&(1<<UDRE0))){};
   // Transmit data
   UDR0 = u8Data;
}

void sendString(char myString[])
{
   uint8_t a = 0;
   while (myString[a])
   {
      USARTSendByte(myString[a]);
      a++;
   }
}
int main(void)
{
   float temperature;
   float voltage;
   uint8_t ADCvalue;
   char t[10] = "";                  // Will hold temperature for sending via
USART
   char a[14] = "Temperature: ";     // Make sure you have " instead of ”
   char b[14] = " degrees C ";       // Make sure you have " instead of ”
   char newline[4] = "\r\n";

DDRD = 0b00000000; // Set PORTD to inputs
   startADC();
   USARTInit();
   while(1)
   {
      // Get reading from TMP36 via ADC
      ADCSRA |= (1 << ADSC);         // Start ADC measurement
      while (ADCSRA & (1 << ADSC) ); // Wait until conversion is complete
      _delay_ms(10);

    ❷ // Get value from ADC register, place in ADCvalue
      ADCvalue = ADC;
    ❸ // Convert reading to temperature value (Celsius)
      voltage = (ADC * 5);
      voltage = voltage / 1024;
      temperature = ((voltage - 0.5) * 100);
```

```
  ❹ // Send temperature to PC via USART
     sendString(a);
     dtostrf(temperature,6,2,t);
     sendString(t);
     sendString(b);
     sendString(newline);
     _delay_ms(1000);
  }
return 0;
}
```

In the code between ❶ and ❷ the microcontroller's ADCs are initialized, then used. Next, we convert the ADC data to a temperature in degrees Celsius ❸. Finally, we convert the temperature to a character array, which we send to the terminal emulator to create a nice output ❹. This is repeated once per second.

At this point, the terminal emulator software can capture the data received from the microcontroller into a text file, which you can open in a text editor or a spreadsheet for further analysis. To enable this in CoolTerm, select **Connection ▸ Capture to Text/Binary File ▸ Start**, as shown in Figure 4-11.

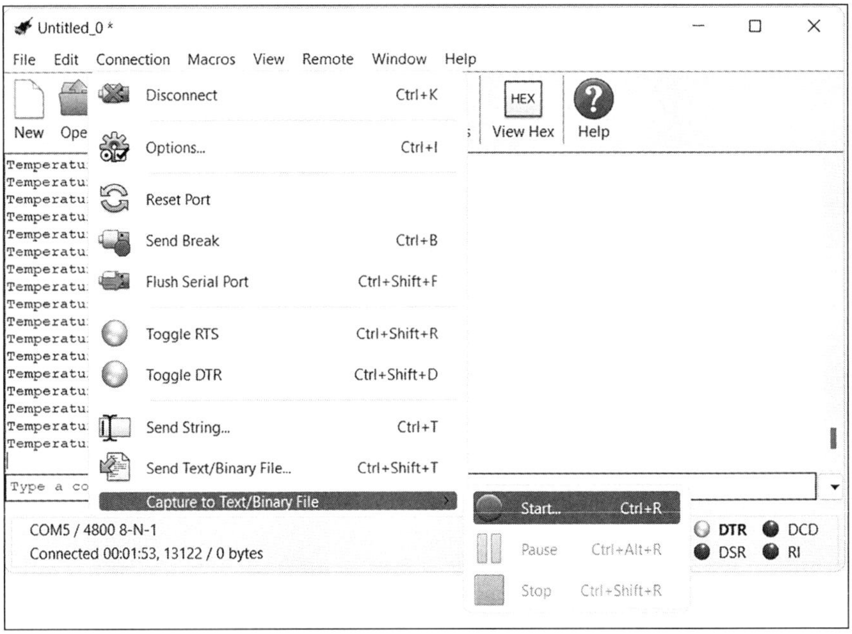

*Figure 4-11: Start recording data from the terminal.*

CoolTerm then asks you to select a location and name for the text file, as shown in Figure 4-12. Once you've done that, click **Save**, and the recording should start. You can pause and finish recording using the Pause and Stop options, respectively.

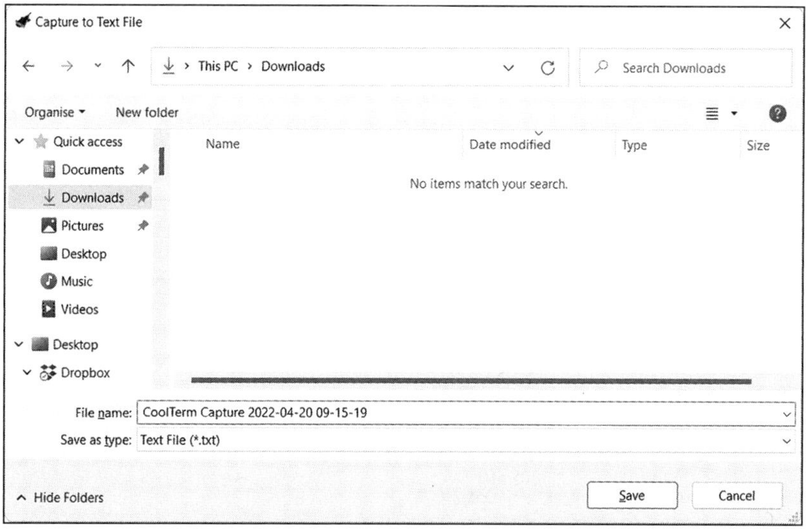

Figure 4-12: Select the filename and location to store the text file.

Once you have captured all the temperature data you want, stop the recording in the terminal emulator software (**Connection ▸ Capture to Text/Binary File ▸ Stop**) and open the resulting text file in your spreadsheet software. For demonstration purposes, I've used Excel. Because this is a text file, you will be prompted to select a *text delimiter*, a single character to insert between data values to allow other software to easily collate the data. Select a space as your delimiter, as shown in Figure 4-13. Note that in the screenshot, "Treat consecutive delimiters as one" is selected; this removes duplicate blank columns if there are double spaces in your file.

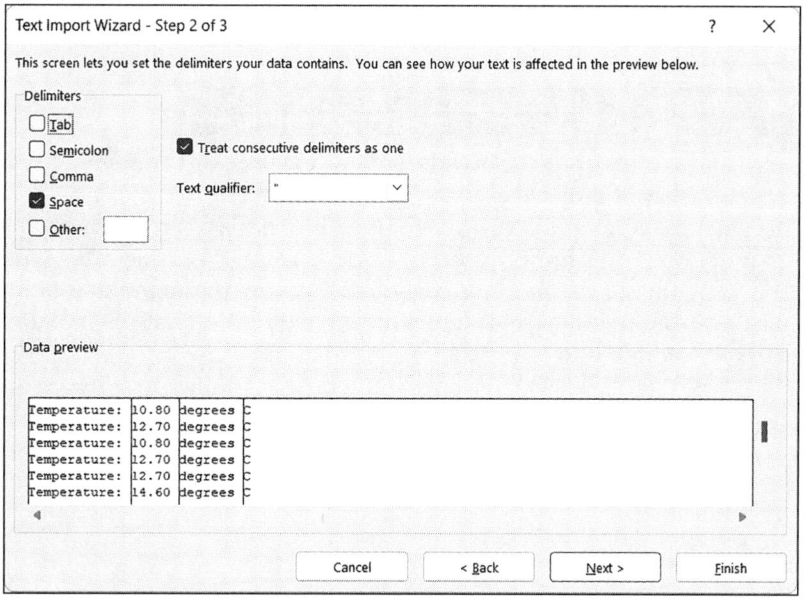

Figure 4-13: Select the delimiter and preview the data.

Click **Next** to import the data. This should create a neat temperature data spreadsheet like the one in Figure 4-14, which you can analyze to your heart's content.

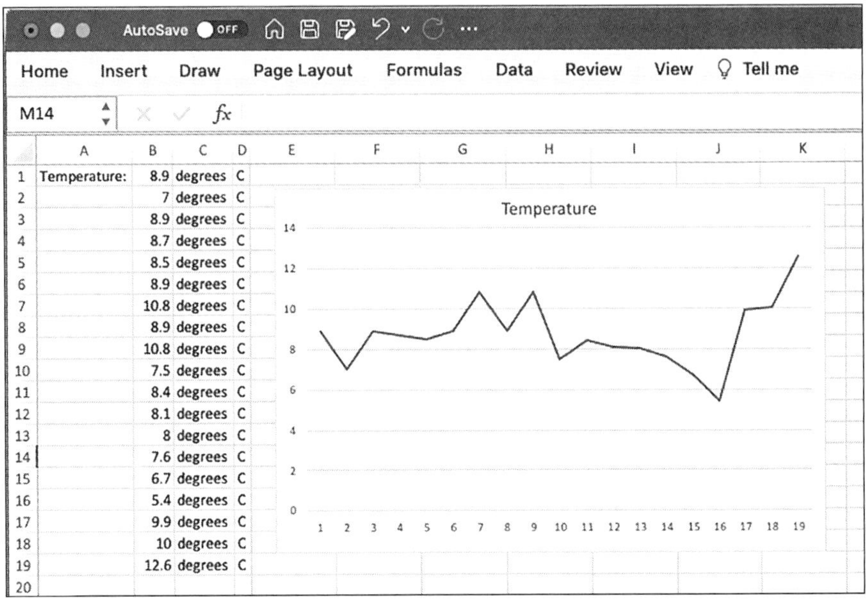

*Figure 4-14: Temperature data ready for analysis*

Although this project uses temperature data, you can log anything sent from the AVR's USART using the method demonstrated here. Keep this in mind if you need to record data in future experiments. For now, we'll turn to our next project: sending data from the terminal software on the computer back to the AVR.

## Project 20: Receiving Data from Your Computer

In this project, you'll learn how to use your computer to control your AVR-based projects or make your own input device by sending data between your computer and the microcontroller in both directions via the USART.

This project uses the same hardware as Project 16. After reproducing that, open a terminal window, navigate to the *Project 20* subfolder of this book's *Chapter 4* folder, and enter the `make flash` command to upload the code for Project 20.

Next, switch over to the terminal emulation software and click the **Options** button, then select **Terminal** from the list on the left-hand side of the window. Set Terminal Mode to **Raw Mode**, as shown in Figure 4-15, then click **OK**, then **Connect**.

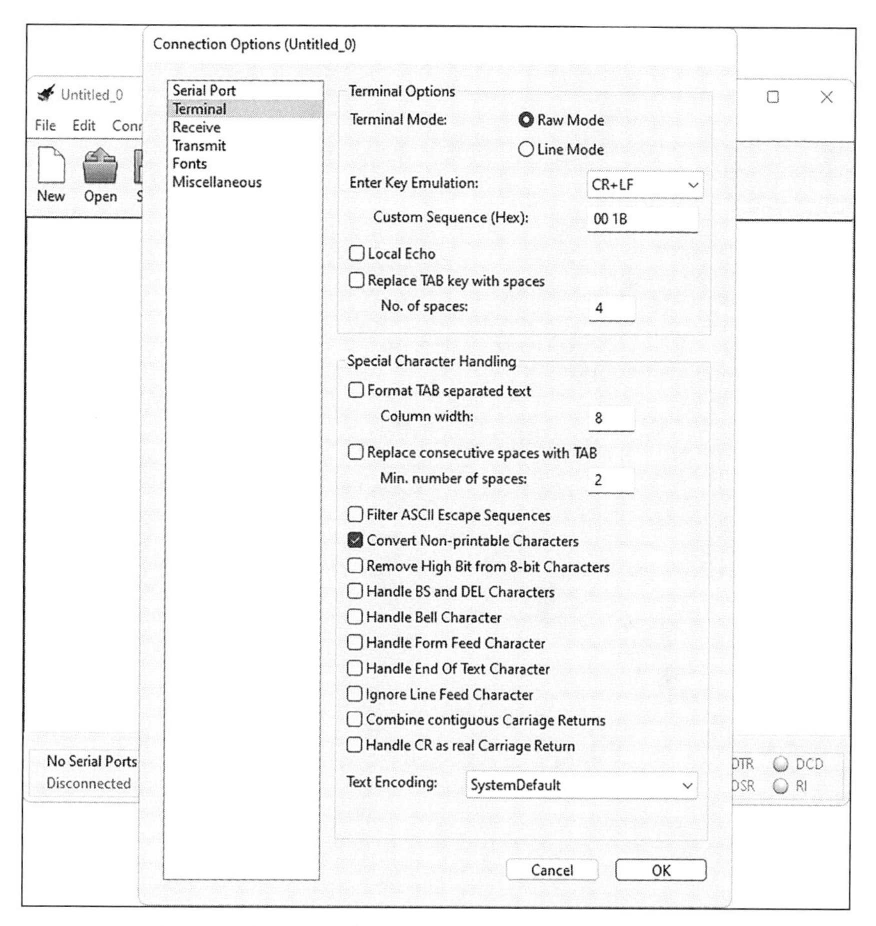


*Figure 4-15: Changing the terminal emulator back to raw mode*

Once you have the terminal emulator in raw mode, type something on your keyboard. Every keystroke you type is sent to the AVR, which then sends it back to the terminal emulator for display. Whatever you type should appear in the terminal window, and the TX and RX indicators should blink.

To see how the USART receives your keystrokes and sends them back out, open the *main.c* file located in the *Project 20* subfolder and take a look at the code:

```
// Project 20 - Receiving Data from Your Computer

#include <avr/io.h>
#include <stdlib.h>
#define USART_BAUDRATE 4800
#define UBRR_VALUE 12

void USARTInit(void)
{
   // Set baud rate registers
   UBRR0H = (uint8_t)(UBRR_VALUE>>8);
   UBRR0L = (uint8_t)UBRR_VALUE;
```

```
    // Set data frame format to 8 data bits, no parity, 1 stop bit
    UCSR0C |= (1<<UCSZ01)|(1<<UCSZ00);

    // Enable transmission and reception
    UCSR0B |= (1<<RXEN0)|(1<<TXEN0);
}

void USARTSendByte(uint8_t sentByte)
{
    // Wait while previous byte is sent
    while(!(UCSR0A&(1<<UDRE0))){};
    // Transmit data
    UDR0 = sentByte;
}

❶ uint8_t USARTReceiveByte()
// Receives a byte of data from the computer into the USART register
{
    // Wait for byte from computer
    while(!(UCSR0A&(1<<RXC0))){};
    // Return byte
    return UDR0;
}

int main(void)
{
    uint8_t tempByte;
    // Initialize USART0
    USARTInit();

    while(1)
    {
        // Receive data from PC via USART
      ❷ tempByte = USARTReceiveByte();

        // Send same data back to PC via USART
      ❸ USARTSendByte(tempByte);
    }
}
```

This code should be quite familiar to you now; it begins with the usual functions to initialize the USART and send bytes to the computer. However, it includes a new function called `USARTReceiveByte()` ❶, which waits for a byte of data to arrive at the USART, then places that data into an integer variable. In this case, the function places the incoming byte into the variable `tempByte` ❷. The `USARTSendByte()` function then sends the same byte of data back to the terminal emulator ❸. It's as simple as that: a byte comes in and is sent back out.

## Project 21: Building a Four-Function Calculator

By now, you've learned to send and receive data between your AVR projects and a computer so that your projects can work with external data and commands. In this project you'll use all the knowledge you've acquired in this and previous chapters to make a simple four-function calculator.

As in the previous projects in this chapter, you'll use the basic AVR and USB-to-serial converter. After reproducing that setup, open a terminal window, navigate to the *Project 21* subfolder of this book's *Chapter 4* folder, and enter the **`make flash`** command to upload this project's code. When you flash the microcontroller with the code, you might receive a warning such as:

```
warning: 'answer' may be used uninitialized in this function
```

That's okay, you can just continue as normal.

Next, open the terminal emulator, make sure the terminal mode is set to raw mode as in Project 20, then click **Connect**. After a moment, the calculator interface should appear in the terminal window, and you should be prompted to enter a command. This calculator can add, subtract, multiply, and divide single-digit numbers. Figure 4-16 shows some examples; have fun and enter your own commands to see the results.

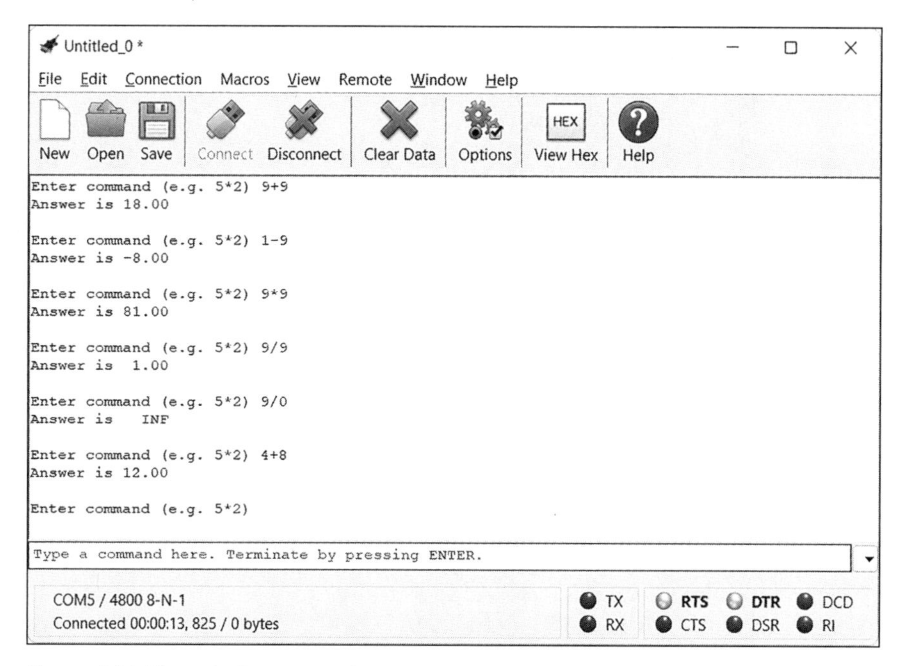


*Figure 4-16: The calculator at work*

Open the *main.c* file in the *Project 21* subfolder. The code is just a sequence of events using functions from previous projects.

```
// Project 21 - Building a Four-Function Calculator

#include <avr/io.h>
#include <stdlib.h>
#include <stdio.h>
#include <string.h>
#include <math.h>
#include <util/delay.h>

#define USART_BAUDRATE 4800
#define UBRR_VALUE 12

void USARTInit(void)
{
   // Set baud rate registers
   UBRR0H = (uint8_t)(UBRR_VALUE>>8);
   UBRR0L = (uint8_t)UBRR_VALUE;
   // Set data frame format to 8 data bits, no parity, 1 stop bit
   UCSR0C |= (1<<UCSZ01)|(1<<UCSZ00);

   // Enable transmission and reception
   UCSR0B |= (1<<RXEN0)|(1<<TXEN0);
}

void USARTSendByte(uint8_t sentByte)
{
   // Wait while previous byte is sent
   while(!(UCSR0A&(1<<UDRE0))){};
   // Transmit data
   UDR0 = sentByte;
}

void sendString(char myString[])
{
   uint8_t a = 0;
   while (myString[a])
   {
      USARTSendByte(myString[a]);
      a++;
   }
}

uint8_t USARTReceiveByte()
// Receives a byte of data from the computer into the USART register
{
   // Wait for byte from computer
   while(!(UCSR0A&(1<<RXC0))){};
   // Return byte
   return UDR0;
}

int main(void)
{
   uint8_t digit1;
```

```
    uint8_t digit2;
    uint8_t operator;
    float answer=0;
    float d1=0;
    float d2=0;

❶  char a[26] = "Enter command (e.g. 5*2) ";
    char b[11] = "Answer is ";
    char answerString[20] = ""; // Holds answer
    char newline[4] = "\r\n";
    USARTInit();
    while(1)
    {
       sendString(newline);
       sendString(a);

    ❷ digit1 = USARTReceiveByte();
    ❸ USARTSendByte(digit1);
    ❹ switch (digit1)          // Convert ASCII code of digit1 to actual
number
       {
          case 48 : digit1 = 0; break;
          case 49 : digit1 = 1; break;
          case 50 : digit1 = 2; break;
          case 51 : digit1 = 3; break;
          case 52 : digit1 = 4; break;
          case 53 : digit1 = 5; break;
          case 54 : digit1 = 6; break;
          case 55 : digit1 = 7; break;
          case 56 : digit1 = 8; break;
          case 57 : digit1 = 9; break;
       }
    ❺ operator = USARTReceiveByte();
    ❻ USARTSendByte(operator);

    ❼ digit2 = USARTReceiveByte();
    ❽ USARTSendByte(digit2);
    ❾ switch (digit2)          // Convert ASCII code of digit2 to actual
number
       {
          case 48 : digit2 = 0; break;
          case 49 : digit2 = 1; break;
          case 50 : digit2 = 2; break;
          case 51 : digit2 = 3; break;
          case 52 : digit2 = 4; break;
          case 53 : digit2 = 5; break;
          case 54 : digit2 = 6; break;
          case 55 : digit2 = 7; break;
          case 56 : digit2 = 8; break;
          case 57 : digit2 = 9; break;
       }
       sendString(newline);

       // Convert entered numbers into float variables
       d1 = digit1;
```

```
        d2 = digit2;

        // Calculate result
        switch (operator)
        {
           case 43 : answer = d1 + d2; break; // Add
           case 45 : answer = d1 - d2; break; // Subtract
           case 42 : answer = d1 * d2; break; // Multiply
           case 47 : answer = d1 / d2; break; // Divide
        }

        // Send result to PC via USART
     ❿ sendString(b);
        dtostrf(answer,6,2,answerString);
        sendString(answerString);
        sendString(newline);
        _delay_ms(1000);
     }
     return 0;
}
```

In this code, we first initialize the variables inside the `main()` function, then initialize the USART with the lines beginning at ❶. The program prompts the user to enter a command consisting of 3 bytes of data: the first digit, the operator, and then the second digit. The USART receives the first digit ❷, the operator ❺, and the second digit ❼ and sends them back to the terminal to give visual feedback at ❸, ❻, and ❽, respectively.

When the user enters a digit, the terminal emulator sends the ASCII code for the digit, not the digit itself, to the AVR. The program then converts the ASCII code into the actual digit ❹ and places it in an integer variable ❾. This same process for converting ASCII codes to digits also determines which operator has been entered (for example, +, -, *, or /).

The program then performs the required calculation on the two digits via the functions in the code's last `switch()` function. Finally, the calculation's result is converted to a character array and sent back to the terminal emulator with `sendString(b)` for the user to read ❿. The calculator is now ready for another calculation.

The projects in this chapter showed you how to use your computer as a terminal with your AVR to send and receive data, preparing you to record and analyze data. In the next chapter I'll show you how to use interrupts, a neat way to let your AVR to respond to input whenever it occurs, instead of at a planned moment in time.

# TAKING CONTROL WITH HARDWARE INTERRUPTS

So far, the code for your projects in this book has been sequential. Any deviations from a linear pattern, such as detecting button presses, required you to monitor digital inputs. However, preplanning button presses in your code isn't always efficient or realistic. *Hardware interrupts* allow your programs to respond to events more efficiently and dynamically.

Hardware interrupts enable the AVR microcontroller to react to a change of state at a digital input pin at any time. In a way, they allow your AVR to multitask: when a button is pressed or when a signal is received at a digital input, the AVR will stop what it is doing and run some other code, called an *interrupt service routine (ISR)*. After the ISR code runs, the AVR picks up execution where it left off before the interrupt.

Interrupts allow you to write more logical code and make your AVR-based projects operate more intuitively. This chapter covers two kinds of hardware interrupts, external interrupts and pin-change interrupts, using the ATmega328P-PU microcontroller. Both interrupts are triggered by state changes in pins (for instance, a change from high to low voltage). Pin-change interrupts can occur on all the pins, while external interrupts can only happen on two pins. Once you've got the basics down, you'll use interrupts to create a counting device with a USART-based display.

## External Interrupts

This section walks you through the basics of initializing external interrupts, which you'll put to use in Project 22. The ATmega328P-PU uses pins 4 and 5, referenced as INT0 and INT1, for external interrupts. These two pins can detect and report changes in the electrical signal connected to them. The interrupts can be set to react to one of the following four possible changes in state at the pin:

**Low level** The voltage at the pin changes to a low state (equivalent to GND).

**Any logic change** The voltage at the pin changes in any way, either from high to low or from low to high.

**Falling edge** The voltage changes from high to low.

**Rising edge** The voltage changes from low to high.

Consider which of these options would allow your code to respond to a button press. For example, if you had a button connected to GND and a microcontroller input with a pullup resistor, then the button press would switch the input from high to low. In this situation, you would typically use the falling edge interrupt option. Conversely, if you had a button connected between 5 V and the microcontroller input with a pulldown resistor, the button press would switch the input from low to high. In this case you could use a rising edge or any logic change interrupt.

Overall, the choice of interrupt type will be determined by the external circuitry connected to the interrupt pin. I'll demonstrate various types of interrupts in action in this chapter, so you can use these examples to help you determine what's right for your own projects.

### *Setting Up Interrupts in Code*

To use interrupts, first add the interrupt library as follows:

```
#include <avr/interrupt.h>
```

Next, you'll need to set up a few registers—we'll step through them one by one below. The first of these is the EICRA register, which we use to determine which of the four state changes the pin will respond to. Here's the template for setting the EICRA register in any AVR program:

```
EICRA = 0b0000abcd;
```

Set interrupt INT0 (pin 4) with bits *c* and *d* and interrupt INT1 (pin 5) with bits *a* and *b*, using the guidelines in Table 5-1.

**Table 5-1**: EICRA Register Option Bits

| Bit a/c | Bit b/d | Interrupt type |
|---|---|---|
| 0 | 0 | Low level |
| 0 | 1 | Any logic change |
| 1 | 0 | Falling edge |
| 1 | 1 | Rising edge |

For example, you could set up INT0 for a rising edge interrupt as follows, using the *c* and *d* bits:

```
EICRA = 0b00000011; // INT0 rising edge
```

Next, set up the EIMSK register, which is used to turn on the interrupt function, as follows:

```
EIMSK = 0b000000ab;
```

Here, bit *a* is INT1 and bit *b* is INT0. Set each bit to 1 for on or 0 for off. For example, to turn the interrupt function on for INT0, use:

```
EIMSK = 0b00000001;
```

Since only the last bit is set to 1, only INT0 will be turned on.

After setting the EICRA and EIMSK registers, enable interrupts in your code using this function call:

```
sei();
```

Don't forget this step—if you skip it, even if you've set your registers properly, the interrupt won't be triggered.

Here's a recap of the code required for interrupts so far:

```
#include <avr/interrupt.h> // Enable the interrupt library
EICRA = 0b0000abcd;        // Determine which state changes the interrupt pin responds to
EIMSK = 0b000000ab;        // Turn on the required interrupt
                sei();     // Enable interrupts in your code
```

Once you've prepared your AVR to respond to interrupts, you must define your ISR—the code that runs when the interrupt is triggered. The ISR is a custom function with the following structure:

```
ISR (INTx_vect)
{
   // Code to be executed when an interrupt is triggered
   EIFR = y;
}
```

The parameter `INTx_vect` specifies the pin that the ISR responds to. When you pass a value to the function, replace the *x* in `INTx_vect` with `0` for INT0 or `1` for INT1. The body of the function is the code that executes when the interrupt is triggered. We always end this ISR code section with this command:

```
EIFR = y;
```

This sets the external interrupt flag registers back to zero, telling the microcontroller that the interrupt code for this particular ISR is complete and the AVR can return to normal operation. If you are using only one interrupt, you can use the following values for *y*: `0b00000001` for INT0 and `0b00000010` for INT1. However, if your project is using both interrupts, if you set the EIFR register with an entire 8-bit value, you will alter both interrupts. Instead, you can use the following, which will just turn off one interrupt:

```
EIFR &= ~(1<<0); // Set interrupt flag register for INT0 to zero
EIFR &= ~(1<<1); // Set interrupt flag register for INT1 to zero
```

This way of addressing register bits individually will be explained in detail in the next chapter. In the meanwhile, let's put pin-change interrupts to the test in the following project.

## Project 22: Experimenting with Rising Edge Interrupts

In this project, you'll first program your microcontroller to rapidly blink an LED, then add logic so that pressing a button interrupts the blinking LED and runs some other code, keeping the LED on for two seconds before it returns to its blinking pattern. You'll accomplish this using a rising edge interrupt.

### *The Hardware*

For this project, you'll need the following hardware:

- USBasp programmer
- Solderless breadboard
- ATmega328P-PU microcontroller
- Jumper wires
- One LED
- One 560 Ω resistor
- One pushbutton
- One 10 kΩ resistor

First, assemble the circuit shown in Figure 5-1.

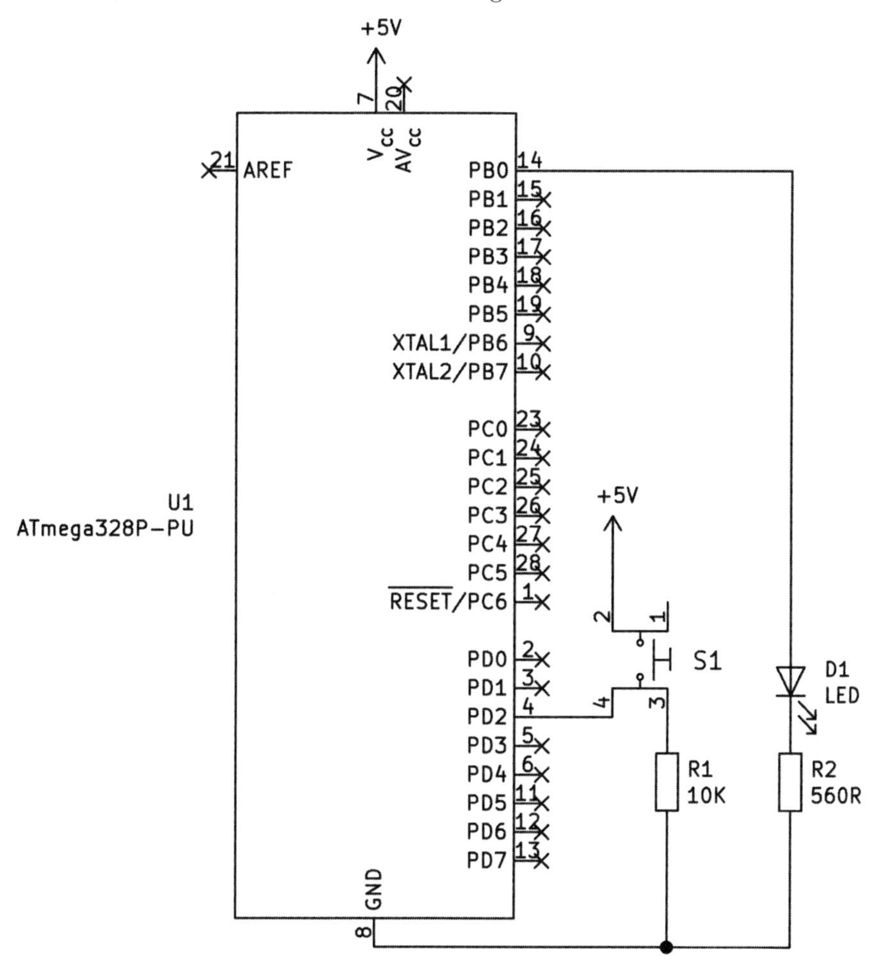


*Figure 5-1: The main circuit for Project 22*

The resistor and button are in the pulldown configuration that I introduced in Chapter 3. When you press the button the current will flow to pin 4, changing its state from low to high. This rising edge state change will trigger the interrupt.

With your circuit assembled, connect the USBasp to your microcontroller via the solderless breadboard, as you did in previous projects.

## The Code

Open a terminal window and navigate to the *Project 22* subfolder of this book's *Chapter 5* folder, then enter the command `make flash`. The toolchain should compile the program file and upload the data to the microcontroller, at which point the LED should begin to blink rapidly, turning on and off every 50 milliseconds. Press the button quickly (don't leave your finger on it for long, since this will trigger switch bouncing), and the LED should stay on for two seconds, then resume blinking.

Open the *main.c* file for Project 22 to see how this code works:

```
// Project 22 - Experimenting with Rising Edge Interrupts

// Blink PORTB. If button pressed, turn on PORTB for 2 seconds.
#include <avr/io.h>
#include <util/delay.h>
#include <avr/interrupt.h>

ISR (INT0_vect)
{                        // Code to be executed when interrupt is triggered
   PORTB = 0b11111111;
   _delay_ms(2000);
❶ PORTB = 0b00000000;
❷ EIFR = 0b00000001; // Clear external interrupt flag register
}

void startInt0()
{
   // Initialize interrupt 0 (PD2/INT0/pin 4)
   // Rising edge (LOW to HIGH at pin 4)
❸ EICRA = 0b00000011;
   // Turn on interrupt INT0
❹ EIMSK = 0b00000001;
// Turn on global interrupt enable flag in order for interrupts to be
processed
   sei();
}

int main(void)
{
   // Declare global variables
   // Set up GPIO pins etc.
❺ DDRB = 0b11111111; // Set PORTB register as outputs
❻ DDRD = 0b00000000; // Set PORTD pins 4 and 5 as inputs

   // Initialize interrupt
   startInt0();

   for(;;)
❼ {
      // Blink LED connected to PB7 (pin 10)
      PORTB = 0b00000001;
      _delay_ms(50);
      PORTB = 0b00000000;
      _delay_ms(50);
   }
   return 0;
}
```

This code defines a `startInt0()` function to initialize the interrupts. Within this function, we first set the EICRA register so that INT0 reacts to a rising edge interrupt ❸, then set the EIMSK register to turn INT0 on ❹, and finally call `sei()` to enable interrupts. In the main section of the code, we set up PORTB pins as outputs to control the LED ❺ and set PORTD pins as inputs ❻. PORTD includes digital pin 4, which will act as input for the interrupt INT0.

Once everything is initialized, the for loop ❼ makes the LED blink on and off. Because there's an interrupt, when you press the button, the resulting rising edge triggers the hardware interrupt INT0. This tells the AVR to stop blinking the LED and run the code in the ISR ❶. When the ISR code has finished, the EIFR register is set to 0 ❷ and the LED returns to rapidly blinking as normal.

Congratulations! You've just seen the most common kind of interrupt in action. Less often, you'll need to detect when current stops flowing to the microcontroller, which you can do using a falling edge interrupt. You'll try this out in the next project.

## Project 23: Experimenting with Falling Edge Interrupts

This project has the same result as Project 22, but this time the circuit uses the pullup configuration introduced in Chapter 3. By default, there will be current at digital pin 4. When you press the button, current should stop flowing to pin 4, changing its state from high to low to trigger the falling edge interrupt.

### *The Hardware*

You'll need the following hardware:

- USBasp programmer
- Solderless breadboard
- ATmega328P-PU microcontroller
- Jumper wires
- One LED
- One 560 Ω resistor
- One pushbutton
- One 10 kΩ resistor

Start by assembling the circuit shown in Figure 5-2.

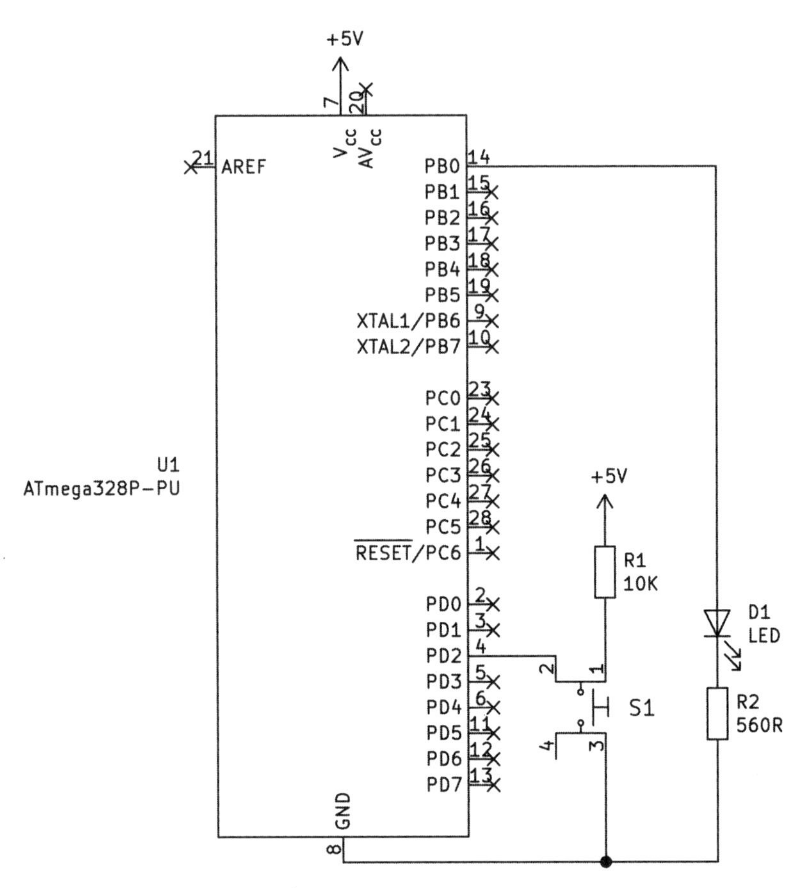


*Figure 5-2: The main circuit for Project 23*

With your circuit assembled, connect the USBasp to your microcontroller via the solderless breadboard as you did in previous projects.

## *The Code*

Open a terminal window, navigate to the *Project 23* subfolder of this book's *Chapter 5* folder, and enter the command **make flash**. Once the project's code has been uploaded, the LED should begin to blink rapidly. Quickly press the button. The LED should stay on for two seconds, then resume blinking.

To see how this works, open the *main.c* file for Project 23:

```
// Project 23 - Experimenting with Falling Edge Interrupts

// Blink PORTB. If button is pressed, turn on PORTB for 2 seconds.
#include <avr/io.h>
#include <util/delay.h>
#include <avr/interrupt.h>

ISR (INT0_vect) {       // Code to be executed when interrupt is triggered
   PORTB = 0b11111111;
```

```
    _delay_ms(2000);
    PORTB = 0b00000000;
    EIFR = 0b00000001;
}

void startInt0()
{
    // Initialize interrupt 0 (PD2/INT0/pin 4)
    // Falling edge (HIGH to LOW at pin 4)
❶   EICRA = 0b00000010;
    // Turn on interrupt INT0
    EIMSK = 0b00000001;
    // Turn on global interrupt enable flag for interrupts to be processed
    sei();
}

int main(void)
{
    // Declare global variables
    // Set up GPIO pins etc.
    DDRB = 0b11111111; // Set PORTB register as outputs
    DDRD = 0b00000000; // Set PORTD pins 4 and 5 as inputs

    // Initialize interrupt
    startInt0();

    for(;;)
    {
      PORTB = 0b00000001;
      _delay_ms(50);
      PORTB = 0b00000000;
      _delay_ms(50);
    }
    return 0;
}
```

The code for this project is identical to that of Project 22, except for one change: we set the EICRA register to `0b00000010` instead of `0b00000011` ❶. Per Table 5-1, the last two bits (`10`) set INT0 to the falling edge type of interrupt. EIMSK stays the same, since we're still using INT0 as the interrupt pin, and as usual we call `sei()` to enable the interrupt.

Play with these projects to familiarize yourself with both rising and falling edge interrupts. Once you feel comfortable using both options for digital inputs, you can choose whether to trigger an interrupt with a high or low signal in your own projects, depending on how the circuit you're working with is built. This may sound like a trivial choice now, but it'll become important as you create more complicated AVR-based projects. In some situations, you can't choose the hardware or the circuit, and you'll have to work around their limitations using code.

Now let's try something more interesting. As I mentioned earlier, the ATmega328P-PU has two interrupt pins. In the next project, you'll use both to make the microcontroller respond to two different interrupts.

## Project 24: Experimenting with Two Interrupts

This project uses two buttons to allow you to trigger two interrupts for different responses. One button triggers a rising edge interrupt and turns on the LED for one second, while the other triggers a falling edge interrupt that turns on the LED for two seconds.

### The Hardware

You'll need the following hardware:

- USBasp programmer
- Solderless breadboard
- ATmega328P-PU microcontroller
- Jumper wires
- One LED
- One 560 Ω resistor
- Two pushbuttons
- Two 10 kΩ resistors

Assemble the circuit shown in Figure 5-3.

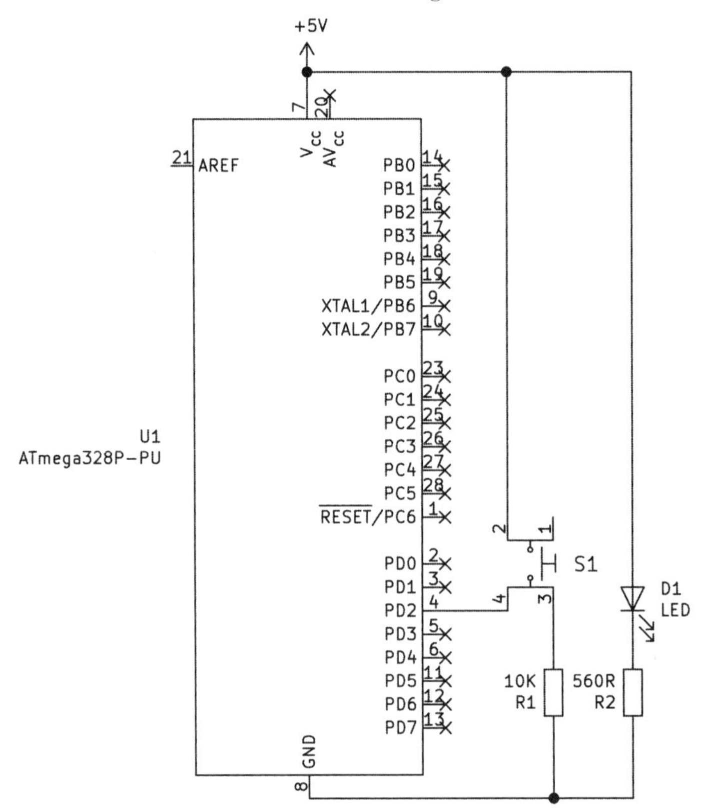


*Figure 5-3: The main circuit for Project 24*

With your circuit assembled, connect the USBasp to your microcontroller via the solderless breadboard as in the previous projects.

### The Code

Open a terminal window, navigate to the *Project 24* subfolder of this book's *Chapter 5* folder, and enter the command **make flash**. As in the previous two projects, once the code has been uploaded the LED should start to blink rapidly. Press the button connected to INT0 (digital pin 4), and the LED should stay on for one second. Press the button connected to INT1 (digital pin 5), and the LED should stay on for two seconds.

To see how the code handles these interrupts, open the *main.c* file for Project 24:

```
// Project 24 - Experimenting with Two Interrupts

// PORTB blinks, INT0 rising interrupt, INT1 falling interrupt
#include <avr/io.h>
#include <util/delay.h>
#include <avr/interrupt.h>

ISR (INT0_vect)
{                       // Code to be executed when interrupt INT0 is triggered
   PORTB = 0b11111111;
   _delay_ms(1000);
   PORTB = 0b00000000;
   EIFR &= ~(1<<0);    // Set interrupt flag register for INT0 to zero
}

ISR (INT1_vect)
{                       // Code to be executed when interrupt INT1 is triggered
   PORTB = 0b11111111;
   _delay_ms(2000);
   PORTB = 0b00000000;
   EIFR &= ~(1<<1);    // Set interrupt flag register for INT1 to zero
}

void startInts()
{
   // Initialize interrupt 0 (PD2/INT0/pin 4)
   // Rising edge (LOW to HIGH at pin 4)
   // Initialize interrupt 1 (PD3/INT1/pin 5)
   // Falling edge (HIGH to LOW at pin 5)
❶ EICRA = 0b00001011;

   // Turn on interrupts INT0 and INT1
❷ EIMSK = 0b00000011;
   // Turn on global interrupt enable flag for interrupts to be processed
   sei();
}

int main(void)
{
   // Declare global variables
```

```
    // Set up GPIO pins etc.
❸ DDRB = 0b11111111; // Set PORTB register as outputs
❹ DDRD = 0b00000000; // Set PORTD pins 4 and 5 as inputs
    // Initialize interrupts
    startInts();

    for(;;)
❺ {
        PORTB = 0b00000001;
        _delay_ms(50);
        PORTB = 0b00000000;
        _delay_ms(50);
    }
    return 0;
}
```

This code is similar to the code for Projects 22 and 23, again with a little modification. As usual, we start by defining a `startInts()` function to initialize the interrupts. Inside this function, we set the EICRA register to respond to each interrupt ❶. Remember that the register is set with the formula `0b0000`*`abcd`*, where bits *a* and *b* correspond to pin 5 and bits *c* and *d* correspond to pin 6. Here, we've set EICRA so that INT0 reacts to a rising edge interrupt and INT1 reacts to a falling edge interrupt (`0b00001011`). Next, we set the EIMSK register to turn on both INT0 and INT1 by setting the last two bits to 1 ❷, then we call `sei()` to enable interrupts.

In the main section of the code, we set up PORTB as outputs so that the code will control the LED ❸. We also set up PORTD as inputs to cover digital pins 4 and 5, which will act as inputs for INT0 and INT1, respectively ❹. Once everything is initialized, the code in the `for` loop ❺ will cause the LED to blink on and off. However, when one of the buttons is pressed, the corresponding interrupt will trigger and the code in the interrupt's ISR will run.

When planning projects with multiple interrupts, remember that an interrupt cannot be called by another interrupt. That is, if one interrupt's ISR code is running, triggering another interrupt will not affect the operation of that ISR. You can't interrupt an interrupt!

What if you need to use more than two interrupt pins, or can't use digital pins 4 and 5 but still need your project to respond to state change triggers? The solution is to use pin-change interrupts.

## Pin-Change Interrupts

While using the external interrupts INT0 and INT1 is simple, straightforward, and gives you lots of control, working with just two interrupts is a huge limitation. Using *pin-change interrupts* is a little trickier, but gives you as many interrupts as you have pins.

Pin-change interrupts can only tell you if a pin has changed state—they can't provide any details about that change of state. While an external interrupt can detect a change from low to high or high to low, a pin-change interrupt can only detect that a change happened. That means any change in state will trigger the interrupt, and you'll have to decide in your code if you want to respond to that change.

The pins used for pin-change interrupts are organized into three banks:

**Bank 0** Includes PCINT0 through PCINT7.

**Bank 1** Includes PCINT8 through PCINT14.

**Bank 2** Includes PCINT16 through PCINT23.

PCINT stands for *pin-change interrupt*. Each bank has its own ISR code, giving you three types of level changes that you can use with their respective pins. This gives you more flexibility, but it means you can't assume that each ISR function corresponds to a single pin. Rather, each ISR function responds to any state change in any of the pins in its corresponding bank. Thus, your code will have to determine not only what the state change was but which pin it came from before it can respond appropriately. Note that there is no PCINT15.

**NOTE** *PCINT numbers are not the same as pin numbers: for example, PCINT8 is pin 23, not pin 8. Refer to the pinout diagram in Figure 5-4 for each PCINT value's pin number, and make sure not to confuse the two numbers in your code. Each bank's byte in binary represents the order of pins in the bank, from highest to lowest.*

You can match the three PCINT*xx* banks to physical pin numbers using the ATmega328P-PU's pinout diagram, shown in Figure 5-4.

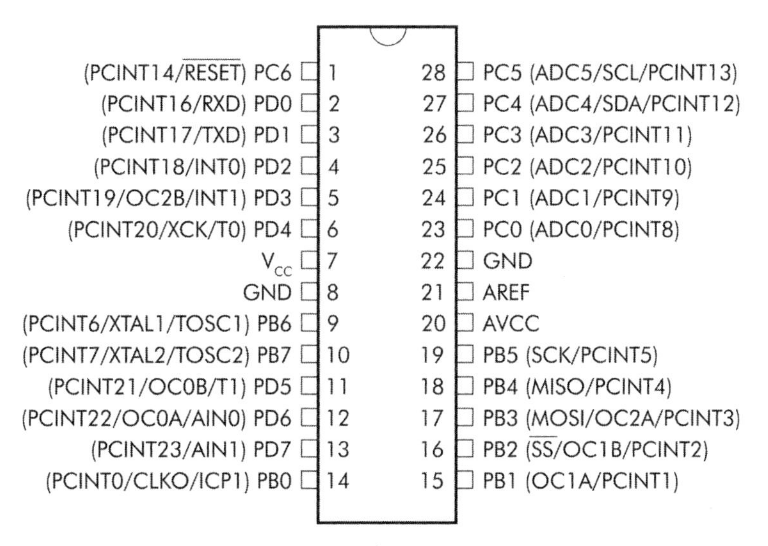

*Figure 5-4: ATmega328P-PU pinout diagram*

For example, Bank 0 responds to PCINT0–PCINT7, meaning you can trigger it with pins 15–19 at the bottom right of Figure 5-4 and pins 9, 10, and 14 at the bottom left.

Using pin-change interrupts in your code is similar to using external interrupts. Once again, first include the interrupt library.

```
#include <avr/interrupt.h>
```

Next, set the PCICR register to turn on the required PCI (pin-change interrupt) banks, using this formula:

```
PCICR = 0b00000xyz;
```

Bank 0 is set with bit *z*, bank 1 with bit *y*, and bank 2 with bit *x*. For example, to turn on banks 0 and 2, you would use:

```
PCICR = 0b00000101;
```

Remember that each bank can respond to several pins, so you'll need to select which pins in each of the banks can be used for an interrupt by turning on the interrupt function corresponding to each pin you select. Each bank has its own interrupt function that you can turn on with the registers PCMSK0, PCMSK1, and PCMSK2. Pins are turned on or off with a 1 or 0 in that pin's spot. To use pin 15 in bank 0, pin 23 in bank 1, and pin 13 in bank 2, you'd set the PCMSK*x* registers like this:

```
PCMSK0 = 0b00000010; // We'll use PCINT1 (pin 15) for bank 0 ...
PCMSK1 = 0b00000001; // and use PCINT8 (pin 23) for bank 1 ...
PCMSK2 = 0b10000000; // and use PCINT23 (pin 13) for bank 2
```

Once you've selected the pins in each bank that should trigger an interrupt, enable the interrupts with the line:

```
sei();
```

Finally, define your ISR. Each PCI has its own ISR with the following structure:

```
ISR (PCINTx_vect)
{
   // Code to be executed when interrupt is triggered
   EIFR &= ~(1<<x); // Set interrupt flag register to zero
}
```

Replace the *x* in `PCINTx_vect` with `0` for bank 0, `1` for bank 1, and `2` for bank 2, and you're ready to add your interrupt code. We always end this ISR code section with the following command:

```
PCIFR = y;
```

This sets the pin-change interrupt flag register back to zero, which tells the microcontroller that the interrupt code for that particular bank has completed and that it can return to running the code in the main loop as normal. You can use the following values for *y* in order to set the pin-change interrupt flag: `0b00000001` for bank 0, `0b00000010` for bank 1, and `0b00000100` for bank 2.

You'll put pin-change interrupts to the test in the following project.

## Project 25: Experimenting with Pin-Change Interrupts

This project expands on the previous ones, using three buttons, in conjunction with a pin on each of the three PCI banks, to demonstrate how to use pin-change interrupts. Pressing each button should trigger a different interrupt, which should turn the LED on for a certain period of time.

### The Hardware

You'll need the following hardware:

- USBasp programmer
- Solderless breadboard
- ATmega328P-PU microcontroller
- Jumper wires
- One LED
- One 560 Ω resistor
- Three pushbuttons
- Three 10 kΩ resistors

To begin, assemble the circuit shown in Figure 5-5.

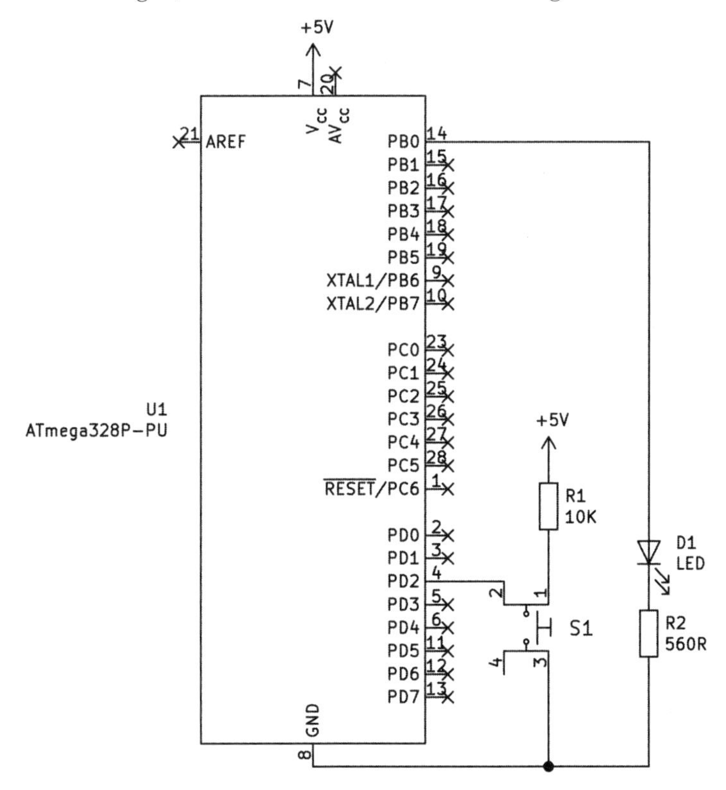


*Figure 5-5: The main circuit for Project 25*

With your circuit assembled, connect the USBasp to your microcontroller via the solderless breadboard as you have for the previous projects.

### The Code

Open a terminal window, navigate to the *Project 25* subfolder of this book's *Chapter 5* folder, and enter the command **make flash** to upload the project's code, as usual. Once the code has uploaded, the LED should start to blink rapidly. Pressing the different buttons should turn the LED on for one, two, or three seconds, as defined in the ISR for each pin-change interrupt bank.

To see how this works, open the *main.c* file for Project 25:

```
// Project 25 - Experimenting with Pin-Change Interrupts

#include <avr/io.h>
#include <util/delay.h>
#include <avr/interrupt.h>

ISR (PCINT0_vect)
{  // Code to be executed when PCI bank 0 PCINT1 pin 15 is triggered
   PORTB = 0b11111111;
   _delay_ms(1000);
   PORTB = 0b00000000;
   PCIFR = 0b00000001;
}

ISR (PCINT1_vect)
{  // Code to be executed when PCI bank 1 PCINT8 pin 23 is triggered
   PORTB = 0b11111111;
   _delay_ms(2000);
   PORTB = 0b00000000;
   PCIFR = 0b00000010;
}

ISR (PCINT2_vect)
{  // Code to be executed when PCI bank 2 PCINT23 pin 13 is triggered
   PORTB = 0b11111111;
   _delay_ms(3000);
   PORTB = 0b00000000;
   PCIFR = 0b00000100;
}

void startInts()
{
❶ PCICR = 0b00000111;  // Activate all three PCIs
   PCMSK0 = 0b00000010; // We'll use PCINT1 (pin 15) for bank 0 ...
   PCMSK1 = 0b00000001; // and use PCINT8 (pin 23) for bank 1 ...
   PCMSK2 = 0b10000000; // and use PCINT23 (pin 13) for bank 2
   sei();
}

int main(void)
{
   // Set up GPIO pins etc.
```

```
❷ DDRB = 0b11111101;  // Set up PORTB register (pin 15 input, rest outputs)
  DDRC = 0b00000000;  // Set up PORTC register (all inputs)
  DDRD = 0b01111111;  // Set up PORTD register (pin 13 input, rest outputs)

  // Initialize interrupts
  startInts();

  for(;;)
  {
     // Blink LED connected to PB7 (pin 10)
  ❸ PORTB = 0b00000001;
     _delay_ms(50);
     PORTB = 0b00000000;
     _delay_ms(50);
  }
  return 0;
}
```

This code has the same structure as the code for the external interrupt projects. First, we define a `startInts()` function to initialize the interrupts. Inside this function, we set the PCICR register to enable all three banks of PCIs ❶; then we set the physical pins to use as each bank's interrupt pin using the three following lines and make a call to `sei()` to enable interrupts.

In the main section of the code, we set up PORTB and PORTD so that the LED pin is an output and PORTC as inputs for the interrupt pins ❷. Once the initializations have taken place, the code in the `for` loop ❸ will make the LED blink on and off. However, when you press one of the buttons, you'll trigger the interrupt for the corresponding PCI bank, running the code in that bank's matching ISR. We end the code run for each interrupt by setting the PCIFR flag to 1 for that particular bank.

At this point, you've seen how to use interrupts with digital inputs to activate code on demand when required by the user. To finish off this chapter's experiments, I'll show you how to use interrupts in a more practical situation.

## Project 26: Creating an Up/Down Counter Using Interrupts

This project combines what you've learned about interrupts with sending data to your computer via the USART (covered in Project 18 in Chapter 4). You'll build a counting device that uses two buttons to accept user input: one button increases the count by one, and the other decreases it. Each button will trigger a rising edge interrupt and call a matching ISR to add to or subtract from the counter's tally.

### The Hardware

You'll need the following hardware:

- USBasp programmer
- Solderless breadboard

- ATmega328P-PU microcontroller
- USB to serial converter
- Jumper wires
- Two pushbuttons
- Two 10 kΩ resistors

Assemble the circuit shown in Figure 5-6.

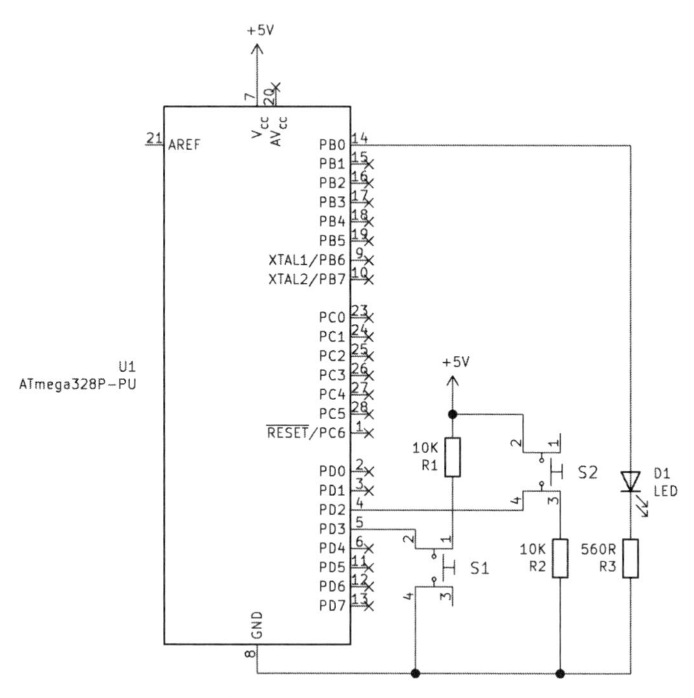


*Figure 5-6: Schematic for Project 26*

With your circuit assembled, connect the USBasp to your microcontroller via the solderless breadboard as in the previous projects. Next, connect your USB-to-serial converter to your computer, just as you did in Chapter 4.

## The Code

Open a terminal window, navigate to the *Project 26* subfolder of this book's *Chapter 5* folder, and enter the command `make flash` to upload the project's code as usual. Now run your terminal software on your computer, as you did in Chapter 4, and click the **Connect** button. After a moment, the terminal software should display the value of the counting variable.

Try pressing each button. The counter value should increase or decrease with each button press, giving you output like that shown in Figure 5-7.

```
Count - 106
Count - 105
Count - 104
Count - 104
Count - 103
Count - 103
Count - 102
Count - 102
Count - 102
Count - 102
Count - 102
Count - 101
Count - 100
Count - 99
Count - 98
Count - 96
Count - 95
Count - 94
Count - 95
Count - 94
Count - 95
Count - 97
Count - 97
Count - 98
Count - 99
Count - 100
Count - 102
Count - 102
Count - 102
```

Figure 5-7: Our counter in action

In this figure, the counter doesn't always appear to increase or decrease by just one. That's because the count is only updated every second, and you can press a button many times in the span of one second.

To see how this is implemented, open the *main.c* file for Project 26:

```
// Project 26 - Creating an Up/Down Counter Using Interrupts

#include <avr/io.h>
#include <stdlib.h>
#include <util/delay.h>
#include <avr/interrupt.h>

#define USART_BAUDRATE 4800
#define UBRR_VALUE 12

❶ volatile uint8_t i = 100; // Initial value for counter
void USARTInit(void)
{
   // Set baud rate registers
   UBRR0H = (uint8_t)(UBRR_VALUE>>8);
   UBRR0L = (uint8_t)UBRR_VALUE;

   // Set data frame format to 8 data bits, no parity, 1 stop bit
```

```
    UCSR0C |= (1<<UCSZ01)|(1<<UCSZ00);

    // Enable transmission and reception
    UCSR0B |= (1<<RXEN0)|(1<<TXEN0);
}

void USARTSendByte(uint8_t u8Data)
{
    // Wait while previous byte is sent
    while(!(UCSR0A&(1<<UDRE0))){};
    // Transmit data
    UDR0 = u8Data;
}

void sendString(char myString[])
{
    uint8_t a = 0;
    while (myString[a])
    {
        USARTSendByte(myString[a]);
        a++;
    }
}

❷ ISR (INT0_vect)
{                     // Code to be executed when interrupt INT0 is triggered
    i = i - 1;        // Subtract one from the counter
    EIFR &= ~(1<<0); // Set interrupt flag register for INT0 to zero
}

❸ ISR (INT1_vect)
{                     // Code to be executed when interrupt INT1 is triggered
    i = i + 1;        // Add one to the counter
    EIFR &= ~(1<<1); // Set interrupt flag register for INT1 to zero
}

❹ void startInts()
{
    // Initialize interrupt 0 (PD2/INT0/pin 4)
    // Rising edge (LOW to HIGH at pin 4)
    // Initialize interrupt 1 (PD3/INT1/pin 5)
    // Rising edge (LOW to HIGH at pin 5)
    EICRA = 0b00001111;

    // Turn on interrupts INT0 and INT1
    EIMSK = 0b00000011;
    // Turn on global interrupt enable flag
    sei();
}

int main(void)
{
    char a[10] = "Count - "; // Make sure you have " instead of ”
    char s[10] = "";          // For our itoa() conversion used in the main loop
    char newline[] = "\r\n";
```

```
    // Set up pins 4 and 5 as inputs for INT0 and INT1
    DDRD = 0b00000000;

    // Initialize interrupts
    startInts();

    // Initialize USART
    USARTInit();

    for(;;)
    {
        // Send the value of our counter to the USART for display on the PC
        itoa(i, s, 10);
        sendString(a);
        sendString(s);
        sendString(newline);
       _delay_ms(1000);
    }
    return 0;
}
```

This project demonstrates how to use interrupts to receive user input, using much less code than if we had to check for a button press in every cycle of the main code. First we set up the library initialization and functions required to use the USART. Then we declare the `i` variable ❶, which stores the value of the counter, followed by the convenient functions used in the last chapter to send text and numbers via the USART to the PC.

Both the main code and the ISR functions need to be able to access the `i` variable, which is why it's defined outside the `int main(void)` section. Declaring a variable outside the main code makes it a *global variable*, which means any part of the code can access it, not just the code in the particular function in which the variable could be declared. When you declare a global variable, you should place the `volatile` keyword before its data type to let the compiler know that it could change at any time, so the microcontroller needs to reload it from memory every time the program uses it.

This project uses pins 4 and 5 as external rising edge interrupts, so next we define the code to run when the INT0 ❷ and INT1 ❸ interrupts are triggered and initialize them using EICRA and EIMSK ❹. Once the main code starts running, it should send the value of the counter variable to your PC via the USART every second. Thanks to the power of the interrupts (and the buttons connected to them), each time you trigger one of the interrupts, the relevant ISR code should add or subtract one to or from the counter variable, depending on which button is pressed. You can test this by pressing either button on and off rapidly; as the variable changes, the updated value should display in your terminal.

## Final Notes on Interrupts

Working with hardware interrupts can give your AVR-based projects more options to complete tasks on demand, instead of during a preprogrammed sequence of events. To simplify this introduction to interrupts, this chapter focused on creating circuits that react to button presses. However, in real-world projects, you'll more often program interrupts to respond to limit switches in machinery or signals from sensors to help your project make a decision.

When using interrupts, always declare the pins used to trigger interrupts as inputs using a `DDRx` function, or the microcontroller won't detect the trigger. Also declare any variables used in both your main code and ISRs as `volatile` and make them global variables.

The next chapter's projects expand on interrupts, showing you how to use them to run functions after a preset period of time.

# 6 USING HARDWARE TIMERS

We use timers to determine when we want a period of time to elapse before an action takes place, which can be incredibly useful. For example, you can set an interrupt to trigger when a timer reaches a certain value. Timers operate in the background; while the microcontroller runs your code, the timers are counting away.

In this chapter, you'll learn about:

- Various timers in your ATmega328P-PU microcontroller
- Timer overflow interrupts
- Clear Timer on Compare Match interrupts

I'll show you how to run parts of code on a regular basis, create longer delays for repetitive actions, and examine the accuracy of internal timers. You'll also learn a more efficient method of addressing individual bits inside a register.

## Introducing Timers

Both our AVR microcontrollers have several timers, each of which contains an incrementing counting variable whose value is stored in a *counter register*. Once a counter reaches its maximum value, a bit in a register changes, and the counter resets to zero and starts over. Beyond using timers to trigger interrupts, you can use them to measure elapsed time with some clever arithmetic based on the incrementing variable's progress.

The ATmega328P-PU has three timers—TIMER0, TIMER1, and TIMER2—with their own counter registers. TIMER0 and TIMER2 are 8-bit counters with a maximum value of 255. TIMER1 is a 16-bit counter and has a maximum value of 65,535. The ATtiny85 also has timers, but as the ATmega is more versatile given its higher number of I/O pins, we'll only discuss the latter in this chapter.

Timers need a *clock source* to count accurate periods of time. A clock source is an oscillator circuit with an output that changes between high and low at a precise frequency. You can use either the internal or an external clock source. In this chapter we'll use the microcontroller's internal clock source, and I'll show you how to use an external clock source when this is necessary in later chapters.

Up to this point, our microcontrollers have been running at a speed of 1 MHz, and you can use their internal clock sources to drive the timers. We determine the period of time between each increment of the timer's counter with this simple formula:

$$T = 1 / f$$

where $f$ is frequency in Hz and $T$ is time in seconds. For example, we calculate the period at 1 MHz as $T = 1 / 1{,}000{,}000$, which results in a value of one millionth of a second, known as one microsecond.

You can adjust the length of the period by using a *prescaler*, a number used to divide the frequency to increase the period time. You use the prescaler when you need to measure amounts of time that exceed the default duration of one of the timers. Five prescalers are available: 1, 8, 64, 256, and 1,024.

To calculate a period altered by a prescaler, we use the following formula:

$$T = 1 / (1{,}000{,}000 / p)$$

where $T$ is time in seconds and $p$ is the prescaler value. We can then determine the length of time before a given register resets. For example, to determine the length of time elapsed before reset for TIMER1, you would multiply the resulting value of *T* for your chosen prescaler by the maximum value of the TIMER1 counter (65,535). If your prescaler is 8, your time per period is 0.00008 seconds, so you'd multiply 65,535 by 0.00008 to get 0.52428 seconds. This means TIMER1 will reset after 0.52428 seconds.

I've already calculated the values for the TIMER1 counter, for your convenience; they're listed in Table 6-1.

**Table 6-1**: Prescaler Values for the TCCR1B Register at 1 MHz and Their Period Times

| Prescaler type | Period (s) | Bit 2 | Bit 1 | Bit 0 |
|---|---|---|---|---|
| /1 (none) | 0.000001 | 0 | 0 | 0 |
| /8 | 0.000008 | 0 | 1 | 0 |
| /64 | 0.000064 | 0 | 1 | 1 |
| /256 | 0.000256 | 1 | 0 | 0 |
| /1024 | 0.001024 | 1 | 0 | 1 |

That's enough theory for now. In the following projects we'll put timers to work, to increase your understanding.

## Project 27: Experimenting with Timer Overflow and Interrupts

In the first of our timer demonstrations, you'll learn how to trigger an ISR once a timer counter overflows, using TIMER1. You'll also experiment with prescalers to alter the length of time before the counter resets.

### *The Hardware*

For this project, you'll need the following hardware:

- USBasp programmer
- Solderless breadboard
- ATmega328P-PU microcontroller
- Jumper wires
- Two LEDs
- Two 560 Ω resistors

Assemble the circuit as shown in Figure 6-1.

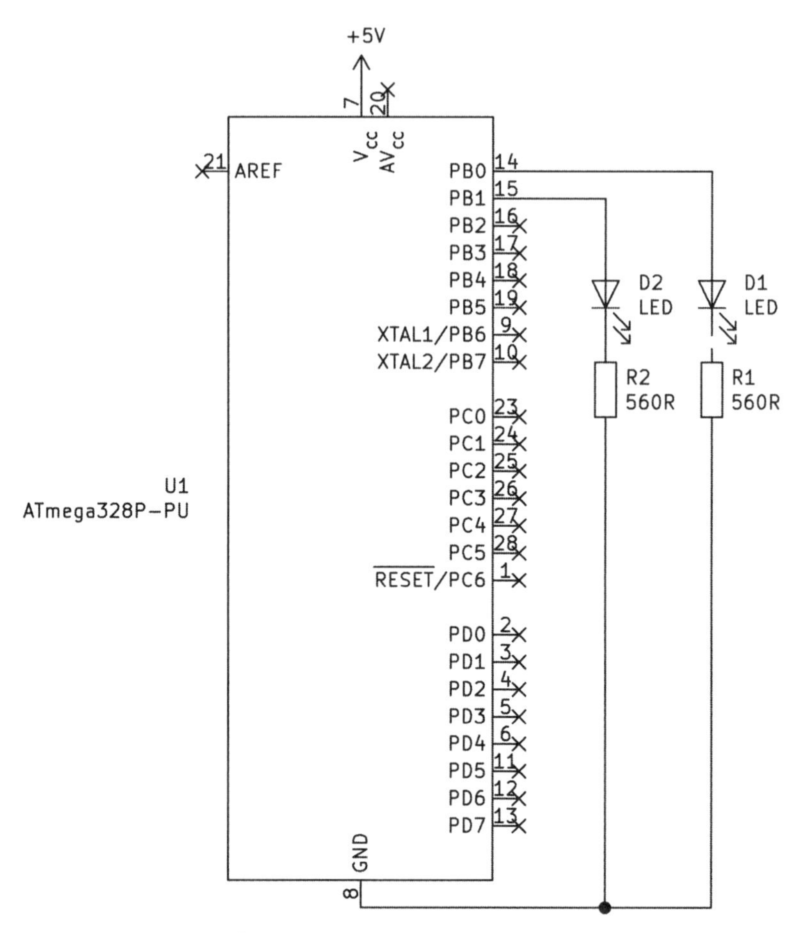


*Figure 6-1: Schematic for Project 27*

With your circuit assembled, connect the USBasp to your microcontroller via the solderless breadboard in the same way you have for the previous projects. Once completed, keep your circuit together, as you'll use it for the next project as well.

## The Code

Open a terminal window, navigate to the *Project 27* subfolder of this book's *Chapter 6* folder, and enter the command `make flash`. The toolchain should compile the program file and upload the data to the microcontroller as usual. At this point the LED connected to PB0 should blink rapidly, and the LED connected to PB1 should quickly flicker on and off about every half-second (every 0.52428 seconds, to be exact!).

Let's see how this works. Open the *main.c* file for Project 27:

```
// Project 27 - Experimenting with Timer Overflows and Interrupts

#include <avr/io.h>
```

```
❶ #include <avr/interrupt.h>
  #include <util/delay.h>

❷ ISR(TIMER1_OVF_vect)
  {
     // Code to be executed when an interrupt is triggered from TIMER1 overflow.
     // For this example, quickly blink LED on PB1.
     PORTB = 0b00000010;
     _delay_ms(5);
     PORTB = 0b00000000;
  }

  void initOVI()
  // Set up overflow interrupt and TIMER1
  {
   ❸ TCCR1B = 0b00000010; // Set CS10 and CS11 for /8 prescaler
   ❹ TIMSK1 = 0b00000001;  // Turn on TIMER1 interrupt on overflow
   ❺ sei();                // Turn on global interrupts
  }

  int main(void)
  {
     DDRB = 0b11111111;    // Set PORTB register as outputs
     initOVI();            // Set up overflow interrupt and TIMER1

     for(;;)               // Do something (such as blink LED on PB0)
     {
        PORTB = 0b00000001;
        _delay_ms(100);
        PORTB = 0b00000000;
        _delay_ms(100);
     }

     return 0;
  }
```

This code includes the function `initOVI()` to initialize TIMER1 for use. First, we include the library for interrupts ❶ and define the timer operations ❷—this is the code that runs when the timer resets. We then set the prescaler to 8 by setting the second bit of the TCCR1B register ❸. This causes the TIMER1 register to reset every 0.52428 seconds. Next, we set the TIMSK1 register with 1 as the first bit ❹ to enable an interrupt to be called every time the TIMER1 counter overflows and resets to initialize the timer operations we defined earlier, and call `sei()` to enable interrupts ❺.

Once operating, the LED should blink on and off as instructed in `int main(void)`, and the TIMER1 counter will count away at 125 kHz (remember, our clock speed is 1 MHz, and we're using a prescaler of 8), so each counter increment takes 0.000008 seconds. With such a tiny length of time for each count, it only takes 0.52428 seconds to count from 0 to 65,535, at which point the TIMER1 counter overflows and the code calls the interrupt code ❷, which blinks the other LED briefly. TIMER1 resets to zero and starts counting again.

Though this code sets the prescaler to 8 using the TCCR1B register, you can also select other prescalers by setting bits 2, 1, and 0 of the register using the values shown in Table 6-1. Spend some time changing the bits for the TCCR1B register with your Project 27 hardware to experiment with the way this impacts the timing.

In the next project, I'll show you how to run a section of code on a recurring, regular basis.

## Project 28: Using a CTC Timer for Repetitive Actions

*Clear Timer on Compare Match (CTC)* is a different method of timing that calls an ISR once a timer's counter has reached a certain value, then resets the timer to zero and starts it counting again. The CTC timing mode is useful when you want to run a section of code on a recurring basis.

In this project, you'll learn how to trigger an ISR every time a counter reaches 15 seconds, again using TIMER1. To determine the duration value, first calculate the number of elapsed periods per second for the timer, using the values in Table 6-1. We'll use the 1,024 prescaler (if you need a longer duration, you can use an appropriate prescaler). This gives us 14,648 periods (rounded down), to which we add 1 to account for the time required for the timer to be reset back to zero. Our code should now check the TIMER1 counter value. Once it reaches 14,649, the code calls the ISR and then resets the counter to zero.

Use the same hardware for this project as for Project 27. With your circuit assembled, connect the USBasp to your microcontroller via the solderless breadboard in the same way you have for the previous projects. Again, keep the circuit together when you're done so you can use it in the next project.

Open a terminal window, navigate to the *Project 28* subfolder of this book's *Chapter 6* folder, and enter the command **make flash**. Once the project's code has been uploaded to the microcontroller the LED connected to PB0 should start to blink rapidly, and the LED connected to PB1 should briefly turn on and off every 15 seconds.

Let's see how this works. Open the *main.c* file for Project 28:

```
// Project 28 - Using a CTC Timer for Repetitive Actions

#include <avr/io.h>
#include <avr/interrupt.h>
#include <util/delay.h>

❶ ISR(TIMER1_COMPA_vect)
{
   // Code to be executed when an interrupt is triggered from TIMER1 overflow.
   // For this example, quickly blink LED on PB1.
 ❷ TCNT1 = 0;
   PORTB = 0b00000010;
   _delay_ms(10);
   PORTB = 0b00000000;
```

```
    // Reset TIMER1 to zero, so counting can start again.
    TCNT1 = 0;
}

❸ void initCTC()
// Set up CTC interrupt and TIMER1
{
 ❹ OCR1A = 14649;       // Number of periods to watch for: 14,649
    // Turn on CTC mode and set CS12 and CS10 for /1024 prescaler
 ❺ TCCR1B = 0b00000101;
 ❻ TIMSK1 = 0b00000010; // Turn on timer compare interrupt
    sei();               // Turn on global interrupts
}

int main(void)
{
    DDRB = 0b11111111;   // Set PORTB register as outputs
    initCTC();           // Set up overflow interrupt and TIMER1
    for(;;)              // Do something (such as blink LED on PB0)
    {
        PORTB = 0b00000001;
        _delay_ms(100);
        PORTB = 0b00000000;
        _delay_ms(100);
    }
    return 0;
}
```

This code includes an `initCTC()` function ❸, which we use to set up our timer. We tell the code to run the ISR when the timer reaches the value 14,649 by setting OCR1A to `14649` ❹. Then we set the prescaler to 1,024 ❺ and turn on the timer compare interrupt feature ❻.

The main code begins by running the `initCTC()` function, then merrily blinks the LED connected to PB0. Once the TIMER1 counter reaches 14,649 (our 15-second mark), the ISR code ❶ will run. Inside the ISR, the code first resets TIMER1 to zero ❷, then blinks the LED connected to PB1.

You should now understand how to execute an ISR after a set period of time. Experiment with prescalers and values for some practice, then we'll move on to using CTC for longer delays in code.

## Project 29: Using CTC Timers for Repetitive Actions with Longer Delays

Sometimes you'll want to set up a recurring event over a longer period than the one in Project 28—perhaps every 15 minutes rather than every 15 seconds. Due to the size of the OCR1A register (65,535) we can't just enter a very large number to count up to a long period of time and expect the CTC timer to work, so we need to use a small workaround. We set up a CTC timer as per Project 28 that triggers the ISR once per second. We then count those seconds, and when the desired delay has elapsed, we call a function to execute the required code.

In more detail, to set up a longer period between recurring events, we do the following:

1. Use a global variable to store the target delay value we wish to use for our period (measured in seconds).
2. Use another global variable to store the number of elapsed seconds in the delay.
3. Set the CTC timer to watch for durations of one second.
4. Have the ISR (called every second) add one to the elapsed seconds variable, then check that it has reached the target delay value—and if so, execute the required code.

With this approach you can implement a variation of multitasking, as you'll see in the following project.

Use the same hardware as in Project 28. With your circuit assembled, connect the USBasp to your microcontroller via the solderless breadboard in the same way you have for the previous projects.

Next, open a terminal window, navigate to the *Project 29* subfolder of this book's *Chapter 6* folder, and enter the command **make flash**. The tool chain should compile the program file and then upload the data to the microcontroller. At this point the LED connected to PB0 should blink rapidly, and the LED connected to PB1 should briefly turn on and off every second.

Let's see how this works. Open the *main.c* file for Project 29:

```
// Project 29 - Using CTC Timers for Repetitive Actions with Longer Delays
#include <avr/io.h>
#include <avr/interrupt.h>
#include <util/delay.h>

❶ uint16_t target = 1;
// Interval in seconds between running function "targetFunction"
❷ uint16_t targetCount = 0;
// Used to track number of seconds for CTC counter resets

❸ void targetFunction()
{
   // Do something once target duration has elapsed
   PORTB = 0b00000010;
   _delay_ms(100);
   PORTB = 0b00000000;
}

ISR(TIMER1_COMPA_vect)
{ // Code to be executed when an interrupt is triggered from CTC
  ❹ targetCount++;       // Add one to targetCount
  ❺ if (targetCount == target)
   // If required period of time has elapsed
   {
    ❻ TCNT1 = 0;        // Reset TIMER1 to zero
    ❼ targetFunction(); // Do something
```

```
    ❽ targetCount = 0;  // Reset targetCount to zero
    }
}

void initCTC()
// Set up CTC interrupt and TIMER1
{
   OCR1A = 15625;       // Number of periods to watch for - 15625 = 1 second
   // Turn on CTC mode and set CS10 and CS11 for /64 prescaler
   TCCR1B = 0b00000011;
   TIMSK1 = 0b00000010; // Turn on timer compare interrupt
   sei();               // Turn on global interrupts
}

int main(void)
{
   DDRB = 0b11111111;   // Set PORTB register as outputs
   initCTC();           // Set up overflow interrupt and TIMER1
   for(;;)              // Do something (blink LED on PB0)
   {
      PORTB = 0b00000001;
      _delay_ms(100);
      PORTB = 0b00000000;
      _delay_ms(100);
   }
   return 0;
}
```

At this stage you should be quite familiar with most of this code, such as the `ISR()` function, but there are a few new components. First, there are two new global variables, `uint16_t target` ❶ and `uint16_t targetCount` ❷. We set `target` to the number of seconds to wait while the required code runs ❸. In this example `target` is set to 1, but you can set it to anything up to 32,767 (for 546.116 . . . minutes), as this is the highest value that can be stored in a 16-bit integer.

The ISR uses the variable `targetCount` to accumulate elapsed seconds, as every time the code calls the ISR (once per second), it increments `targetCount` by 1 ❹. When the code calls the ISR, it checks to see if `targetCount` matches `target` ❺. If so, the code resets TIMER1 to zero ❻, then runs the required code via the function `targetFunction()` ❼, and finally resets `targetCount` back to zero ❽, allowing the process to start again.

Although our example runs the `targetFunction()` code once per second, remember that you can easily increase the duration by changing the value for `target`. For example, to run `targetFunction()` every 5 minutes, change `target` to 300 (5 minutes × 60 seconds = 300 seconds).

Now that you've had some opportunities to experiment with the AVR's ATmega's timers, I'd like to briefly discuss the accuracy of the internal timers themselves.

## Examining the Accuracy of the Internal Timer

You may be wondering how accurately the internal timer can keep time. You can easily check this by running the Project 29 code (with a one-second interval) and measuring the results with a digital storage oscilloscope, as shown in Figure 6-2.

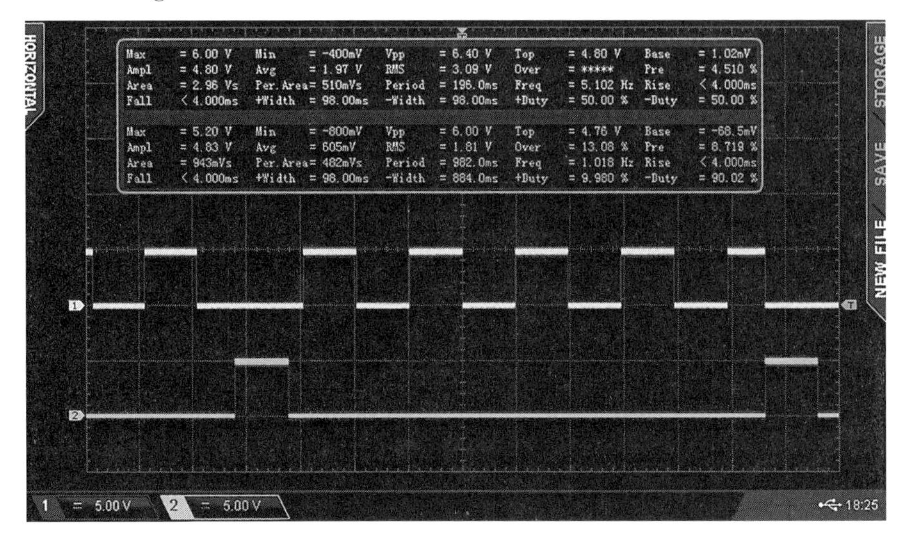


*Figure 6-2: Measuring the outputs of Project 29 using a DSO*

The blinking LED on PB0 is connected to channel 1 (the upper signal) of the DSO, while the LED on PB2 that `targetFunction()` controls is connected to channel 2 (the lower signal). You should see the signal rise and fall as the LED on PB0 turns on then off every 100 milliseconds. The signal for the first LED stays low while the code controlling the second LED on PB1 operates, as the microcontroller cannot operate two things at the same time. After the one second has elapsed, the other LED on PB0 connected to channel 1 turns on and off as directed by `targetFunction()`, and the whole process repeats.

In this case the DSO has measured the frequency for the second LED as 1.018 Hz, or 1.018 times per second—this is awfully close to the required 1 second. Considering we're not using any external timing hardware on the circuit, this is a good outcome. However, if you'd like to run much longer delay periods, you'll need to take such slight variances into account. For example, that 0.018 Hz imbalance from 1 second can equate to 5.4 seconds over a 5-minute period (the actual time of 5.4 seconds is calculated by multiplying 0.018 Hz by 300 seconds). Keep this in mind in your future timing projects.

## Addressing Registers with Bitwise Operations

From the beginning of this book, we've been addressing various registers using the binary number format. For example, with an ATmega328P-PU, we use the following two lines of code to set pin PB0 as an output and turn it on:

```
DDRB = 0b11111111;  // Set PORTB to outputs
. . .
PORTB = 0b00000001; // Set PB0 on
```

This method has worked well for us so far, but it requires us to consider every bit in the register every time we address one or more bits inside the entire register. In this section I'll show you how to use bitwise operations, which allow us to change just a particular bit (or bits) in a register, leaving the others as they were. This will be useful for future projects in the book and beyond, as it allows you to easily set individual or multiple bits without worrying about changing all the bits in a register at once.

### Addressing Individual Bits in a Register

By default, all bits in a register are set to 0 (low, or off) when we reset or turn on the microcontroller. We then set bits to 1 (high, or on) or 0 when required, using the following operations:

**Turning a bit to high**

Use the following code to turn a bit to high (on) by setting it to 1:

```
registername |= (1 << bitname);
```

For example, to turn on the output from PORTB's PB7, you would use the following line, since 7 is the bit number in the PORTB register matching pin PB7:

```
PORTB |= (1 << 7);
```

This allows you to turn on PB7 without needing to concern yourself with the status of the other bits in the register.

**Toggling a bit between high and low**

To *toggle* a bit is to change it from its current state to the alternate state (from off to on, or vice versa). Use the following code to do so:

```
registername ^= (1 << bitname); // Toggle bit "bitname"
```

For example, to toggle the output from PORTB's PB3 you would use the following line, as the bit number in the PORTB register matching pin PB3 is 3:

```
PORTB ^= (1 << 3); // Toggle bit PB3
```

To demonstrate this, you can blink an LED connected to PB3 with a 250 ms delay by using the following two lines in your code's `for (;;)` loop:

```
PORTB ^= (1 << 3); // Toggle PB3
_delay_ms(250);    // Wait 250 milliseconds
```

By using bitwise operations you also save space, as the LED blinking example only requires two lines of code instead of four.

### Turning a bit to low

Use the following code to turn a bit to low (off) by setting it to 0:

```
registername &= ~(1 << bitname); // Set bit "bitname" to 0
```

For example, to turn off the output from PORTB's PB7 you would use the following line, as 7 is the bit number in the PORTB register matching pin PB7:

```
PORTB &= ~(1 << 7);
```

Using bitwise operations in this way is more efficient than the binary number format we used to work with register bits earlier. However, you can improve on this method even further by using the name of the register bit instead of the number, which makes it easier to determine which bit is being altered. For example, say you want to set bit 0 of TIMSK1 on to enable the TIMER1 overflow interrupt used in Project 27. Instead of using `TIMSK1 = 0b00000001`, you can use `TIMSK1 |= (1 << TOIE1)`.

To determine which register bit name to use for the corresponding bit number (`TOIE1` rather than `0b00000001` in the previous example), check the data sheets for your microcontrollers. If you have not already done so, you can download the full data sheets in Adobe PDF format from the Microchip website:

**ATtiny85 data sheet** *https://www.microchip.com/wwwproducts/en/ATtiny85/*

**ATmega data sheet** *https://www.microchip.com/wwwproducts/en/ATmega328p/*

You can then learn the bit names to match a given register, such as the TIMSK1 register, which appears in section 16.11.8 of the ATmega328P-PU's data sheet, as shown in Figure 6-3.

| Bit | 7 | 6 | 5 | 4 | 3 | 2 | 1 | 0 | |
|---|---|---|---|---|---|---|---|---|---|
| (0x6F) | – | – | ICIE1 | – | – | OCIE1B | OCIE1A | TOIE1 | TIMSK1 |
| Read/write | R | R | R/W | R | R | R/W | R/W | R/W | |
| Initial value | 0 | 0 | 0 | 0 | 0 | 0 | 0 | 0 | |

*Figure 6-3: The ATmega328P-PU TIMSK1 register*

Now that you can address a single bit inside a register, next I'll show you how to address two or more bits in the same register without affecting other bits.

### Addressing Multiple Bits in a Register

You can also change more than one bit in a single register at the same time (again, without worrying about the bits that you're not changing) using bitwise operations. Note, however, that you have to perform the same operation on all the bits—for example, you can turn three bits on in one line, but you can't turn two on and one off in one line. If you need to do the latter, use the binary number method instead. Here's how this works:

**Turning multiple bits high**

To turn two bits on at once, use the following code:

```
registername |= (1 << bitname)|(1 << bitname);
```

For example, to turn on the output from PORTB's PB0 and PB7, you could use:

```
PORTB |= (1 << PORTB7)|(1 << PORTB0);
```

You can add additional addressing to the same line. For example, you could turn on PB0, PB3, and PB7 as follows:

```
PORTB |= (1 << PORTB7)|(1 << PORTB3)|(1 << PORTB0);
```

You could use this method to address up to seven bits. If you want to change all bits, then use the usual `PORTx` function.

**Toggling multiple bits between high and low**

To toggle multiple bits high or low at once, use the following code:

```
registername ^= (1 << bitname)|(1 << bitname);
```

For example, to toggle the output from PORTB's PB0 and PB3 you could use:

```
PORTB ^= (1 << PORTB3)|(1 << PORTB0);
```

Again, you can add additional addressing to the same line. For example, to toggle PB0, PB3, and PB7, you could use:

```
PORTB ^= (1 << PORTB7)|(1 << PORTB3)|(1 << PORTB0);
```

**Turning multiple bits low**

To turn multiple bits off at once, use the following code.

```
registername &= ~(1 << bitname)&~(1 << bitname);
```

In this case we use the &~ characters between the bracketed operations, rather than the pipe (|) character. For example, to turn off the output from PORTB's PB7 and PB0, you could use:

```
PORTB &= ~(1 << PORTB7)&~(1 << PORTB0);
```

And to turn off PB0, PB3, and PB7 all at once you could use:

```
PORTB &= ~(1 << PORTB7)&~(1 << PORTB3)&~(1 << PORTB0);
```

Now that we've reviewed all these bitwise operations for addressing registers, let's see how we could use them to improve some of our earlier code.

## Project 30: Experimenting with Overflow Timers Using Bitwise Operations

This project has the same results as Project 27, but I've rewritten the code to take advantage of bitwise operations for addressing registers.

Use the same hardware used in Project 27. With your circuit assembled, connect the USBasp to your microcontroller via the solderless breadboard in the usual way. Open a terminal window, navigate to the *Project 30* subfolder of this book's *Chapter 6* folder, and enter the command **make flash**. Once the code has been uploaded to the microcontroller the LED connected to PB0 should start to blink rapidly, and the LED connected to PB1 should quickly flicker on and off every 0.52428 seconds.

To see how the updated code works, open the *main.c* file for Project 30:

```
// Project 30 - Experimenting with Overflow Timers Using Bitwise Operations

#include <avr/io.h>
#include <avr/interrupt.h>
#include <util/delay.h>

ISR(TIMER1_OVF_vect)
{
   // Code to be executed when an interrupt is triggered from TIMER1 overflow.
   // For this example, quickly blink LED on PB1.
❶ PORTB |= (1 << PORTB1);          // PB1 on
   _delay_ms(5);
❷ PORTB &= ~(1 << PORTB1);         // PB1 off
}

void initOVI()
// Set up overflow interrupt and TIMER1
{
❸ TCCR1B |= (1 << CS11);           // Set prescaler to /8
❹ TIMSK1 |= (1 << TOIE1);          // Enable TIMER1 overflow interrupts
   sei();                           // Turn on global interrupts
}
```

```
int main(void)
{
❺ DDRB |= (1 << DDB0)|(1 << DDB1); // Set PORTB pins 0 and 1 as outputs
  initOVI();                       // Set up overflow interrupt and TIMER1

  for(;;)                          // Do something (blink LED on PB0)
  {
  ❻ PORTB ^=(1 << PORTB0);         // Toggle PB0
     _delay_ms(100);
  }
  return 0;
}
```

In this code, the LED blinks on and off in turn when the code calls the ISR ❶❷. When setting the prescaler, we only need to set `CS11` to 1 ❸, as `CS10` remains 0 for a prescaler of 8, as you'll recall from Table 6-1. (Remember, bits are 0 by default.)

We enable the TIMER1 overflow interrupts by setting bit `TOIE1` of the TIMSK1 register to 1 ❹, then set the PB0 and PB1 pins of PORTB to outputs for our LEDs ❺. Finally, we toggle PB0 on and off in order to flash the LED ❻.

I encourage you to spend some time getting comfortable with the registers we have used so far in the book—PORTB, DDRB, and so on—and experimenting with previous projects to familiarize yourself with bitwise methods of addressing registers. These methods will come in handy in the next chapter, as we begin using pulse-width modulation to experiment with LEDs, motors, and more.

# USING PULSE-WIDTH MODULATION

When you need a digital output to simulate an analog signal, such as operating an LED at partial brightness, you can use *pulse-width modulation (PWM)* to adjust the amount of time between each high and low signal from a digital output pin. PWM can generate various effects, such as adjusting the brightness of an LED, controlling the speed of an electric motor, and creating sounds using tools that convert electricity into vibrations.

In this chapter, you will:

- Learn how pulse-width modulation works and is generated by AVRs.
- Use PWM with ATtiny85 and ATmega328P-PU microcontrollers.
- Make varying tones of sound using piezo elements with PWM.
- Learn how to use PWM to create colorful effects using RGB LEDs.

## Pulse-Width Modulation and Duty Cycles

PWM allows us to control the perceived brightness of an LED, instead of simply switching it on and off as we've done in previous chapters. The LED's brightness is determined by the *duty cycle*, or the length of time the PORT*x* pin is on (meaning the LED is lit) versus the length of time it is off (the LED is unlit). Duty cycle is expressed as the percentage of "on" time. The greater the duty cycle—that is, the longer the PORT*x* pin is on compared to off in each cycle—the greater the perceived brightness of the LED connected to the pin.

Furthermore, the higher the frequency of the PWM signal is—that is, the faster the signal is turned on and off—the smoother the visual effect is. If you're controlling a motor, a higher PWM frequency will make the rotational speed a closer approximation of the actual speed required.

Figure 7-1 shows four possible PWM duty cycles. The filled-in gray areas represent the amount of time that the LED is on; as you can see, this increases with the duty cycle.

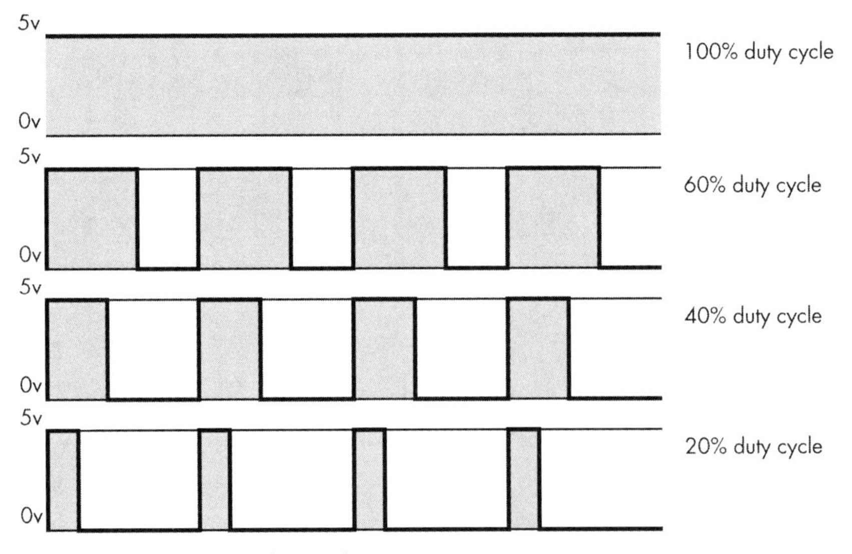


*Figure 7-1: Various PWM duty cycles*

We can only use certain pins on our AVR microcontrollers for PWM. For the ATtiny85, we use PB0, PB1, and PB4; for the ATmega328P-PU, we use PB1 to PB3 and PD3, PD5, and PD6. To create a PWM signal, we need to set the required registers depending on the microcontroller used. I'll demonstrate this for both microcontrollers in this chapter. Let's begin with the ATtiny85.

## Project 31: Demonstrating PWM with the ATtiny85

In this project, you'll learn how to trigger the available PWM outputs offered by the ATtiny85 microcontroller. We trigger each output slightly differently, but the process is always simple.

### *The Hardware*

For this project, you'll need the following hardware:

- USBasp programmer
- Solderless breadboard
- ATtiny85 microcontroller
- Jumper wires
- Three LEDs
- Three 560 Ω resistors

Assemble the circuit as shown in Figure 7-2.

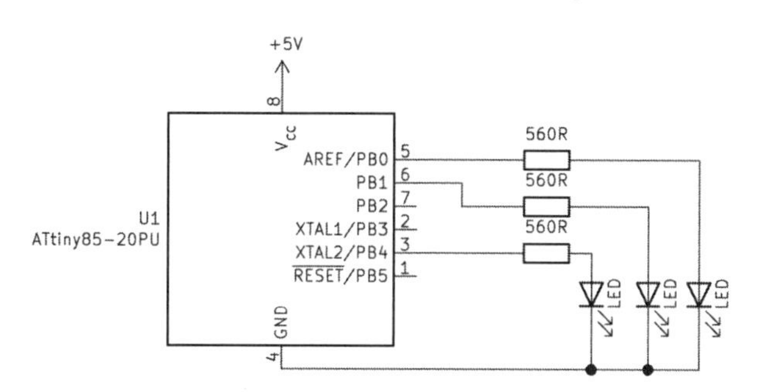


*Figure 7-2: Schematic for Project 31*

With your circuit assembled, connect the USBasp to your microcontroller in the usual way.

### *The Code*

Open a terminal window, navigate to the *Project 31* subfolder of this book's *Chapter 7* folder, and enter the command `make flash` as usual. Once the project's code has been uploaded to the microcontroller the LEDs connected to PB4, PB1, and PB0 should all start to fade on and off at the same time, repeatedly displaying rising then falling levels of brightness.

Let's see how this works. Open the *main.c* file for Project 31.

```
// Project 31 - Demonstrating PWM with the ATtiny85

#include <avr/io.h>
#include <util/delay.h>

void initPWM(void)
{
   // Set PB4, PB1, PB0 as outputs
❶ DDRB |= (1 << PORTB4)|(1 << PORTB1)|(1 << PORTB0);

❷ // PB0
   // Set timer mode to FAST PWM
   TCCR0A |= (1 << WGM01)|(1 << WGM00);
   // Connect PWM signal to pin (OC0A => PB0)
   TCCR0A |= (1 << COM0A1);
   // No prescaler
   TCCR0B |= (1 << CS00);

❸ // PB1
   // Connect PWM signal to pin (OC0B => PB0)
   TCCR0A |= (1 << COM0B1);

❹ // PB4
   // Connect PWM signal to pin (OCR0B => PB4)
   TCCR1 |= (1 << PWM1A)|(1 << COM1A0);
   // Toggle PB4 when timer reaches OCR1B (target)
   GTCCR |= (1 << COM1B0);
   // Clear PB4 when timer reaches OCR1C (top)
   GTCCR |= (1 << PWM1B);
   // No prescaler
   TCCR1 |= (1 << CS10);
}

int main(void)
{
   uint8_t duty = 0;
   initPWM();
   while (1)
   {
    ❺ for (duty = 1; duty <100; duty++)
      {
         OCR0A = duty;                    // PB0
         OCR0B = duty;                    // PB1
         OCR1B = duty;                    // PB4
         _delay_ms(10);
      }
    ❻ for (duty = 100; duty >0; --duty)
      {
         OCR0A = duty;                    // PB0
         OCR0B = duty;                    // PB1
         OCR1B = duty;                    // PB4
         _delay_ms(10);
      }
   }
}
```

This code defines the function initPWM(). The function operates every pin at the same time, but we'll go over how to initialize and operate each pin one by one.

We first set the required pins, PORTB0, PORTB1, and PORTB4, to outputs ❶. Next, in turn we address the required three registers to enable PWM on PORTB0. To set the timer to fast PWM mode, we allocate the timer signal to pin PORTB0—note that we don't use a prescaler, so the PWM can operate at its maximum frequency ❷. We only need to address one register to allow PWM over PORTB1 ❸, but using PWM on PORTB4 requires a different timer, so we have to address different registers ❹.

Now it's time to assign values to the PWM pins to set their duty cycle. The microcontrollers require a value between 1 and 254, which maps out to a duty cycle of just over 0 to just under 100. (If you use 0, this is a 0 percent duty cycle—that is, the pin will be off. If you use 255, this is a 100 percent duty cycle, so the pin will be on continuously.)

Three registers store the duty cycle value for our three PWM pins:

- OCR0A for PORTB0
- OCR0B for PORTB1
- OCR1B for PORTB4

Next, we add a simple loop that causes the duty value to rise incrementally, increasing the brightness of the LEDs over time ❺. This process is then reversed by using another loop that decreases the brightness of the LEDs over time ❻.

Experiment with adjusting the value in the _delay_ms() functions to alter the speed of the change in brightness. You may notice that there's little to no difference in brightness between higher duty cycle values. That's because during high-frequency PWM operations (anything faster than 50 cycles per second), the LED blinks too rapidly for the average human eye to perceive when it's off.

## Individual PWM Pin Control for the ATtiny85

Now that you've assembled and tested all the PWM pins using the ATtiny85, it's time to learn how to use each PWM pin so you can apply them in your own projects.

To activate all the PWM pins at once, simply use the initPWM() function as used in Project 31. (You'll learn how to deactivate them all later on.) To activate and deactivate each pin individually, follow the instructions in this list:

**Activating PWM on ATtiny85 pin PORTB0**

To activate PWM on PORTB0, use the following:

```
DDRB |= (1 << PORTB0);                    // Set PB0 as output
TCCR0A |= (1 << WGM01)|(1 << WGM00);   // Set timer mode to FAST PWM
```

```
// Connect PWM signal to pin (OCOA => PBO)
TCCROA |= (1 << COMOA1);
TCCROB |= (1 << CS00);                    // No prescaler
```

You can then set the duty cycle by allocating a value between 1 and 254 inclusive to the OCR0A register.

### Activating PWM on ATtiny85 pin PORTB1

To activate PWM on PORTB1, use the following:

```
DDRB |= (1 << PORTB1);                    // Set PB1 as output
TCCROA |= (1 << WGM01)|(1 << WGM00);     // Set timer mode to FAST PWM
// Connect PWM signal to pin (OCOB => PB1)
TCCROA |= (1 << COMOB1);
TCCROB |= (1 << CS00);                    // No prescaler
```

You can then set the duty cycle by allocating a value between 1 and 254 to the OCR0B register.

### Activating PWM on ATtiny85 pin PORTB4

To activate PWM on PORTB4, use the following:

```
DDRB |= (1 << PORTB4);                    // Set PB4 as output
// Connect PWM signal to pin (OCROB => PB4)
TCCR1 |= (1 << PWM1A)|(1 << COM1A0);
// Toggle PB4 when timer reaches OCR1B (target)
GTCCR |= (1 << COM1B0);
// Clear PB4 when timer reaches OCR1C (top)
GTCCR |= (1 << PWM1B);
TCCR1 |= (1 << CS10);                     // No prescaler
```

You can then set the duty cycle by allocating a value between 1 and 254 to the OCR1B register.

### Deactivating ATtiny85 PWM

If your project needs to use a pin for both PWM and output on or off, then you must deactivate PWM mode before using `PORTx |=` commands. You'll need to define `initPWM()` and `disablePWM()` functions to switch PWM on and off when required. Use the following code to disable PWM for all the pins (PORTB0, PORTB1, and PORTB4):

```
TCCROA &= ~(1 << WGM01)&~(1 << WGM00); // Turn off fast PWM for PORTBO/1
TCCROA &= ~(1 << COMOA1);              // Disconnect PWM from PORTBO
TCCROA &= ~(1 << COMOB1);              // Disconnect PWM from PORTB1
TCCR1 &= ~(1 << PWM1A)&~(1 << COM1A0); // Turn off PWM for PORTB4
// Disconnect PWM from PORTB4 off timer/counter
TCCR1 &= ~(1 << CS10);
GTCCR &= ~(1 << PWM1B);                // Disable PWM for PORTB4
GTCCR &= ~(1 << COM1B0);               // Disconnect PWM from PORTB4
```

In general, it's a good idea to keep these lines of code required to set up PWM in their own function, as in Project 31. Now that we've reviewed the ins and outs of PWM on the ATtiny85, let's have some fun making noise with a piezo element.

## Project 32: Experimenting with Piezo and PWM

A *piezo element* is a device that converts an electrical charge into a different form of energy. It can convert electrical energy into physical movement in the form of vibration, which generates sound waves that you can hear. By applying an electrical current and varying it using PWM, you can change the piezo's tone. For this project you can use a small, prewired piezo like the one shown in Figure 7-3.

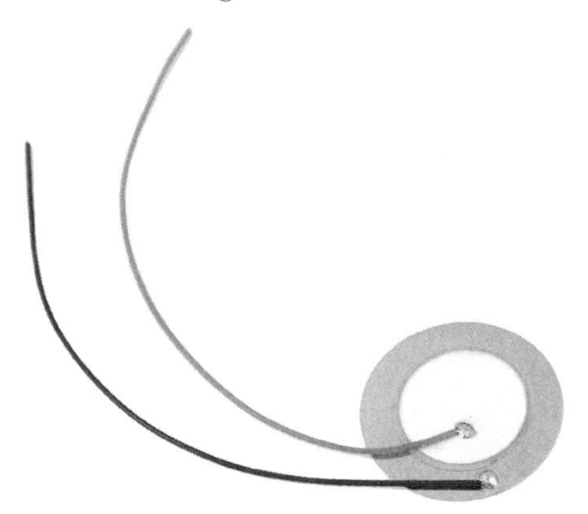

*Figure 7-3: A prewired 27 mm piezo element*

Figure 7-4 shows the schematic symbol for our piezo element.

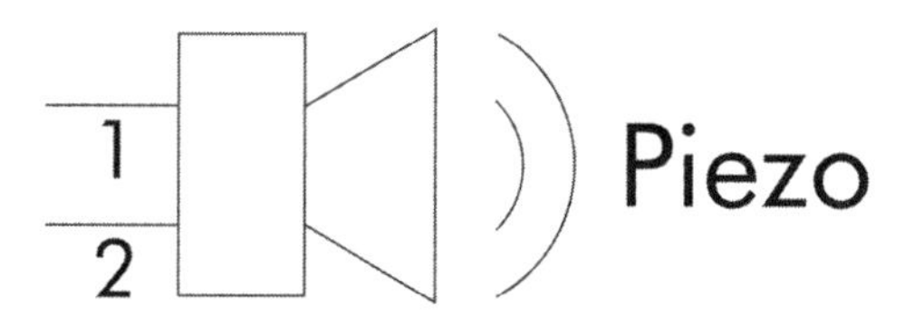


*Figure 7-4: Schematic symbol for piezo element*

In this project, you'll learn to change the pitch of sound from a piezo element by adjusting a trimpot. We'll use an ADC to read the trimpot value, then use that value to determine the duty cycle for PWM control of the piezo element.

### The Hardware

For this project, you'll need the following hardware:

- USBasp programmer
- Solderless breadboard

- ATtiny85 microcontroller
- Jumper wires
- Prewired piezo
- 10 kΩ breadboard-compatible linear trimpot

Assemble the circuit as shown in Figure 7-5.

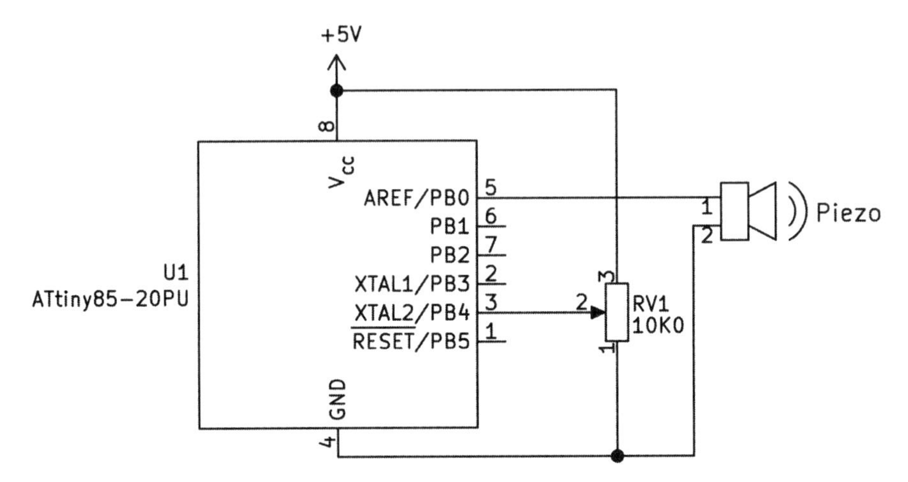


*Figure 7-5: Schematic for Project 32*

With your circuit assembled, connect the USBasp to your microcontroller via the solderless breadboard in the same way you have for previous projects.

## The Code

Open a terminal window, navigate to the *Project 32* subfolder of this book's *Chapter 7* folder, and enter the command **make flash**. Once the code has been uploaded to the microcontroller, slowly turn the trimpot in different directions, altering the piezo's tone.

To see how this works, open the *main.c* file for Project 32:

```
// Project 32 - Experimenting with Piezo and PWM

#include <avr/io.h>
#include <util/delay.h>

void startADC()
❶ // Set up the ADC
{
   ADMUX |= (1 << ADLAR)|(1 << MUX1);
   ADCSRA |= (1 << ADEN)|(1 << ADPS1)|(1 << ADPS0);
}

❷ void initPWM(void)
{
   DDRB |= (1 << PORTB0);              // Set PB0 as output
```

```
    TCCR0A |= (1 << WGM01)|(1 << WGM00);
    TCCR0A |= (1 << COM0A1);
    TCCR0B |= (1 << CS00);
}

int main(void)
{
❸ startADC();
❹ initPWM();
    for(;;)
    {
     ❺ ADCSRA |= (1 << ADSC);          // Start ADC measurement
        while (ADCSRA & (1 << ADSC) ); // Wait until conversion completes
        _delay_ms(5);

     ❻ OCR0A = ADCH;                   // Set PWM duty cycle with ADC value
    }
    return 0;
}
```

This code recalls that from Project 31 and the other PWM examples in this chapter. It accomplishes its goal of setting the duty cycle, as the code takes the 8-bit value of the ADC register and places it in the PWM register OCR0A.

The code initializes the ADC and uses pin PORTB4 for the input ❶. It then initializes the PWM output on PB0 ❷, as in Project 31, starts the ADC ❸, and initializes the PWM ❹. Next, it reads the analog input ❺ and then finally assigns the ADC value (which falls between 0 and 255) to the PWM duty cycle register, thereby driving the piezo ❻.

## Individual PWM Pin Control for the ATmega328P-PU

Now it's time to move on to the PWM functions available for the ATmega328P-PU. Table 7-1 lists the six ATmega328P-PU pins that can be used with PWM.

**Table 7-1**: ATmega328P-PU PWM Pins

| Port register bit | Physical pin | Duty cycle register |
|---|---|---|
| PORTB1 | 15 | OCR1A |
| PORTB2 | 16 | OCR1B |
| PORTB3 | 17 | OCR2A |
| PORTD3 | 5 | OCR2B |
| PORTD5 | 11 | OCR0B |
| PORTD6 | 12 | OCR0A |

Let's go over how to activate (and deactivate) these pins for use with PWM.

### Activating PWM on ATmega328P-PU pins PORTD5/6

To activate PWM on PORTD5/6, use the following:

```
TCCR0A |= (1 << WGM01)|(1 << WGM00);
TCCR0B |= (1 << CS01);
```

You can then connect the PWM output to pins PORTD5/6 with:

```
TCCR0A |= (1 << COM0A1);          // PWM to OCR0A - PD6
TCCR0A |= (1 << COM0B1);          // PWM to OCR0B - PD5
```

Set the duty cycle by allocating a value between 1 and 254 to the duty cycle registers. If you want to control the pins directly, to use them as regular inputs or outputs, you will need to disconnect them from the PWM output. You can do this as follows:

```
TCCR0A &= ~(1 << COM0A1);         // Disconnect PWM from OCR0A - PD6
TCCR0A &= ~(1 << COM0B1);         // Disconnect PWM from OCR0B - PD5
```

### Activating PWM on ATmega328P-PU pins PORTB1/2

To activate PWM on PORTB1/2, use the following:

```
TCCR1A |= (1 << WGM10);
TCCR1B |= (1 << WGM12);
TCCR1B |= (1 << CS11);
```

You can then connect the PWM output to pins PORTB1/2 with:

```
TCCR1A |= (1 << COM1A1);          // PWM to OCR1A - PB1
TCCR1A |= (1 << COM1B1);          // PWM to OCR1B - PB2
```

To disconnect the pins from PWM, use the following:

```
TCCR1A &= ~(1 << COM1A1);         // Disconnect PWM from OCR1A - PB1
TCCR1A &= ~(1 << COM1B1);         // Disconnect PWM from OCR1B - PB2
```

### Activating PWM on ATmega328P-PU pins PORTB3 and PORTD3

To activate PWM on PORTB3 and PORTD3, use the following:

```
TCCR2A |= (1 << WGM20);
TCCR2A |= (1 << WGM21);
TCCR2B |= (1 << CS21);
```

You can then connect the PWM output to pins PORTB3 and PORTD3 like so:

```
TCCR2A |= (1 << COM2A1);          // PWM to OCR2A - PB3
TCCR2A |= (1 << COM2B1);          // PWM to OCR2B - PD3
```

To disconnect the pins from PWM, use the following:

```
TCCR2A &= ~(1 << COM2A1);       // Disconnect PWM from OCR2A - PB3
TCCR2A &= ~(1 << COM2B1);       // Disconnect PWM from OCR2B - PD3
```

Remember that you can connect and disconnect pins from PWM when required only if you've run the activation code first.

You've already seen a few ways to use PWM in this chapter, and next I'll give you one more example: generating colors using RGB LEDs.

## The RGB LED

An RGB LED is simply three LED elements—one red, one green, one blue—in a single enclosure, as shown in Figure 7-6. These LEDs are great for saving space, and you can also use them to create your own colors by changing the brightness of the individual elements. RGB LEDs are available in many sizes; one common option is 10 mm in diameter.

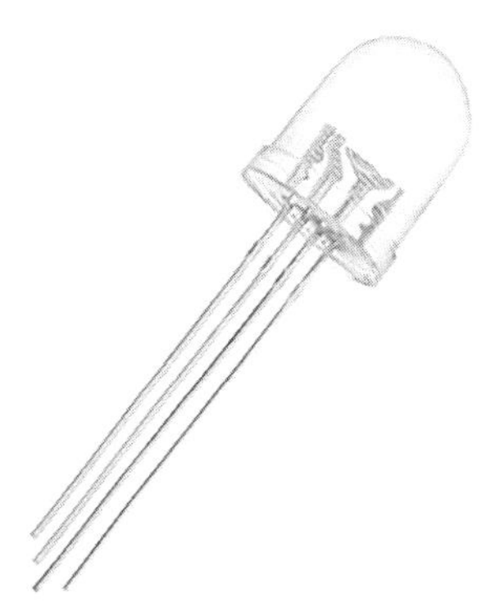

Figure 7-6: Typical RGB LED

As you can see, RGB LEDs have four legs. As with the single-color LEDs discussed in Chapter 3, there two types of these LEDs available: *common anode* and *common cathode.* In the common-anode configuration, all three LED anodes are connected, while the cathodes are separate. The common-cathode configuration has three separate anodes, with all three cathodes connected. Figure 7-7 shows the schematics for both.

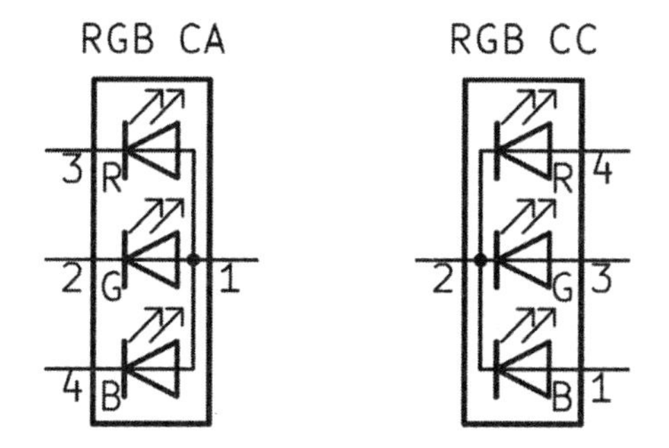


Figure 7-7: Schematic symbols for RGB LEDs: common anode (left) and common cathode (right)

The pinouts for each LED unfortunately may vary, so check with your supplier. However, if you cannot find this information, the longest pin is usually the common anode or cathode pin. You can order clear (where the body of the LED is transparent) or diffused (with a clouded LED body) RGB LEDs. I recommend the latter, as diffused LEDs work better at blending their three primary color elements to make color combinations.

For the projects in this book, starting with the next one, we'll use common-cathode RGB LEDs.

## Project 33: Experimenting with RGB LEDs and PWM

This project allows you to generate various colors by mixing two of the three primary colors in an RGB LED with varying brightness caused by PWM. The diagram in Figure 7-8 shows which color combinations will result in a given hue.

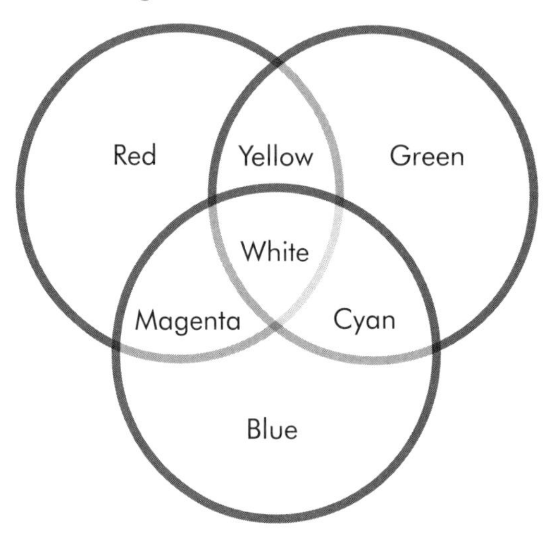


*Figure 7-8: Results of mixing red, green, and blue in various combinations*

For example, mixing red and green light will create a yellow glow. By increasing the brightness of one color while decreasing the brightness of the other color, you can run through multiple shades of yellow. This project mixes only two colors, but you can mix three at once to create a white light if you wish in your own projects. You'll use the ATmega328P-PU microcontroller for this project; it has many more pins than the ATtiny85, so you can use it to work with the RGB LED and still have pins left over for other purposes.

### The Hardware

For this project, you'll need the following hardware:

- USBasp programmer
- Solderless breadboard
- ATmega328P-PU microcontroller
- Jumper wires
- One diffused common-cathode RGB LED
- Three 560 Ω resistors

Assemble the circuit as shown in Figure 7-9.

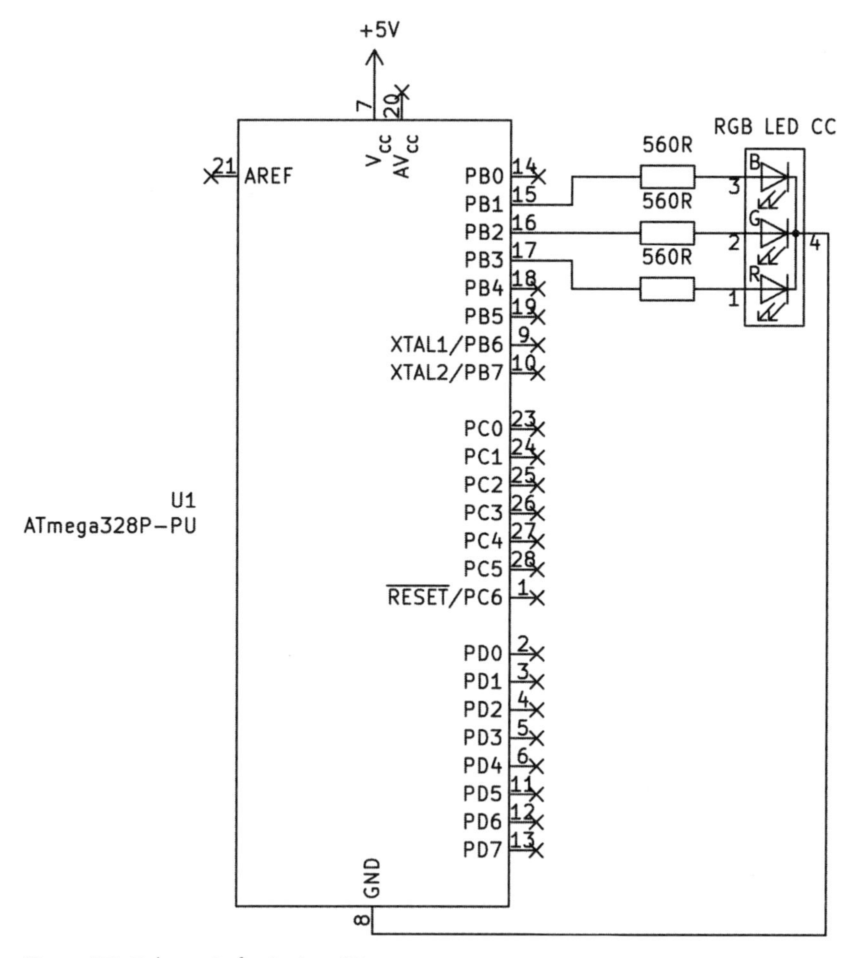


*Figure 7-9: Schematic for Project 33*

Open a terminal window, navigate to the *Project 33* subfolder of this book's *Chapter 7* folder, and enter the command **make flash**. Once you've uploaded the code to the microcontroller, your RGB LED should begin to glow with constantly changing colors. (If your LED has a clear body, you can easily diffuse the output for a better effect by placing some white paper over the LED.)

To see how this works, open the *main.c* file for Project 33:

```
// Project 33 - Experimenting with RGB LEDs and PWM

#include <avr/io.h>
#include <util/delay.h>

❶ #define wait 10
❷ void initPWM(void)
{
   // Timers 1A and 1B
```

```
    TCCR1A |= (1 << WGM10);  // Fast PWM mode
    TCCR1B |= (1 << WGM12);  // Fast PWM mode
    TCCR1B |= (1 << CS11);
    TCCR1A |= (1 << COM1A1); // PWM to OCR1A - PB1
    TCCR1A |= (1 << COM1B1); // PWM to OCR1B - PB2

    // Timer 2
    TCCR2A |= (1 << WGM20);  // Fast PWM mode
    TCCR2A |= (1 << WGM21);  // Fast PWM mode
    TCCR2B |= (1 << CS21);
    TCCR2A |= (1 << COM2A1); // PWM to OCR2A - PB3
}

❸ void PWMblue(uint8_t duty)
// Blue LED is on PB1
{
    OCR1A = duty;
}

❹ void PWMred(uint8_t duty)
// Red LED is on PB3
{
    OCR2A = duty;
}

❺ void PWMgreen(uint8_t duty)
// Green LED is on PB2
{
    OCR1B = duty;
}

int main(void)
{
    // Set PORTB1, PORTB2, and PORTB3 as outputs
  ❻ DDRB |= (1 << PORTB1)|(1 << PORTB2)|(1 << PORTB3);
    initPWM();
    uint8_t a;

    while(1)
    {
        // Red to green
      ❼ for (a=1; a<255; a++)
        {
            PWMred(255-a);
            PWMgreen(a);
            _delay_ms(wait);
        }

        // Green to blue
      ❽ for (a=1; a<255; a++)
        {
            PWMgreen(255-a);
            PWMblue(a);
            _delay_ms(wait);
        }
```

```
      // Blue to red
   ❾ for (a=1; a<255; a++)
      {
         PWMblue(255-a);
         PWMred(a);
         _delay_ms(wait);
      }
   }
}
```

We begin with our usual initPWM() function, which sets up the PORTB1, PORTB2, and PORTB3 pins as PWM outputs ❷. This is followed by three simple functions, one for each LED ❸❹❺, that pass on the required duty cycle, to make controlling each of the primary colors in the LED easy when required.

The main code sets up PORTB1, PORTB2, and PORTB3 as output pins, then initializes the PWM outputs ❻. Finally, to mix the colors we use the three functions we defined to mix two of the LED colors by starting one color at a high level of brightness with the other color at a low level, then gradually decreasing and increasing the levels of the two colors, respectively ❼❽❾. You can adjust the speed of the color transition by changing the value of wait ❶.

I hope you found making sound effects with the piezo element and light effects with the RGB LED enjoyable. You're just getting started: there are many more uses for PWM, including controlling motors for robotics and electric fans, as well as learning how to use MOSFETs to controller larger currents. We'll explore all of these in the next chapter.

# CONTROLLING MOTORS WITH MOSFETS

AVRs cannot directly control motors. In order to enable this, we need to use external components: *metal-oxide-semiconductor field-effect transistors (MOSFETs)*, transistors that can switch or amplify voltages in circuits.

In this chapter, you'll learn how to:

- Use PWM with MOSFETs to control DC motors.
- Use MOSFETs to control larger currents.
- Use motor driver ICs to interface larger motors with your AVR microcontrollers.

Along the way, you'll build a temperature-controlled fan and a two-wheel-drive robot vehicle, building on prior knowledge to complete more interesting and complex projects. By the end of the chapter, you'll have the skills to begin using MOSFETs in your own projects both for fun and for more serious applications, such as robotics, automation, or toys.

## The MOSFET

We use MOSFETs when we need to control large currents and voltage using a small signal, such as that from a microcontroller's digital output pin. MOSFETs are available in various sizes, such as those shown in Figure 8-1, to match different projects' requirements.

*Figure 8-1: Various MOSFETs*

We will be using the small 2N7000 version shown at the bottom left in Figure 8-1, which has three leads. When you're looking at the front of the 2N7000—the flat-faced side—the pins are, from left to right, the source, gate, and drain pins (I'll explain their functions momentarily).

Figure 8-2 shows the schematic symbol for the 2N7000 MOSFET.

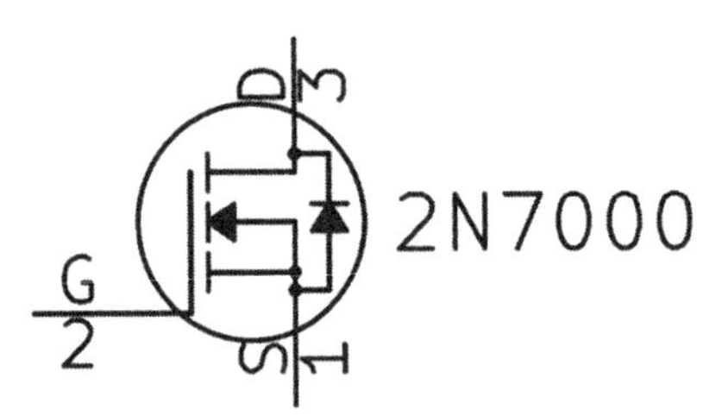


*Figure 8-2: Schematic symbol for 2N7000 MOSFET*

It's easy to operate a MOSFET. When you apply a small current to the gate pin, a large current can flow in through the drain pin and out through the source pin. You can also connect a PWM signal to the gate pin of a MOSFET, allowing you to control lights, motors, and more in various ways. That's what we'll focus on in this chapter.

Our 2N7000 MOSFET can handle up to 60 V DC at 200 mA continuously, or 500 mA in bursts. When choosing a MOSFET for your projects, be sure to check the voltage and current maximums against the signal you want to switch.

We connect a 10 kΩ resistor between the MOSFET's gate and the source pin every time we use a MOSFET, as you'll see in the following project. This acts to keep the gate switched off when a current is not applied to it, in the same way a resistor pulls down a button, as shown in Chapter 3; it stops the MOSFET from turning slightly on or off at random.

## Project 34: DC Motor Control with PWM and MOSFET

This project demonstrates how to control a small DC motor using PWM and a MOSFET. As the microcontroller cannot provide enough current for the motor on its own, we use an external power supply and a MOSFET to handle the motor's requirements.

### *The Hardware*

For this project, you'll need the following hardware:

- USBasp programmer
- Solderless breadboard
- ATmega328P-PU microcontroller
- Jumper wires
- Small DC motor and matching power
- 2N7000 MOSFET
- 10 kΩ resistor

A small DC motor model like the one shown in Figure 8-3 with a maximum of 12 V DC will suffice.

*Figure 8-3: Small DC motor*

You will also need external power, such as a battery pack that holds several AA cells. A 6 AA cell pack like the one shown in Figure 8-4 will provide up to 9 V DC, enough power to run a 12 V DC motor nicely.

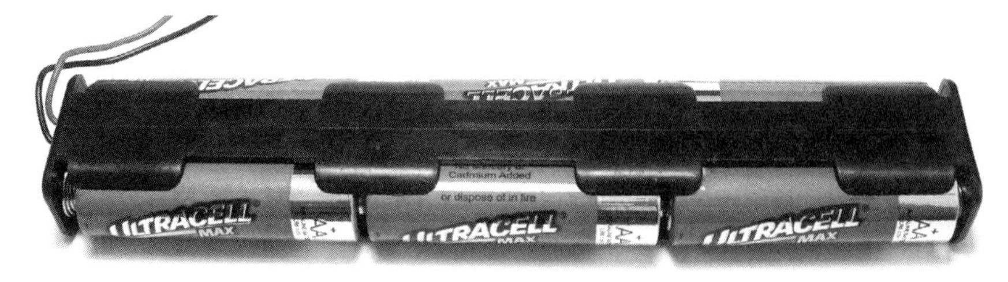

*Figure 8-4: AA battery pack*

Assemble the circuit as shown in Figure 8-5. Note that the black/negative lead from the battery pack will be connected to GND.

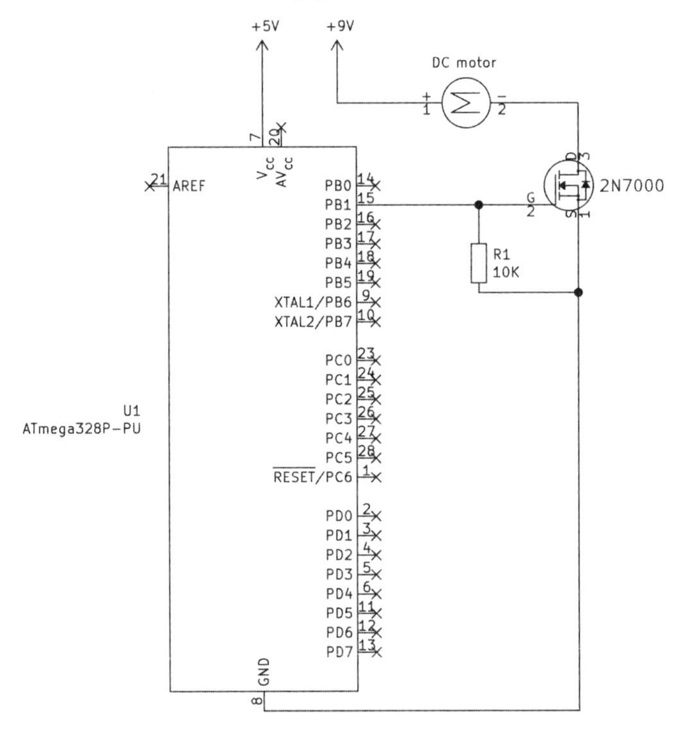


*Figure 8-5: Schematic for Project 34*

Don't disassemble the circuit once you've finished using it, as you will use it as part of the circuit in the following project.

## The Code

Open a terminal window, navigate to the *Project 34* subfolder of this book's *Chapter 8* folder, and enter the command `make flash`. The DC motor should start from 0, increase speed to a complete on state, then reduce speed back to a complete off state, then repeat the procedure.

Let's see how this works:

```
// Project 34 - DC Motor Control with PWM and MOSFET

#include <avr/io.h>
#include <util/delay.h>

❶ #define wait 10
❷ void initPWM(void)
{
   // Timers 1A and 1B
   TCCR1A |= (1 << WGM10);        // Fast PWM mode
   TCCR1B |= (1 << WGM12);        // Fast PWM mode
   TCCR1B |= (1 << CS11);
}

void motorOn(void)
{

 ❸ TCCR1A &= ~(1 << COM1A1);      // Disconnect PWM from PB1
   PORTB |= (1 << PORTB1);        // Set PB1 on
}

void motorOff(void)
{
 ❹ TCCR1A &= ~(1 << COM1A1);      // Disconnect PWM from PB1
   PORTB &= ~(1 << PORTB1);       // Set PB1 off
}

void motorPWM(uint8_t duty)
{
 ❺ TCCR1A |= (1 << COM1A1);       // Connect PWM to OCR1A-PB1
   OCR1A = duty;
}

int main(void)
{
   DDRB |= (1 << PORTB1);         // Set PORTB pin 1 as output
 ❷ initPWM();
   uint8_t a;

   while(1)
   {
      motorOff();                 // Motor off
      _delay_ms(3000);

      for (a = 1; a <255; a++)    // Slowly increase motor speed
      {
         motorPWM(a);
         _delay_ms(wait);
      }

      motorOn();                  // Motor full on
      _delay_ms(1000);
```

```
      for (a = 254; a > 0;-a)   // Slowly decrease motor speed
      {
         motorPWM(a);
         _delay_ms(wait);
      }
   }
}
```

You should be familiar with the code used in this project by now as you learned about using PWM in Chapter 7, but let's go through it together. First, we set the required registers to initialize PWM operation ❷. To make control easier, we use three functions: `motorOn()`, `motorOff()`, and `motorPWM()`. The `motorOn()` function turns the motor completely on by first disconnecting PORTB1 from PWM ❸ and then setting it to high. This gives 100 percent power to the motor via the MOSFET at all times.

We use the `motorOff()` function to completely turn the motor off by disconnecting PORTB1 from PWM ❹ and setting it to low. This turns off the MOSFET gate pin, so the motor has no power. Again, this is necessary as you can't send a 0 percent duty cycle to the OCR1A register and expect it to be off 100 percent of the time. Even with a duty cycle of 0 percent, every time the hardware timer resets the output is turned on briefly during the reset.

Finally, the function `motorPWM()`, which accepts the required duty cycle value, is used to set the motor speed with PWM. It connects PORTB1 to PWM ❺ and then loads the OCR1A register with the required value.

Our main code repeatedly turns the motor on and increases the speed to 100 percent, then reduces the speed back to 0, then turns the motor off for 3 seconds. We turn the motor off at the start of the code, to allow the end user a moment's notice before spinning it up. You can change the delay time in the PWM loops by altering the value of `wait` ❶.

Now that you know how to control a DC motor, let's apply this skill to a practical example by building a temperature-controlled fan system.

## Project 35: Temperature-Controlled Fan

In this project, you'll combine your existing knowledge of motor control with your newfound MOSFET skills, using temperature sensors to make a temperature-controlled fan.

### The Hardware

For this project, you'll need the following hardware:

- USBasp programmer
- Solderless breadboard
- ATmega328P-PU microcontroller
- Jumper wires
- Small DC motor and matching power

- 2N7000 MOSFET
- TMP36 temperature sensor
- 0.1 µF ceramic capacitor
- 10 kΩ resistor

You can use the small DC motor from the previous project just to see how this works, or you can pick up a DC motor–powered cooling fan from an electrical retailer, such as the unit from PMD Way (part number 59119182) shown in Figure 8-6. Some fans may have four wires, but only two of these are required (power and GND). Once again, we'll need to use external power for the fan.

Figure 8-6: DC cooling fan

Assemble the circuit as shown in Figure 8-7.

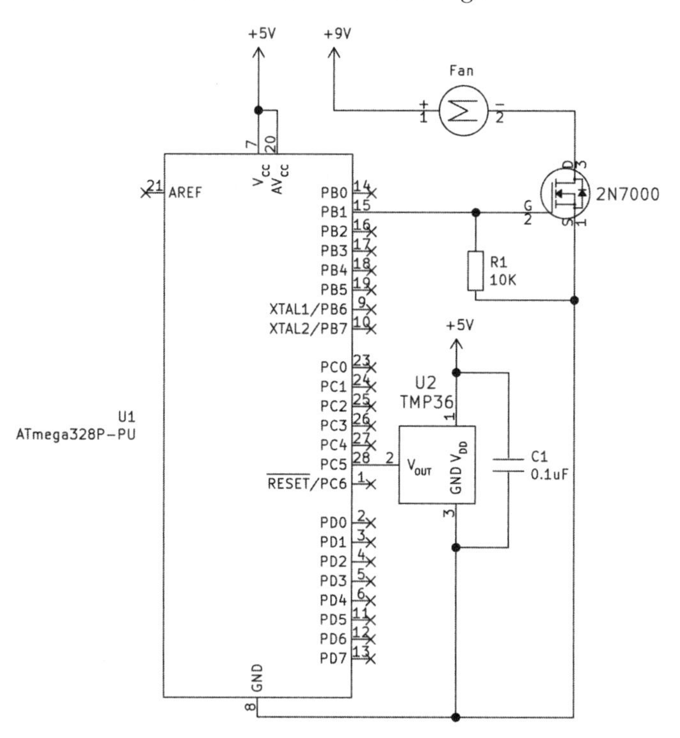


Figure 8-7: Schematic for Project 35

As you assemble the project, note that the black/negative lead from the battery pack or fan power supply will be connected to GND. Also, don't forget to connect $AV_{CC}$ to 5 V.

## The Code

Open a terminal window, navigate to the *Project 35* subfolder of this book's *Chapter 8* folder, and enter the command **make flash**. Once you've applied power, the project should wait three seconds before taking the temperature and operating the fan, depending on the current temperature.

To see how this works, take a look at the code:

```
// Project 35 - Temperature-Controlled Fan

#include <avr/io.h>
#include <util/delay.h>

void startADC()
{
   ADMUX |= (1 << REFS0);                   // Use AVcc pin with ADC
   ADMUX |= (1 << MUX2) | (1 << MUX0);      // Use ADC5 (pin 28)
   ADCSRA |= (1 << ADPS1) | (1 << ADPS0); // Prescaler for 1 MHz (/8)
   ADCSRA |= (1 << ADEN);                   // Enable ADC
}

void initPWM(void)
{
   // Timers 1A and 1B
   TCCR1A |= (1 << WGM10);                  // Fast PWM mode
   TCCR1B |= (1 << WGM12);                  // Fast PWM mode
   TCCR1B |= (1 << CS11);
}

void motorOff(void)
{
   TCCR1A &= ~(1 << COM1A1);                // Disconnect PWM from PB1
   PORTB &= ~(1 << PORTB1);                 // Set PB1 off
}

void motorOn(void)
{
   TCCR1A &= ~(1 << COM1A1);                // Disconnect PWM from PB1
   PORTB |= (1 << PORTB1);                  // Set PB1 on
}

void motorPWM(uint8_t duty)
{
   TCCR1A |= (1 << COM1A1);                 // Connect PWM to OCR1A—PB1
   OCR1A = duty;
}

int main(void)
{
   DDRB |= (1 << PORTB1);                   // Set PORTB pin 1 as output
```

```
❶ startADC();
   initPWM();
❷ uint8_t ADCvalue;
   float voltage;
   float temperature;

   // Delay motor action for a few moments on start
❸ _delay_ms(3000);

   while(1)
   {
      // Get reading from TMP36 via ADC
    ❹ ADCSRA |= (1 << ADSC);              // Start ADC measurement
      while (ADCSRA & (1 << ADSC) );      // Wait until conversion complete
      _delay_ms(10);

      // Get value from ADC register, convert to 8-bit value
      ADCvalue = ADC >> 2;

      // Convert reading to temperature value (Celsius)
      voltage = (ADCvalue * (5000 / 256));
    ❺ temperature = (voltage-500) / 10;

      // Now you have a temperature value, take action
    ❻ if (temperature<25)
      {
         // Under 25 degrees, turn motor off
         motorOff();
      }
    ❼ else if ((temperature>=25) & (temperature <35))
      {
         // At or above 25 and below 35 degrees, set motor to 50% PWM
         motorPWM(127);
      }
    ❽ else if (temperature>=35)
      {
         // 35 degrees and over, turn motor full on
         motorOn();
      }
    ❾ _delay_ms(500); // Prevent rapid motor speed changes
   }
}
```

This code builds upon the ADC and temperature sensor from Project 19 in Chapter 4 and the PWM motor control used in Project 34. First, we activate the ADC to read the TMP36 temperature sensor and activate PWM for variable-speed motor control ❶ (the `startADC()` and `initPWM()` functions are defined at the beginning of the program). We introduce the variables required to calculate the temperature for the thermostat ❷, and then we introduce a delay at startup so the motor doesn't jump into life straight after a reset or power-up ❸.

In the main loop, we take the value from the ADC ❹ and convert it to degrees Celsius ❺. The code can now use this temperature value to determine whether to operate the motor. In this project, the motor is switched off if the temperature is below 25 degrees ❻. If the temperature is between 25 and 34 degrees inclusive, the fan runs at half speed ❼. If the temperature is 35 degrees or over, the fan runs at full speed ❽.

Finally, after checking the temperature, there is a short delay ❾ to avoid *hysteresis*—that is, rapid changes in the characteristics of the circuit. For example, if the sensor were in the path of a breeze or a fluttering curtain, the temperature might fluctuate rapidly between 24.99 and 25 degrees, causing the motor to turn continuously on and off. The delay allows us to avoid this.

At this point, I hope you're beginning to see how we can combine basic AVR code and tools to solve new problems. Building on prior knowledge, we've started to move beyond the simpler projects in earlier chapters to more complex, practical applications.

Now that we've experimented with basic motor control using the MOSFET, we'll move on to controlling the direction of rotation as well as the speed of a DC motor. To do this, we'll use the L293D motor driver IC.

## The L293D Motor Driver IC

To control the speed and direction of one or two small DC motors, we'll use the L293D motor driver IC from STMicroelectronics, shown in Figure 8-8. This is in the same type of package as a microcontroller, and thus we can easily use it in a solderless breadboard for experimenting.

*Figure 8-8: L293D motor driver IC*

You can use small motor driver ICs like the L293D for robotics or small toys that run from 4.5 to 36 V DC at up to 600 mA, with some restrictions with regard to heat that I'll explain later. The L293D saves you a lot of time, as it takes care of distributing power to the motors and spares you from building a bunch of external circuitry. It is known as an *H-BRIDGE IC* because it has an internal circuit of MOSFETs and other components configured in the shape of the letter H, as shown in Figure 8-9.

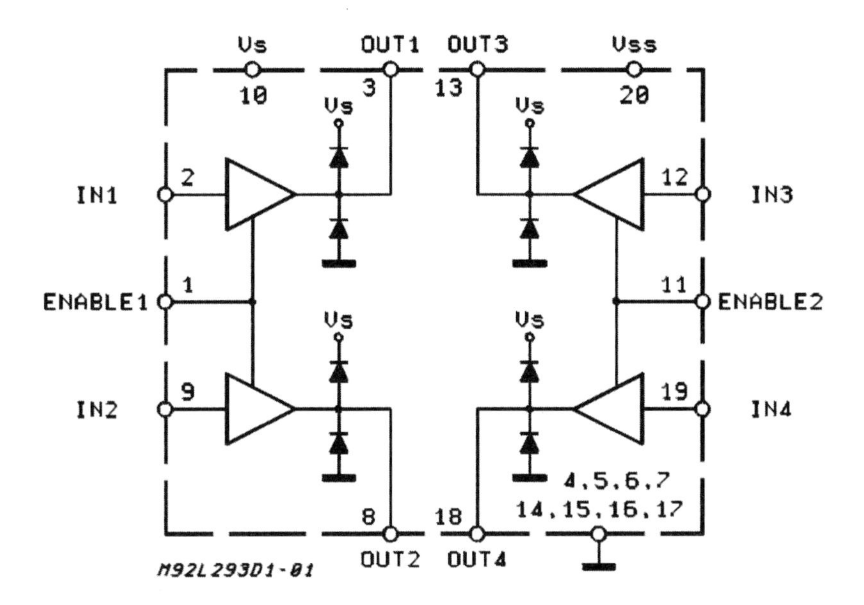

*Figure 8-9: L293D IC block diagram*

Thankfully, we don't need to build the L293D IC's circuitry ourselves; it's already set up and ready for us to connect the motors, control logic, and power. Instead, we just connect motors, power, GND, and outputs from a microcontroller. To see how to wire up the L293D to one DC motor, take a look at the pinouts in Figure 8-10.

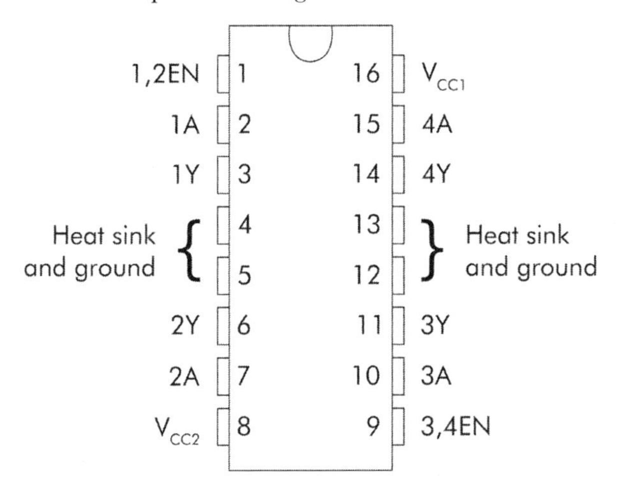


*Figure 8-10: L293D IC pinouts*

There are four GND pins: 4, 5, 12, and 13. Connect those to GND. Next, locate the two power pins. Connect the first one—$V_{CC}1$, which is the logic (or control) power pin—to the 5 V, as you did with your microcontroller in our earlier projects. Then connect the second power pin, $V_{CC}2$, to the positive of the motor power supply (up to 36 V DC). Finally, connect the motor: one wire to pin 3 and the other to pin 6.

Controlling the motor requires three signals from digital outputs on our microcontroller. First, we set the ENABLE pin: either to high, so that the driver IC sends power to the motor, or to low, so that the motor stops. The signals from the next two pins, 1A and 2A, control the polarity of the power to the motor, and thus the rotational direction. With ENABLE set high, the motor will rotate in one direction with 1A high and 2A low and rotate in the other direction with 1A low and 2A high. Table 8-1 summarizes all this for easy reference.

**Table 8-1:** L293D Single Motor Control

| ENABLE pin/EN1 (pin 1) | 1A pin/out 1 (pin 2) | 2A pin/out 2 (pin 7) | Motor action |
|---|---|---|---|
| High | High | Low | Forward |
| High | Low | High | Backward |
| Low | High or Low | High or Low | Stop |

There's no way to tell from the outside whether your motor will run forward or backward; you will need to do a test run to determine which of the two 1A/2A combinations is which for your motor. You can alter the speed of the motor by applying a PWM signal to the ENABLE pin.

### A FEW WORDS ABOUT HEAT

The L293D can become warm (or hot) when running toward the higher end of its capacity. It shouldn't be used in a solderless breadboard in these situations, as the four GND pins are also used as a heatsink. This means they might melt the plastic around the pins, leaving the L293D stuck in the breadboard. If you're going to control larger motors, build your circuit using your own PCB, use a breakout board for the motor control, or solder the circuit onto a stripboard.

Now that you're familiar with the theory of the L293D, let's put it into practice in the next project.

## Project 36: DC Motor Control with L293D

This project demonstrates how you can control a small DC motor using PWM and the L293D motor driver IC, operating the motor in either direction and at various speeds. This will give you the remaining skills you need to build your first moving robot vehicle in the next project.

### The Hardware

For this project, you'll need the following hardware:

- USBasp programmer
- Solderless breadboard
- ATmega328P-PU microcontroller
- Jumper wires
- Small DC motor and matching power
- L293D motor driver IC

Use the same DC motor and matching power supply you used for Project 34. Assemble the circuit as shown in Figure 8-11.

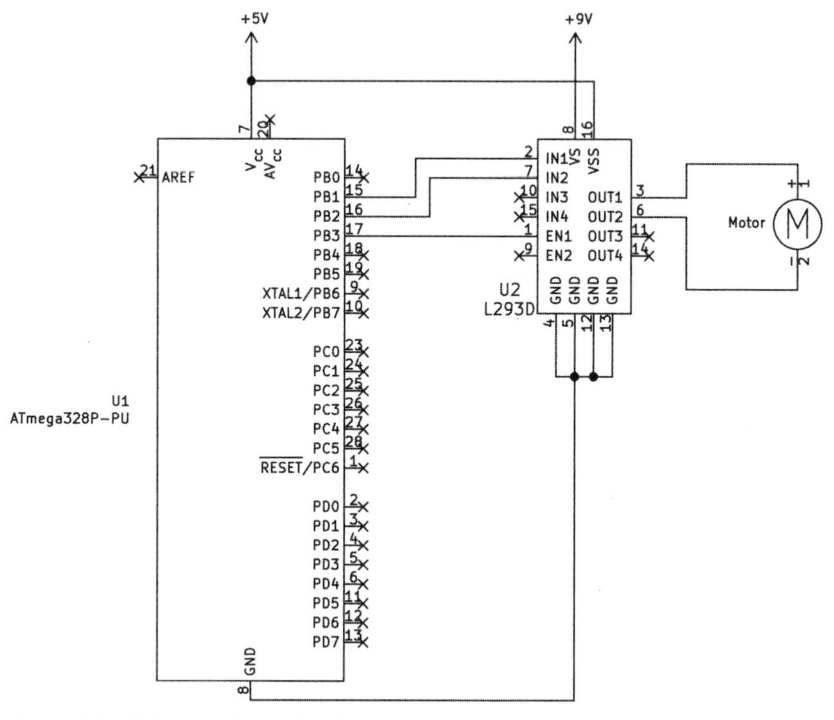


*Figure 8-11: Schematic for Project 36*

As you assemble the circuit, connect the black/negative lead from the battery pack or external power to GND once again.

### The Code

Open a terminal window, navigate to the *Project 36* subfolder of this book's *Chapter 8* folder, and enter the command `make flash`. Once you've applied power, the project should wait three seconds before operating the motor at two different speeds successively, both forward and backward.

Let's see how this works.

```
// Project 36 - DC Motor Control with L293D

#include <avr/io.h>
#include <util/delay.h>

❶ void initPWM(void)
{
   TCCR2A |= (1 << WGM20);     // Fast PWM mode
   TCCR2A |= (1 << WGM21);     // Fast PWM mode, part 2
   TCCR2B |= (1 << CS21);      // PWM Freq = F_CPU/8/256
}
❷ void motorForward(uint8_t duty)
{
   // Set direction
 ❸ PORTB |= (1 << PORTB1);     // PB1 HIGH
   PORTB &= ~(1 << PORTB2);    // PB2 LOW

   // Set speed
 ❹ if (duty == 255)
   {
      PORTB |= (1 << PORTB3);  // Set PORTB3 to on
   } else if (duty < 255)
   {
    ❺ TCCR2A |= (1 << COM2A1); // PWM output on OCR2A—PB3
      OCR2A = duty;            // Set PORTB3 to PWM value
   }
}

❻ void motorBackward(uint8_t duty)
{
   // Set direction
   PORTB &= ~(1 << PORTB1);    // PB1 LOW
   PORTB |= (1 << PORTB2);     // PB2 HIGH

   // Set speed
   if (duty == 255)
   {
      PORTB |= (1 << PORTB3);  // Set PORTB3 to on
   } else if (duty < 255)
   {
      TCCR2A |= (1 << COM2A1); // PWM output on OCR2A—PB3
      OCR2A = duty;            // Set PORTB3 to PWM value
   }
}

❼ void motorOff(void)
{
   // Disconnect PWM output from OCR2A—PB3
   TCCR2A &= ~(1 << COM2A1);
   // Set ENABLE to zero for brake
   PORTB &= ~(1 << PORTB3);
}

int main(void)
```

```
{
   // Set PORTB3, 2, and 1 as outputs
   DDRB |= (1 << PORTB3)|(1 << PORTB2)|(1 << PORTB1);
❽ initPWM();
   _delay_ms(3000);          // Wait a moment before starting
   while(1)
   {
    ❾ motorForward(64);
      _delay_ms(2000);
      motorOff();
      _delay_ms(2000);
      motorForward(255);
      _delay_ms(2000);
      motorOff();
      _delay_ms(2000);
      motorBackward(64);
      _delay_ms(2000);
      motorOff();
      _delay_ms(2000);
      motorBackward(255);
      _delay_ms(2000);
      motorOff();
      _delay_ms(2000);
   }
}
```

This code builds on that of previous motor control projects in this chapter, with the required additions for the L293D. We set up PWM at points ❶ and ❽. The first of the motor control functions, `motorForward()` ❷, rotates the motor in one direction and accepts a duty cycle value of between 1 and 255. Per Table 8-1, we set the outputs as high and low for motor directional control ❸. The code then checks if the required duty cycle value is 255 ❹, and if so simply switches the ENABLE pin to high for full-speed motor running instead of using PWM. However, if it's less than 255, then PWM is enabled for the output pin controlling the L293D ENABLE pin ❺ and the required duty cycle value is dropped into OCR2A.

The motor control method used in `motorForward()` is repeated with the function `motorBackward()` ❻, except with the outputs for motor control set to low and high for reverse rotation. Finally, the `motorOff()` function ❼ turns off the motor by first disabling PWM for the output pin controlling the L293D ENABLE pin and then setting it to low. With all this complete, you can now use the motor control functions to control the speed and direction of motor rotation, as demonstrated in the main loop of the code ❾.

Now that you know how to control the speed and direction of a DC motor, let's use two motors to control a small robot vehicle.

## Project 37: Controlling a Two-Wheel-Drive Robot Vehicle

In this project, you'll learn to control a small two-wheel-drive robot vehicle. The suggested hardware includes two DC motors and a *castor* (a small, swiveling wheel fixed to the bottom of your robot vehicle), allowing you to easily control the speed and direction of travel. I hope this inspires you to create your own more complex robotic creations!

## The Hardware

For this project, you'll need the following hardware:

- USBasp programmer
- Solderless breadboard
- ATmega328P-PU microcontroller
- Jumper wires
- Two small DC motors and matching power
- 2WD robot vehicle chassis (such as PMD Way part number 72341119)
- Four AA battery cells
- 1N4004 power diode
- L293D motor driver IC

### The Chassis

The foundation of any robot vehicle is a solid chassis containing the motors, drivetrain, and power supply. You can choose from many chassis models available on the market. To keep things simple, this project relies on an inexpensive robot chassis with two small DC motors that operate at around 6 V DC and two matching wheels, as shown in Figure 8-12.

*Figure 8-12: Two-wheel-drive robot vehicle chassis (PMD Way part number 72341119)*

The task of physically assembling the robot chassis varies between models, but most require a few additional tools beyond those included in the kit, such as screwdrivers. If you haven't settled on a final design and wish to get your robot moving in a temporary configuration, you can attach the electronics to the chassis with a reusable putty adhesive like Blu-Tack.

### The Power Supply

The motors included with the robot chassis typically operate at around 6 V DC, so we'll use the 4 AA cell battery holder included with the example chassis in Figure 8-12. We can't use 6 V to power the microcontroller circuit, so we place a 1N4004 diode between the power supply positive and the 5 V connection on the microcontroller. The diode will cause a 0.7 V drop in voltage, bringing the microcontroller supply to around 5.3 V DC. The voltage will again drop as the battery life decreases.

Assemble the circuit as shown in Figure 8-13.

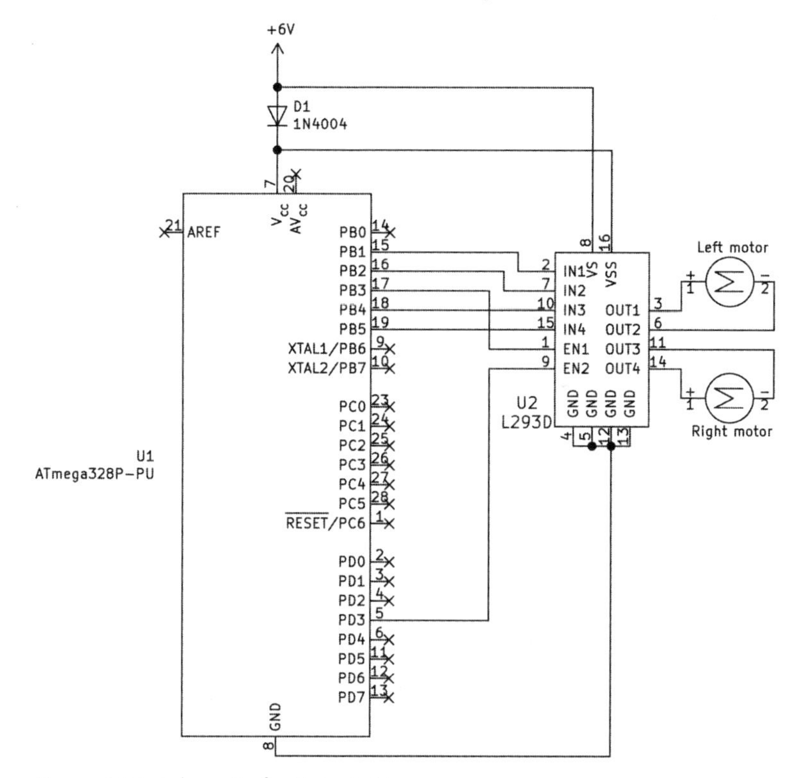


*Figure 8-13: Schematic for Project 37*

Again, the black/negative lead from the battery pack or external power connects to GND, and the red/positive lead runs to both the L293D $V_{CC}2$ pin and the 1N4004 diode.

## The Code

Open a terminal window, navigate to the *Project 37* subfolder of this book's *Chapter 8* folder, and enter the command **`make flash`**. Once you remove the AVR programmer and the vehicle starts up, it should wait three seconds before moving off forward and then turning left, right, and so on as directed by the sequence of functions in the main loop of the code.

This code is the culmination of our experiments with DC motor control using the L293D motor controller IC and PWM. Let's see how it works:

```
// Project 37 - Controlling a Two-Wheel-Drive Robot Vehicle

#include <avr/io.h>
#include <util/delay.h>

void initPWM(void) ❶
{
   TCCR2A |= (1 << WGM20);                 // Fast PWM mode
   TCCR2A |= (1 << WGM21); );              // Fast PWM mode, part 2
   TCCR2B |= (1 << CS21); );               // PWM Freq = F_CPU/8/256
}

void moveForward(uint8_t duty)
{
   // Set direction
   PORTB |= (1 << PORTB4)|(1 << PORTB1); ❷ // PB4,1 HIGH
   PORTB &= ~(1 << PORTB5)&~(1 << PORTB2); // PB5,2 LOW

   // Set speed
   if (duty == 255) ❸
   {
      PORTB |= (1 << PORTB3);              // Set PORTB3 to on
      PORTD |= (1 << PORTD3);              // Set PORTD3 to on
   } else if (duty < 255)
   {
      TCCR2A |= (1 << COM2A1); ❹           // PWM output on OCR2A-PB3
      TCCR2A |= (1 << COM2B1);             // PWM to OCR2B-PD3
      OCR2A = duty;                        // Set PORTB3 to PWM value
      OCR2B = duty;                        // Set PORTD3 to PWM value
   }
}

void moveBackward(uint8_t duty)
{
   // Set direction
   PORTB &= ~(1 << PORTB4)&~(1 << PORTB1); // PB4,1 LOW
   PORTB |= (1 << PORTB5)|(1 << PORTB2);   // PB5,2 HIGH

   // Set speed
   if (duty == 255)
   {
      PORTB |= (1 << PORTB3);              // Set PORTB3 to on
      PORTD |= (1 << PORTD3);              // Set PORTD3 to on
   } else if (duty < 255)
   {
      TCCR2A |= (1 << COM2A1);             // PWM output on OCR2A-PB3
      TCCR2A |= (1 << COM2B1);             // PWM to OCR2B-PD3
      OCR2A = duty;                        // Set PORTB3 to PWM value
      OCR2B = duty;                        // Set PORTD3 to PWM value
   }
}

void moveLeft(uint8_t duty)
{
```

```
   // Set direction
   PORTB |= (1 << PORTB4)|(1 << PORTB2);    // PB4,2 HIGH
   PORTB &= ~(1 << PORTB5)&~(1 << PORTB1); // PB5,1 LOW

   // Set speed
   if (duty == 255)
   {
      PORTB |= (1 << PORTB3);              // Set PORTB3 to on
      PORTD |= (1 << PORTD3);              // Set PORTD3 to on
   } else if (duty < 255)
   {
      TCCR2A |= (1 << COM2A1);             // PWM output on OCR2A-PB3
      TCCR2A |= (1 << COM2B1);             // PWM to OCR2B-PD3
      OCR2A = duty;                        // Set PORTB3 to PWM value
      OCR2B = duty;                        // Set PORTD3 to PWM value
   }
}

void moveRight(uint8_t duty)
{
   // Set direction
   PORTB |= (1 << PORTB5)|(1 << PORTB1);    // PB5,1 HIGH
   PORTB &= ~(1 << PORTB4)&~(1 << PORTB2); // PB4,2 LOW

   // Set speed
   if (duty == 255)
   {
      PORTB |= (1 << PORTB3);              // Set PORTB3 to on
      PORTD |= (1 << PORTD3);              // Set PORTD3 to on
   } else if (duty < 255)
   {
      TCCR2A |= (1 << COM2A1);             // PWM output on OCR2A-PB3
      TCCR2A |= (1 << COM2B1);             // PWM to OCR2B-PD3
      OCR2A = duty;                        // Set PORTB3 to PWM value
      OCR2B = duty;                        // Set PORTD3 to PWM value
   }
}

void motorsOff(void) ❺
{
TCCR2A &= ~(1 << COM2A1); // Disconnect PWM from OCR2A-PB3
TCCR2A &= ~(1 << COM2B1); // Disconnect PWM from OCR2B-PD3
PORTB &= ~(1 << PORTB3);  // Set ENABLE pins to zero for brake
PORTD &= ~(1 << PORTD3);
}

int main(void)
{
    // Set PORTB5, 4, 3, 2, and 1 as outputs
    DDRB |= (1 << PORTB5)|(1 << PORTB4)|(1 << PORTB3)|(1 << PORTB2)|(1 << PORTB1); ❻
    DDRD |= (1 << PORTD3); ❼                // Set PORTD3 as output
    initPWM(); ❽
    _delay_ms(3000);                       // Wait a moment before starting
    while(1)
    {
```

```
moveForward(128);
_delay_ms(2000);
moveLeft(128);
_delay_ms(2000);
moveRight(128);
_delay_ms(2000);
motorsOff();
moveBackward(128);
_delay_ms(2000);
}
        }
```

At ❶ and ❽, the code initiates PWM for two digital outputs so it can control two motors. After PWM initiation comes `moveForward()`, the first of five functions to control the motors. You might need to switch the wires on each motor if they appear to work opposite to the code. Four of these functions—`moveForward()`, `moveBackward()`, `moveLeft()`, and `moveRight()`—are identical, except in the order of motor rotation. They all accept a value for the duty cycle to control the speed of the motors. The function `motorsOff()` cuts the power off to both motors.

In this case, we set the direction of the motors forward by making digital outputs high or low, depending on required rotation type ❷. Refer to Table 8-1 for the requisite output configurations. The motor movement functions check if the user requires full speed (a duty cycle of 100 percent, represented by 255) ❸. If so, it simply sets the ENABLE pins of the L293D to on. However, if you pass a lower value for the duty cycle through a motor movement function, the program activates the PWM output to the ENABLE pins ❹ and fills the PWM registers OCR2A and B with the required duty cycle.

The other three movement functions operate similarly, except that the motor rotations are set up to allow for turning left or right or moving backward. The `motorsOff()` function stops movement by turning off PWM and setting both L293D ENABLE pins to low ❺. Finally, the program sets the six required pins to outputs to control the L293D ❻❼.

You can use the functions used in the main loop of the code to change the direction of movement, the speed via the duty cycles, and the duration with the delay functions, and stop the motors when required.

We have used a single timer with two PWM outputs for both motors (OCR2A and OCR2B) so that they share the same PWM generation and will thus synchronize with each other. If you use two different timers for two motors that need to operate together, the PWM signals will differ slightly and the two motors will operate slightly differently from one other.

Now that we have experimented with DC motors, in the next chapter we'll examine another useful tool of the AVR system: the internal EEPROM.

# USING THE INTERNAL EEPROM

When you define and use a variable in your AVR code, the stored data only lasts until the hardware is reset or the power is turned off. But what if you need to keep some values for future use? That's where we turn to the microcontroller's *electrically erasable programmable read-only memory (EEPROM)*, a special type of memory that holds information even when power is disconnected.

In this chapter, you will:

- Learn how to store byte, word, and floating-point variables in the microcontroller's EEPROM, and retrieve them.
- Build an ATtiny85 EEPROM storage and retrieval unit and a simple EEPROM datalogger with an ATmega328P-PU.
- Create a program to log temperatures to the ATmega328P-PU EEPROM for later retrieval.

## Storing Bytes in EEPROM

An EEPROM is a microcontroller component that doesn't need electricity to retain the contents of its memory. The concept originates from read-only memory (ROM) ICs, such as those found in gaming console cartridges, where the game code stays in the IC even when it's not connected to the console. Taking this concept further, an EEPROM lets the host controller write over old information with new information, which the EEPROM can still remember when the power is disconnected—that's what "electrically erasable" signifies.

Different AVR microcontrollers have EEPROMs of various sizes. For example, our ATtiny85 can store 512 bytes of data, while the ATmega328P-PU can store 1,024 bytes. In this chapter I'll show you how to store and retrieve data in both microcontrollers' EEPROMs so you can do so for your own projects. There are several different methods of storing data in the EEPROM, depending on the type of data. We'll begin by discussing how to store bytes.

Before we go any further, however, there are two things you'll need to keep in mind. First, the EEPROM has a lifespan of around 100,000 read/write cycles. Tests have shown that they may last longer, but be aware of the approximate lifespan when building your own projects. Second, remember that when you upload new code to your AVR, the data in the EEPROM is erased.

To use the EEPROM in our code for either microcontroller, we first include the EEPROM library:

```
#include <avr/eeprom.h>
```

Then, to write a byte of data (for example, a number between 0 and 255), we use the following function:

```
eeprom_write_byte((uint8_t*)a, b);
```

where *a* is the location inside the EEPROM—between 0 and 511 for the ATtiny85, and between 0 and 1023 for the ATmega328P-PU—and *b* is the byte of data to store, between 0 and 255. We prefix the location variable *a* with `(uint8_t*)` as the EEPROM functions require the parameter to be an 8-bit integer.

You can also *update* an EEPROM location to change the value stored in it, as follows:

```
eeprom_update_byte((uint8_t*)a, b);
```

where again *a* is the location inside the EEPROM—between 0 and 511 for the ATtiny85, and between 0 and 1023 for the ATmega328P-PU—and *b* is the byte of data to store, between 0 and 255. Before writing a byte of data to a location, an update command first checks the value currently at that location. If the value to be written is the same as the current value, no write occurs. Though this check adds processing time, it saves unnecessarily writing to the EEPROM, thus extending its lifespan.

To retrieve the byte of data stored in a location, use the following:

```
uint8_t i = eeprom_read_byte((uint8_t*)a);
```

This allocates the value stored in EEPROM location *a* to the variable `i`. You'll test a few of these functions out in the next project.

## Project 38: Experimenting with the ATtiny85's EEPROM

This project will demonstrate writing and retrieving bytes of data from the ATtiny85's EEPROM. It uses four LEDs as a quick way to display the numbers 0 to 15 in binary form, with diode D1 being the least significant bit (for 0) and diode D4 being the most significant bit (for 15).

You will need the following hardware:

- USBasp programmer
- Solderless breadboard
- ATtiny85–20PU microcontroller
- Four LEDs
- Four 560 Ω resistors
- Jumper wires

Assemble the circuit as shown in Figure 9-1.

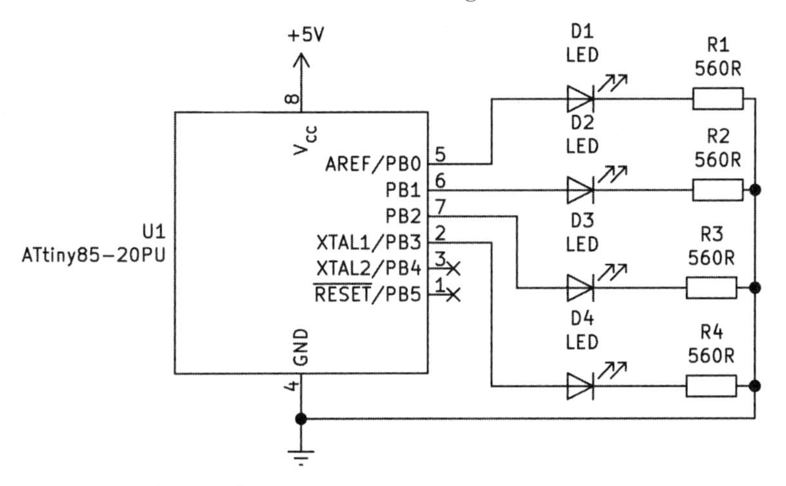

*Figure 9-1: Schematic for Project 38*

Open a terminal window, navigate to the *Project 38* subfolder of this book's *Chapter 9* folder, and enter the command **`make flash`**. After a moment, the LEDs should display the numbers 0 to 15 in binary and then repeat.

To see how this works, take a look at the *main.c* file for Project 38.

```
// Project 38 - Experimenting with the ATtiny85's EEPROM

#include <avr/io.h>
#include <util/delay.h>
❶ #include <avr/eeprom.h>

int main(void)
{
   DDRB = 0b00001111; // Set PORTB3-0 to output for LEDs
   int a;
   while (1)
   {
      // Write 0-15 to locations 0-15
      for (a=0; a<16; a++)
      {
       ❷ eeprom_update_byte((uint8_t*)a, a);
      }
      // Read locations 0-15, display data on LEDs
      for (a=0; a<16; a++)
      {
       ❸ PORTB = eeprom_read_byte((uint8_t*)a);
         _delay_ms(250);
      }
      // Turn off LEDs
      PORTB = 0b00000000;
   }
}
```

As mentioned earlier, we need to include the EEPROM library ❶ to take advantage of the functions to update and read bytes in the EEPROM. The first `for` loop repeats 16 times, updating the values in EEPROM locations 0 . . . 15 to 0 . . . 15 ❷. The second loop retrieves the data from EEPROM locations 0 . . . 15 and sets the PORTB register to the number retrieved from the EEPROM ❸. This activates the LEDs connected to the matching pins of PORTB, thus displaying each value in binary.

Now that you know how to store small numbers in the microcontroller's EEPROM, I'll show you how to store larger numbers using words of data.

## Storing Words

A *word* of data uses 16 bits, or 2 bytes, to represent 16-bit signed or unsigned integers. As you learned in Chapter 2, these can be in the range of –32,768 to 32,767 for signed integers or 0 to 65,535 for unsigned integers. For example, a word could represent 12,345 or –23,567. To write words, we again use functions from the EEPROM library, which we include as follows:

```
#include <avr/eeprom.h>
```

To write a word of data, we use the following function:

```
eeprom_write_word((uint16_t*)a, b);
```

where *a* is the location inside the EEPROM and *b* is the word of data to store. While a word of data is 2 bytes in size, an EEPROM location is 1 byte in size. This means that when you write a word of data it will fill two EEPROM locations. Consequently, if you want to write two words of data at the start of the EEPROM, you'll need to write the first word to location 0 and the second word to location 2.

As with bytes, you can also update words. You do this with the following function:

```
eeprom_update_word((uint16_t*)a, b);
```

To retrieve the word of data stored in a location, use one of the following:

```
uint16_t i = eeprom_read_word((uint16_t*)a);
// For values between 0 and 65535

int16_t i = eeprom_read_word((int16_t*)a);
// For values between -32768 and 32767
```

This allocates the value stored in EEPROM location *a* to the variable `i`. Note that *a* should be the first of the two locations where the word is stored, not the second (so, 0 or 2 in our previous example). You'll test these functions out in the next project.

## Project 39: A Simple EEPROM Datalogger

In this project, you'll create a basic data-logging device. It not only demonstrates writing words of data to and reading them from the ATmega328P-PU's EEPROM but also incorporates the USART and custom functions. Instead of writing arbitrary numbers to the EEPROM, this project repeatedly reads the status of digital input pin PORTB0, writing a 0 or 1 to the specified EEPROM location (for low or high, respectively). We'll use the USART to create a basic text-based interface control system to log, retrieve, and erase EEPROM data.

### The Hardware

You will need the following hardware:

- USBasp programmer
- Solderless breadboard
- ATmega328P-PU microcontroller
- USB-to-serial converter
- Jumper wires

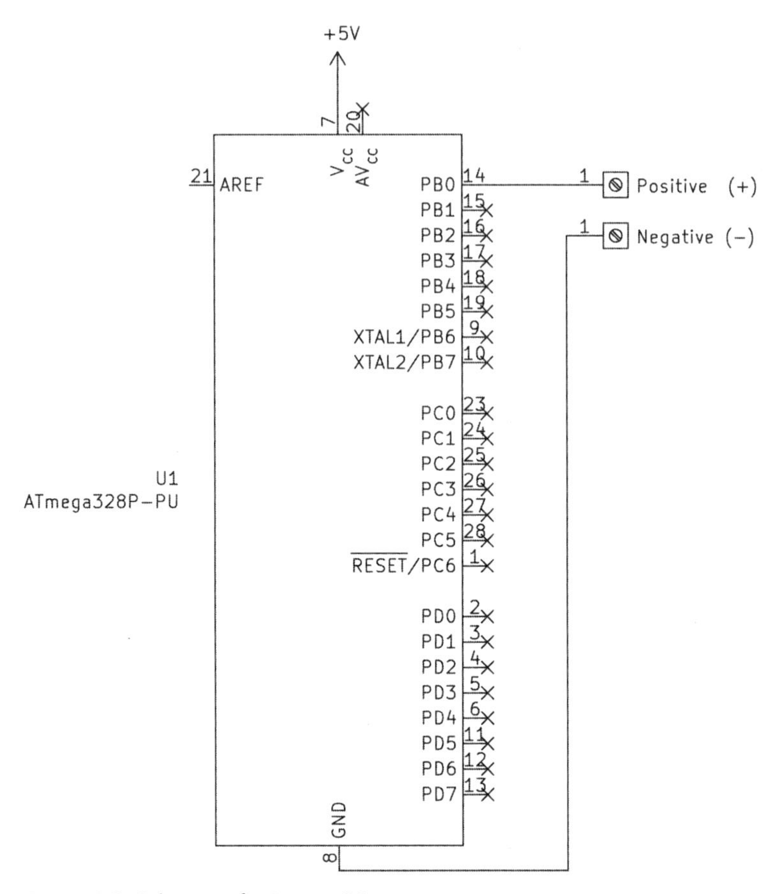


*Figure 9-2: Schematic for Project 39*

Don't forget to connect the USB-to-serial converter as demonstrated in Chapter 4. The positive and negative points shown in the schematic are used to log the low and high signals. To experiment, you could run tests connecting only the positive to 5 V or GND (the negative must always go to GND).

## The Code

Open a terminal window, navigate to the *Project 39* subfolder of this book's *Chapter 9* folder, and enter the command `make flash`. Then open the terminal software you installed in Chapter 4. After a moment you will be prompted to "Enter 1 to start, 2 to dump, 3 to erase." Press **1** on your computer's keyboard to run the data-logging function, **2** for the microcontroller to read the EEPROM and send the data back to the terminal window, or **3** to erase the data by writing all the EEPROM locations back to 0.

Figure 9-3 shows an example of this sequence. (In the interest of saving space, I've altered the code that produced this figure to use only the first 10 EEPROM locations. When you run the code, your sequences will be much longer.)

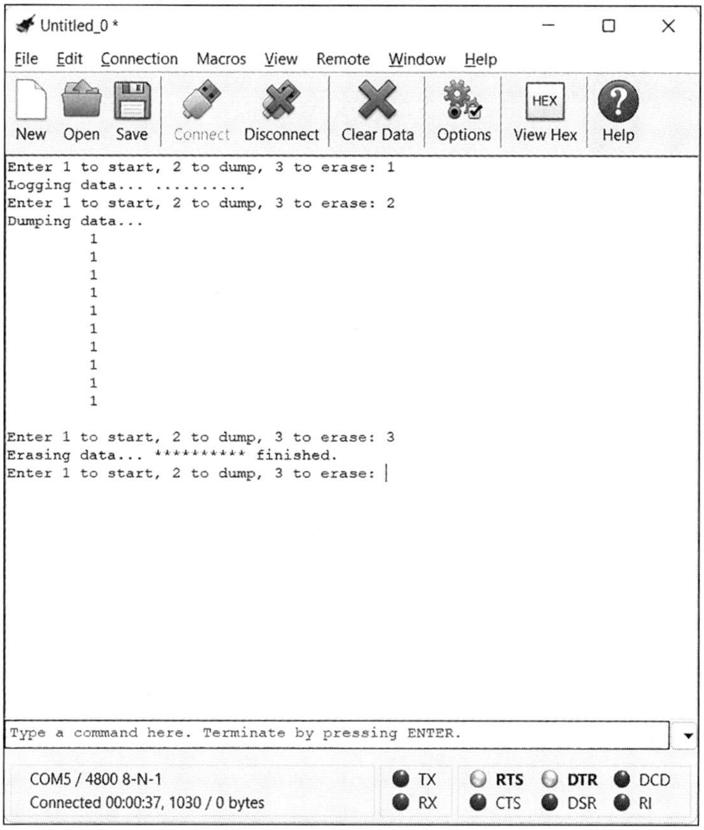

```
Enter 1 to start, 2 to dump, 3 to erase: 1
Logging data... ..........
Enter 1 to start, 2 to dump, 3 to erase: 2
Dumping data...
         1
         1
         1
         1
         1
         1
         1
         1
         1
         1

Enter 1 to start, 2 to dump, 3 to erase: 3
Erasing data... ********** finished.
Enter 1 to start, 2 to dump, 3 to erase:
```

*Figure 9-3: Example output for Project 39*

Let's see how this works:

```
// Project 39 - A Simple EEPROM Datalogger

#include <avr/io.h>
#include <util/delay.h>
#include <avr/eeprom.h>
#include <math.h>
#include <stdio.h>
#include <stlib.h>

#define USART_BAUDRATE 4800
#define UBRR_VALUE 12
❶ #define logDelay 1000

char newline[4] = "\r\n";

❷ void USARTInit(void)
{
   // Set baud rate registers
   UBRR0H = (uint8_t)(UBRR_VALUE>>8);
```

```
    UBRR0L = (uint8_t)UBRR_VALUE;
    // Set data type to 8 data bits, no parity, 1 stop bit
    UCSR0C |= (1<<UCSZ01)|(1<<UCSZ00);
    // Enable transmission and reception
    UCSR0B |= (1<<RXEN0)|(1<<TXEN0);
}

❸ void USARTSendByte(unsigned char u8Data)
{
    // Wait while previous byte is sent
    while(!(UCSR0A&(1<<UDRE0))){};
    // Transmit data
    UDR0 = u8Data;
}

❹ uint8_t USARTReceiveByte()
// Receives a byte of data from the computer into the USART register
{
    // Wait for byte from computer
    while(!(UCSR0A&(1<<RXC0))){};
    // Return byte
    return UDR0;
}

❺ void sendString(char myString[])
{
    uint8_t a = 0;
    while (myString[a])
    {
       USARTSendByte(myString[a]);
       a++;
    }
}

❻ void logData()
{
    uint16_t portData = 0;
    uint16_t location = 0;
    char z1[] =  "Logging data . . .";
    sendString(z1);
    for (location=0; location<1024; location++)
    {
       if (PINB == 0b0000001) // If PORTB0 is HIGH
       {
          eeprom_update_word((uint16_t*)location, 1);
       } else
       {
          eeprom_update_word((uint16_t*)location, 0);
       }
       location++;             // Skip an EEPROM location as we're using words
       USARTSendByte('.');
       _delay_ms(logDelay);
     }
     sendString(newline);
}
```

```
❼ void dumpData()
  {
      uint8_t portData = 0;
      uint16_t location = 0;
      char t[10] = "";          // For our dtostrf
      char z1[] = "Dumping data . . .";
      sendString(z1);
      sendString(newline);
      for (location=0; location<1024; location++)
         {
           // Retrieve data from EEPROM location
            portData=eeprom_read_word((uint16_t*)location);
            dtostrf((float)portData,12,0,t);
            sendString(t);
            sendString(newline);
            location++;         // Skip an EEPROM location as we're using words
         }
         sendString(newline);
  }

❽ void eraseData()
  {
     uint16_t location = 0;
     char msg2[] = "Erasing data . . .";
     char msg3[] = " finished.";
     sendString(msg2);
     for (location=0; location<1024; location++)
     {
        eeprom_write_byte((uint16_t*)location, 0); // Write 0 to EEPROM location
        USARTSendByte('*');
     }
     sendString(msg3);
     sendString(newline);
  }

  int main(void)
  {
     DDRB = 0b11111111;         // Set PORTB0 as input
     char msg1[44] = "Enter 1 to start, 2 to dump, 3 to erase: ";
     uint8_t option;
     USARTInit();
     while (1)
     {
      ❾ sendString(msg1);
        option = USARTReceiveByte();
        USARTSendByte(option);
        sendString(newline);
        switch (option)
        {
           case 49 : logData(); break;
           case 50 : dumpData(); break;
           case 51 : eraseData(); break;
        }
     }
  }
```

First we import all the required libraries and set the data speed for the USART, as usual. We also set `logDelay` to 1,000 ❶, specifying the delay in milliseconds between each logging event (you can change this value to suit your own needs).

Next, we declare the functions required to initialize the USART ❷, send bytes from the USART to the computer ❸ and receive bytes coming in the other direction ❹, and send strings to the terminal emulator ❺. When needed, the user can call the data-logging function `logData()` ❻. This function reads the value of PORTB0 and writes a 1 for high or 0 for low to the EEPROM locations from 0 to 1,022 in turn. The function skips writing to every second location, as we need two locations per byte. If you want to increase the time between logging events, you can adjust the speed, as mentioned earlier, by altering the value of `logDelay` ❶.

The function `dumpData()` ❼ sends the value of each EEPROM location to the USART and thus the PC for viewing. As with the `logData()` function, it skips every second location, since we use two locations per word. Before running this function, you can set the terminal software to capture output for further analysis with a spreadsheet, as demonstrated in Project 19 in Chapter 4.

The `eraseData()` function ❽ writes a 0 in every EEPROM location, thus writing over any previously stored data. While not really necessary here, this function might be useful in your own future projects that require erasing data in the EEPROM.

The main code loop provides a user interface of sorts, by prompting the user for their selection ❾ and then calling the required function using the `switch...case` statement, following the displayed menu options.

Now that you can store and retrieve bytes and words of data in the EEPROM, we'll move on to our final type of data: floating-point variables.

## Storing Floating-Point Variables

A *floating-point variable* represents a floating-point number (as described in Chapter 3) that falls in the range of $-3.39 \times 10^{38}$ to $3.39 \times 10^{38}$. These variables require 32 bits of storage, or 4 bytes. To write a floating-point (float) variable, we again need to include the EEPROM library:

```
#include <avr/eeprom.h>
```

We then use the following function to write a word of data (for example, a number between 0 and 65,535):

```
eeprom_write_float((float*)a, b);
```

where *a* is the location inside the EEPROM and *b* is the float of data to store.

To update a float stored in the EEPROM, we use this function:

```
eeprom_update_float((float*)a, b);
```

As a float takes up 4 bytes and an EEPROM location can contain only 1 byte, you'll need to allocate four EEPROM locations when storing a float. For example, if you were writing two floats of data at the start of the EEPROM, you'd write the first one to location 0 and the second to location 4.

To retrieve a float stored in a given location, use the following function:

```
float i = eeprom_read_float((float*)a);
```

This allocates the value stored in EEPROM location *a* to the variable *i*. Remember that you always need to use the first location when using words. In the next project, you'll put the ability to store floats in EEPROM to use.

## Project 40: Temperature Logger with EEPROM

This project combines your knowledge of capturing data with a TMP36 temperature sensor (introduced in Chapter 3) and writing floating-point variable data to and reading it from the EEPROM, again using the USART and custom functions. The project code samples and stores the temperature in the EEPROM 256 times, so you can retrieve and view the readings with your terminal software or capture the data for spreadsheet analysis.

### *The Hardware*

You will need the following hardware:

- USBasp programmer
- Solderless breadboard
- 5 V breadboard power supply
- ATmega328P-PU microcontroller
- One TMP36 temperature sensor
- 0.1 μF ceramic capacitor
- Jumper wires
- USB-to-serial converter

Assemble your circuit as shown in Figure 9-4, using the external power supply and wiring up the USB-to-serial converter.

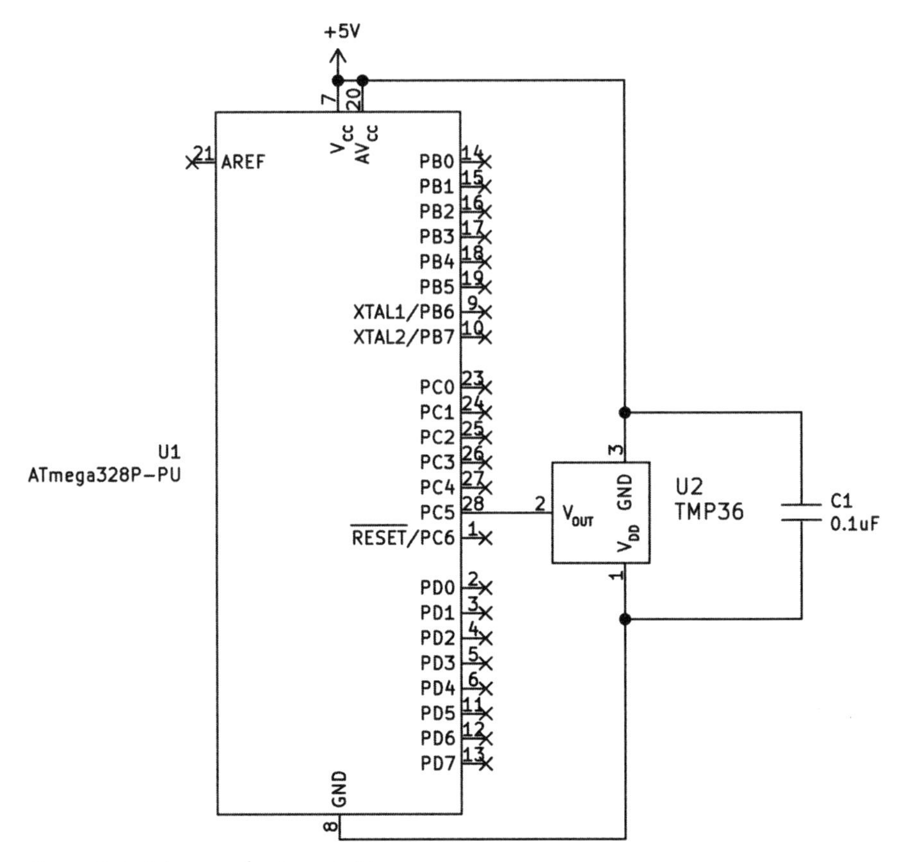


*Figure 9-4: Schematic for Project 40*

Open a terminal window, navigate to the *Project 40* subfolder of this book's *Chapter 9* folder, and enter the command `make flash`. Next, open the terminal application, just as you did for the previous project. After a moment you should be presented with the option to enter 1 to start, 2 to dump, or 3 to erase. Pressing 1 on your computer's keyboard should run the temperature data logging function; pressing 2 tells the microcontroller to read the EEPROM and send the temperature data back to the terminal software for display, and pressing 3 should erase the data by writing all the EEPROM locations back to 0.

Figure 9-5 shows an example of this sequence. (Again, to save space, I've altered the code to use only the first six EEPROM locations.)

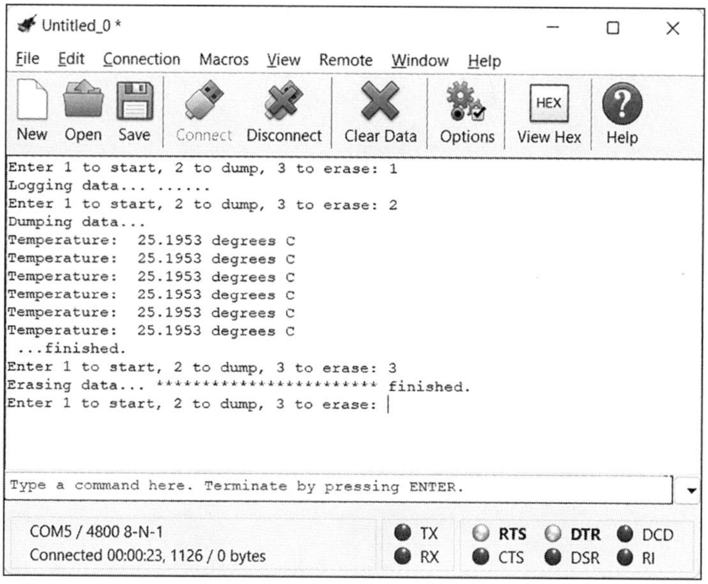

Figure 9-5: Example output for Project 40

Let's look at the code to see how this works:

```
// Project 40 - Temperature Logger with EEPROM

#include <avr/io.h>
#include <util/delay.h>
#include <avr/eeprom.h>
#include <math.h>
#include <stdio.h>

#define USART_BAUDRATE 4800
#define UBRR_VALUE 12
❶ #define logDelay 1000

char newline[4] = "\r\n";

❷ void startADC()
// Set up the ADC
{
   ADMUX |= (1 << REFS0);                      // Use AVcc pin with ADC
   ADMUX |= (1 << MUX2) | (1 << MUX0);         // Use ADC5 (pin 28)
   ADCSRA |= (1 << ADPS1) | (1 << ADPS0); // Prescaler for 1MHz (/8)
   ADCSRA |= (1 << ADEN);                      // Enable ADC
}

❸ void USARTInit(void)
{
   // Set baud rate registers
   UBRR0H = (uint8_t)(UBRR_VALUE>>8);
   UBRR0L = (uint8_t)UBRR_VALUE;
```

```
    // Set data type to 8 data bits, no parity, 1 stop bit
    UCSR0C |= (1<<UCSZ01)|(1<<UCSZ00);
    // Enable transmission and reception
    UCSR0B |= (1<<RXEN0)|(1<<TXEN0);
}

void USARTSendByte(unsigned char u8Data)
{
    // Wait while previous byte is sent
    while(!(UCSR0A&(1<<UDRE0))){};
    // Transmit data
    UDR0 = u8Data;
}

uint8_t USARTReceiveByte()
// Receives a byte of data from the computer into the USART register
{
    // Wait for byte from computer
    while(!(UCSR0A&(1<<RXC0))){};
    // Return byte
    return UDR0;
}

void sendString(char myString[])
{
    uint8_t a = 0;
    while (myString[a])
    {
        USARTSendByte(myString[a]);
        a++;
    }
}

❹ float readTemperature()
{
    float temperature;
    float voltage;
    uint8_t ADCvalue;
    // Get reading from TMP36 via ADC
    ADCSRA |= (1 << ADSC);         // Start ADC measurement
    while (ADCSRA & (1 << ADSC) ); // Wait until conversion is complete
    _delay_ms(10);

    // Get value from ADC register, place in ADCvalue
    ADCvalue = ADC;

    // Convert reading to temperature value (Celsius)
    voltage = (ADCvalue * 5);
    voltage = voltage / 1024;
    temperature = ((voltage - 0.5) * 100);
    return temperature;
}

❺ void logData()
{
```

```
    float portData = 0;
    uint16_t location = 0;
    char z1[] =  "Logging data . . .";
    sendString(z1);
    for (location=0; location<1021; location=location+4)
    {
       portData=readTemperature();
       eeprom_update_float((float*)location,portData);
       USARTSendByte('.');
    }
    sendString(newline);
    _delay_ms(logDelay);
}

❻ void dumpData()
{
    float portData = 0;
    uint16_t location = 0;
    char t[10] = "";                    // For our dtostrf
    char msg1[14] = "Temperature: "; // Make sure you have " instead of "
    char msg2[12] = " degrees C ";
    char msg3[] = "Dumping data . . .";
    char msg4[] = ". . . finished.";
    sendString(msg3);
    sendString(newline);
    for (location=0; location<1021; location=location+4)
    {
       sendString(msg1);
       portData=eeprom_read_float((float*)location); // HERE
       dtostrf(portData,8,4,t);
       sendString(t);
       sendString(msg2);
       sendString(newline);
    }
    sendString(msg4);
    sendString(newline);
}

❼ void eraseData()
{
    int16_t location = 0;
    char msg1[] = "Erasing data . . .";
    char msg2[] = " finished.";
    sendString(msg1);
    for (location=0; location<1024; location++)
    {
       eeprom_write_byte((uint8_t*)location, 0);
       USARTSendByte('*');
    }
    sendString(msg2);
    sendString(newline);
}

int main(void)
{
```

```
❽ char msg1[44] = "Enter 1 to start, 2 to dump, 3 to erase: ";
   uint8_t option;
   DDRD = 0b00000000;              // Set PORTD to inputs
   startADC();
   USARTInit();
   while (1)
   {
      sendString(msg1);
      option = USARTReceiveByte();
      USARTSendByte(option);
      sendString(newline);
      switch (option)
      {
         case 49 : logData(); break;
         case 50 : dumpData(); break;
         case 51 : eraseData(); break;
      }
   }
}
```

This project again pulls together knowledge from previous chapters to bring a new idea to life. First we import all the required libraries and set the data speed for the USART, as usual. I've again set the delay in milliseconds between each logging event to 1,000, but you can adjust the speed by altering the `logDelay` value ❶. Next, we provide the functions required to initialize and operate the ADC ❷ and the USART ❸, as in the previous project.

The `readTemperature()` function ❹ takes the temperature readings from the TMP36; we'll call this from the `logData()` function ❺, which stores these readings in the EEPROM locations from 0 to 1020 in turn, skipping three each time as we need four locations per float variable.

The `dumpData()` function ❻ sends the value of each EEPROM location to the USART and thus to the PC for viewing. Like `logData()`, this function skips to every fourth EEPROM location so that we have space to store our float variables. Before running this function, you can set the terminal software to capture the output in a text file, which you can open in a spreadsheet; see Project 19 in Chapter 4 if you need a refresher on this.

The `eraseData()` function ❼ writes a 0 in every EEPROM location, thus erasing any previously stored data. As mentioned in Project 39, you may find it useful in your own projects.

The main code loop provides a user interface of sorts, prompting the user to choose whether to log, dump, or erase the data ❽ and then calling the required function using the `switch...case` statement, following the displayed menu options.

Along with learning these EEPROM functions, in this chapter you've taken one step further toward developing complex projects that could inspire your own projects later in your microcontroller journey. In the next chapter, you'll learn how to make your own libraries to save time and write more useful code.

# WRITING YOUR OWN AVR LIBRARIES

Cast your mind back to Project 15 in Chapter 3, which required us to convert the voltage measured by the TMP36 temperature sensor to degrees Celsius. To complete those calculations, we called the math library and used the functions within it to perform operations on floating-point numbers. Using this library meant we didn't have to create our own mathematical functions or include their code in the project, saving time and effort.

In this chapter, you'll learn to create your own libraries, allowing you to reuse tested functions in multiple projects to increase your efficiency. You'll build a simple library for a repetitive task, a library that accepts values to perform a function, and a library that processes data from a sensor and returns values in an easy-to-use form. These examples will equip you with the skills required to make your own custom libraries.

## Creating Your First Library

In this section you'll create your first library, which you'll then use in Project 41. First, consider the functions defined in Listing 10-1, `blinkSlow()` and `blinkFast()`. These two functions blink an LED (connected via a resistor between PORTB0 and GND) at a slow or fast rate, respectively.

```
#include <avr/io.h>
#include <util/delay.h>

void blinkSlow()
{
  uint8_t i;
  for (i = 0; i < 5; i++)
  {
    PORTB |= (1 << PORTB0);
    _delay_ms(1000);
    PORTB &= ~(1 << PORTB0);
    _delay_ms(1000);
  }
}

void blinkFast()
{
  uint8_t i;
  for (i = 0; i < 5; i++)
  {
    PORTB |= (1 << PORTB0);
    _delay_ms(250);
    PORTB &= ~(1 << PORTB0);
    _delay_ms(250);
  }
}

int main(void)
{
    DDRB |= (1 << PORTB0); // Set PORTB0 as outputs
    while (1)
    {
        blinkSlow();
       _delay_ms(1000);
        blinkFast();
       _delay_ms(1000);
    }
}
```

*Listing 10-1: Example code that demonstrates two functions that blink LEDs slow and fast*

The custom functions to blink the LED slowly or rapidly are convenient, but it's not very efficient to enter them into your project code every time you want to use them. However, if you offload the code that describes the functions into a library, you can simply call the library with one line in future projects, then use the functions as needed without rewriting them. Let's create such a library now.

## *Anatomy of a Library*

A library consists of two files: *library.h*, the header file, and *library.c*, the source file, where "library" is a placeholder for an individual library's name. We'll call our first example library the *blinko* library, so our two files will be *blinko.h* and *blinko.c*.

A header file contains the definitions of the functions, variables, or other components the library contains. The following is our header file:

```
// blinko.h

void blinkSlow();
// Blinks PORTB0 slowly, five times

void blinkFast();
// Blinks PORTB0 rapidly, five times
```

The file declares the names of the two functions inside the library, `void blinkSlow()` and `void blinkFast()`. Each of these lines is followed by a comment describing the function's purpose. Get into the habit of including comments like these about the custom functions in your library.

Our source file contains the code that will be made available to the main code in Project 41 when we include this library:

```
  // blinko.c

❶ #include <avr/io.h>
  #include <util/delay.h>

❷ void blinkSlow()
  {
    uint8_t i;
    for (i = 0; i < 5; i++)
    {
      PORTB |= (1 << PORTB0);
      _delay_ms(1000);
      PORTB &= ~(1 << PORTB0);
      _delay_ms(1000);
    }
  }

❸ void blinkFast()
  {
    uint8_t i;
    for (i = 0; i < 5; i++)
    {
      PORTB |= (1 << PORTB0);
      _delay_ms(250);
      PORTB &= ~(1 << PORTB0);
      _delay_ms(250);
    }
  }
```

The *blinko.c* file is identical to the first section of Listing 10-1. We first include the other libraries required by the code in our own library ❶—this allows us to use the functions for I/O and `_delay_ms()`. We then add the `blinkSlow()` ❷ and `blinkFast()` ❸ custom functions we want to include in the library.

## Installing the Library

To make the library available to the main code for Project 41, we have to do two things. First, we copy the header and source files into the same project directory as the *main.c* file and Makefile, as shown in the directory listing screen capture in Figure 10-1. You can find these in the *Project 41* subfolder of this book's *Chapter 10* folder.

```
C:\Users\john\Dropbox\AVRcode\AVRW\c10\p41>ls -la
total 7
drwxr-xr-x    2 john     Administ     4096 Jul 20 11:06 .
drwxr-xr-x    7 john     Administ     4096 Jul 20 11:04 ..
-rw-r--r--    1 john     Administ     2549 Jan  6  2021 Makefile
-rw-r--r--    1 john     Administ      418 Jan  6  2021 blinko.c
-rw-r--r--    1 john     Administ      128 Jan  6  2021 blinko.h
-rw-r--r--    1 john     Administ      322 Jul 20 11:04 main.c
```

*Figure 10-1: Place the library files in the same directory as the project files.*

Second, we edit the project's Makefile so that the toolchain knows to look for the library when compiling the code to upload to the microcontroller. To do so, we add `blinko.c` after `main.o` in the Makefile's `OBJECTS` line, as shown in Figure 10-2.

```
19  PROGRAMMER = -c USBasp
20
21  ## added "blinko.c" to objects list in line 22
22  OBJECTS     = main.o blinko.c
23  FUSES       = -U lfuse:w:0x62:m -U hfuse:w:0xdf:m -U
24
```

*Figure 10-2: Adding the* blinko.c *library to the Makefile for Project 41*

Now that we've installed the library, let's put it to the test by using it to program a simple circuit.

## Project 41: Your First Library

You will need the following hardware for this project:

- USBasp programmer
- Solderless breadboard
- ATmega328P-PU microcontroller
- One LED
- One 560 Ω resistor

Assemble the circuit shown in Figure 10-3 on your breadboard.

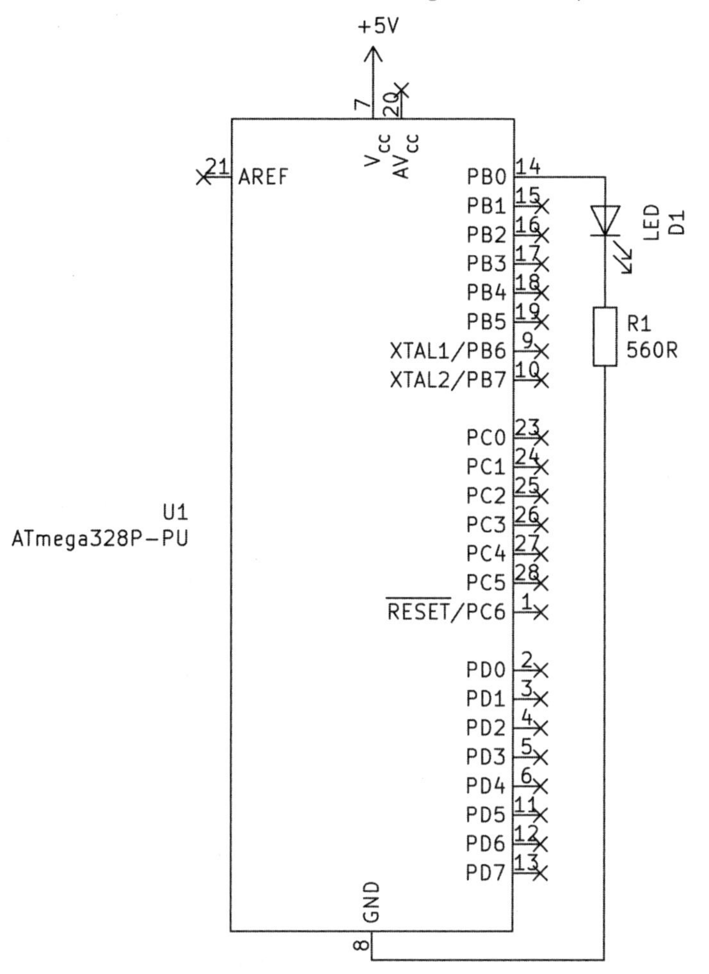


*Figure 10-3: Schematic for Project 41*

Keep this circuit together, as you'll use it again in the following project.

Next, open a terminal window, navigate to the *Project 41* subfolder of this book's *Chapter 10* folder, and enter the command **make flash**. The LED should blink rapidly five times, then blink slowly five times, then repeat.

The code for this project is quite simple:

```
// Project 41 - Your First Library

#include <avr/io.h>
#include <util/delay.h>
❶ #include "blinko.h"   // Use our new library

int main(void)
{
   DDRB = 0b11111111; // Set PORTB as outputs
   for(;;)
```

```
   {
    ❷ blinkSlow();
      _delay_ms(1000);
      blinkFast();
      _delay_ms(1000);
   }
   return 0;
}
```

First, we include our new library ❶. (Note that the custom library name is surrounded by quotes, not left and right angle brackets.) We then take advantage of our library functions to blink the LED ❷.

Although this project is a somewhat minimal demonstration, it illustrates the basic process for creating and using your own AVR libraries. Next, we'll look at some more complex examples.

## Creating a Library That Accepts Values to Perform a Function

Now that you know how to create a basic library, you're ready for the next level: creating a library that can accept values and act on them. Again, we'll begin with an example function and convert that into a library.

Consider the code in Listing 10-2, which uses the `blinkType()` function to set the number of times to blink the LED connected to PORTB0, as well as the on/off period.

```
#include <avr/io.h>
#define __DELAY_BACKWARD_COMPATIBLE__ // Required for macOS users
#include <util/delay.h>

void blinkType(int blinks, int duration)
// blinks - number of times to blink the LED
// duration - blink duration in milliseconds
{
   uint8_t i;
   for (i = 0; i < blinks; i++)
   {
      PORTB |= (1 << PORTB0);
      _delay_ms(duration);
      PORTB &= ~(1 << PORTB0);
      _delay_ms(duration);
   }
}

int main(void)
{
   DDRB |= (1 << PORTB0);              // Set PORTB0 as outputs
   while (1)
   {
      // Blink LED 10 times, with 150 ms duration
      blinkType(10, 150);
      _delay_ms(1000);
```

```
        // Blink LED 5 times, with 1 s duration
        blinkType(5, 1000);
        _delay_ms(1000);
    }
}
```

*Listing 10-2: An example sketch that demonstrates our LED blinking function, which will be converted into a library*

As you can see, `blinkType()` accepts two values and then acts on them. The `blinks` value is the number of times you'd like to turn the onboard LED on and off, and the `duration` value is the delay in milliseconds for each blink. Let's turn this into a library named *blinko2*.

First we need to create the *blinko2.h* header file, which contains the definitions of the functions and variables used inside the library:

```
// blinko2.h

void blinkType(int blinks, int duration);
// blinks - number of times to blink the LED
// duration - blink duration in milliseconds
```

As before, we declare the name of the function inside the library, followed by a comment describing its purpose. In this case, we provide comments describing the function's parameters.

Next, we build our *blinko2.c* source file:

```
// blinko2.c

❶ #include <avr/io.h>
❷ #define __DELAY_BACKWARD_COMPATIBLE__
  #include <util/delay.h>

❸ void blinkType(int blinks, int duration)
  // blinks - number of times to blink the LED
  // duration - blink duration in milliseconds
  {
    uint8_t i;
    for (i = 0; i < blinks; i++)
    {
      PORTB |= (1 << PORTB0);
      _delay_ms(duration);
      PORTB &= ~(1 << PORTB0);
      _delay_ms(duration);
    }
  }
```

The source file includes the necessary libraries for operating our library ❶, then our library's function ❸, as usual. The line at ❷ is required for those of you using Apple computers, as the version of the compiler is slightly different.

The next step is to edit the Makefile of the project in which we'll use this library, which you can find in the *Project 42* subfolder of the book's *Chapter 10* folder. Add the library name `blinko2.c` after `main.o` in the `OBJECTS` line, as shown in Figure 10-4.

```
20
21 ## added "blinko2.c" to objects list in line 22
22 OBJECTS     = main.o blinko2.c
23 FUSES       = -U lfuse:w:0x62:m -U hfuse:w:0xdf:m -U efuse:w:0xff:m
24
```

*Figure 10-4: Adding the* blinko2.c *library to the Makefile for Project 42*

Now that you have the library set up, let's test it out.

## Project 42: Using the blinko2.c Library

You can use the hardware you assembled for Project 41 for this project as well. Open a terminal window, navigate to the *Project 42* subfolder of this book's *Chapter 10* folder, and enter the command **make flash**. The LED should blink rapidly 10 times, then blink slowly 5 times, then repeat.

Let's see how this works:

```
// Project 42 - Using the blinko2.c Library

#include <avr/io.h>
#include <util/delay.h>
❶ #include "blinko2.h"  // Use our new library

int main(void)
{
   DDRB = 0b11111111; // Set PORTB as outputs
   for(;;)
   {
      // Blink LED 10 times, with 150 ms duration
    ❷ blinkType(10, 150);
      _delay_ms(1000);

      // Blink LED 5 times, with 1 s duration
    ❸ blinkType(5, 1000);
      _delay_ms(1000);
   }
   return 0;
}
```

As before, we include our new library ❶, then use that library's function to blink the LED rapidly with the short duration ❷, then slowly with a longer duration ❸. Again, this is intended as a simple demonstration that gives you the framework for creating your own AVR libraries with functions that can accept values. If you'd like a challenge, you can try creating your own PWM library based on the example code from Chapter 7.

# Creating a Library That Processes Data and Returns Values

For this chapter's final project, you'll learn how to create a library that can return values back to the main code. We'll create a "thermometer" library that not only returns values from an Analog Devices TMP36 temperature sensor but also has a function to simplify displaying numbers on a seven-segment LED display.

Our library source code, which you'll find in the *Project 43* subfolder of this book's *Chapter 10* folder, contains two functions: one to return the value in degrees Celsius from the temperature sensor as a float variable, and another that accepts an integer between 0 and 99 to display on the single-digit LED display from Project 15. Let's take a look at the *temperature.h* header file defining the functions and variables used inside the library:

```
// thermometer.h

void displayNumber(uint8_t value);
// Displays a number between 0 and 99 on the seven-segment LED display

float readTMP36();
// Returns temperature from TMP36 sensor using ATmega328P-PU pin PC5
```

As usual, we declare the names of the functions inside the library and provide comments describing their use. Note that the type of the `readTMP36()` function is `float`, not `void`, as this function will return a floating-point value for the temperature to our project's main code.

Next, let's examine our *thermometer.c* source file:

```
  // thermometer.c

❶ #include <avr/io.h>
  #include <math.h>
  #include <util/delay.h>

❷ float readTMP36()
  // Returns temperature from TMP36 sensor using ATmega328P-PU pin PC5
  {
     float temperatureC;
     float voltage;
     uint16_t ADCvalue;

     ADCSRA |= (1 << ADSC);                  // Start ADC measurement
     while (ADCSRA & (1 << ADSC) );          // Wait for conversion to finish
     _delay_ms(10);

     // Get value from ADC (which is 10-bit) register, store in ADCvalue
     ADCvalue = ADC;

     // Convert reading to temperature value (Celsius)
     voltage = (ADCvalue * 5);
     voltage = voltage / 1024;
     temperatureC = ((voltage - 0.5) * 100);
   ❸ return temperatureC;
  }

❹ void displayNumber(uint8_t value)
```

```
// Displays a number between 0 and 99 on the seven-segment LED display
{
   uint8_t tens=0;
   uint8_t ones=0;
   uint8_t delayTime=250;
   tens = value / 10;
   ones = value % 10;
   switch(tens)
   {
      case 0 : PORTB = 0b00111111; break; // 0
      case 1 : PORTB = 0b00000110; break; // 1
      case 2 : PORTB = 0b01011011; break; // 2
      case 3 : PORTB = 0b01001111; break; // 3
      case 4 : PORTB = 0b01100110; break; // 4
      case 5 : PORTB = 0b01101101; break; // 5
      case 6 : PORTB = 0b01111101; break; // 6
      case 7 : PORTB = 0b00000111; break; // 7
      case 8 : PORTB = 0b01111111; break; // 8
      case 9 : PORTB = 0b01101111; break; // 9
   }
   _delay_ms(delayTime);
   switch(ones)
   {
      case 0 : PORTB = 0b00111111; break; // 0
      case 1 : PORTB = 0b00000110; break; // 1
      case 2 : PORTB = 0b01011011; break; // 2
      case 3 : PORTB = 0b01001111; break; // 3
      case 4 : PORTB = 0b01100110; break; // 4
      case 5 : PORTB = 0b01101101; break; // 5
      case 6 : PORTB = 0b01111101; break; // 6
      case 7 : PORTB = 0b00000111; break; // 7
      case 8 : PORTB = 0b01111111; break; // 8
      case 9 : PORTB = 0b01101111; break; // 9
   }
   _delay_ms(delayTime);
   PORTB = 0; // Turn off display
}
```

All the code in the library should be familiar to you by now. First, we include the required libraries for use in our library ❶. The `readTMP36()` function ❷ sends the temperature back in degrees Celsius, using the return function ❸ in the same way as the custom function explained in Project 11 in Chapter 3. The `displayNumber(uint8_t value)` function displays an integer between 0 and 99 on the single-digit LED display ❹.

As before, to make this library available for use in Project 43, we add it to line 22 in the Makefile, as shown in Figure 10-5.

```
21  ## added "thermometer.c" to objects list in line 22
22  OBJECTS    = main.o thermometer.c
```

*Figure 10-5: Adding the* thermometer.c *library to the Makefile for Project 43*

You're now ready to build your digital thermometer using this library.

## Project 43: Creating a Digital Thermometer with the thermometer.c Library

In this project, you'll read an analog temperature sensor (the TMP36) with your microcontroller, which will use the seven-segment LED from Project 15 to display the temperature one digit at a time.

You'll need the following hardware:

- USBasp programmer
- Solderless breadboard
- 5 V breadboard power supply
- ATmega328P-PU microcontroller
- One TMP36 temperature sensor
- One common-cathode seven-segment LED display
- Seven 560 Ω resistors (R1–R7)
- 0.1 µF ceramic capacitor
- Jumper wires

Assemble your circuit as shown in Figure 10-6.

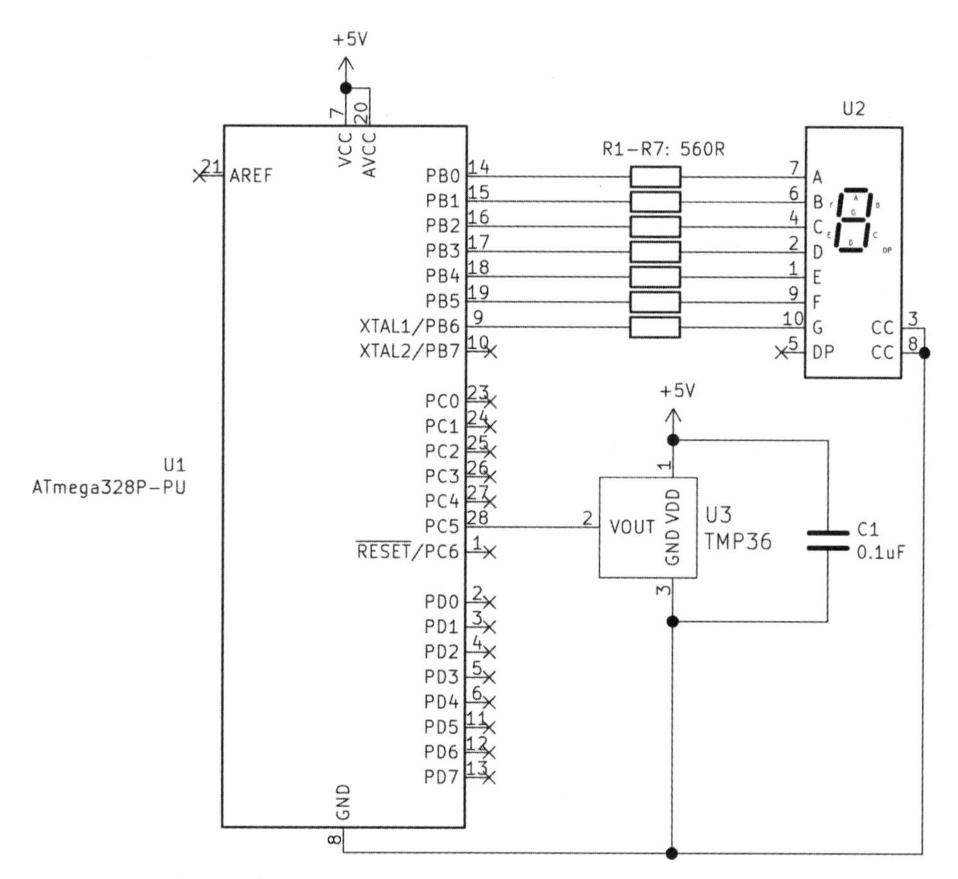


*Figure 10-6: Schematic for Project 43*

Note that the power supply to the project must be as close to 5 V as possible, since the TMP36 is an analog sensor whose output is a function of the supply voltage and the temperature.

Now open a terminal window, navigate to the *Project 43* subfolder in the *Chapter 10* folder, and upload the code for Project 43 as usual. Once you've completed this, you should be presented with the temperature in degrees Celsius on the LED display—first the left-hand digit, then the right.

Let's see how this works:

```
// Project 43 - Creating a Digital Thermometer with the thermometer.c Library

❶ #include <avr/io.h>
#include <util/delay.h>
#include "thermometer.h"                    // Use our new library

❷ void startADC()
// Set up the ADC
{
   ADMUX |= (1 << REFS0);                   // Use AVcc pin with ADC
   ADMUX |= (1 << MUX2) | (1 << MUX0);      // Use ADC5 (pin 28)
   ADCSRA |= (1 << ADPS1) | (1 << ADPS0); // Prescaler for 1MHz (/8)
   ADCSRA |= (1 << ADEN);                   // Enable ADC
}

int main(void)
{
 ❸ float temperature;
   uint8_t finalTemp;

 ❹ startADC();
 ❺ DDRB = 0b11111111;                       // Set PORTB as outputs
   for(;;)
   {
    ❻ temperature = readTMP36();            // Get temperature from sensor
      finalTemp = (int)temperature;

    ❼ displayNumber(finalTemp);
      _delay_ms(1000);
   }
   return 0;
}
```

Now that we have placed the measurement and calculation code in the thermometer library, you can see how simple the main code can be. We first include the required libraries ❶ and define the function to start the ADC ❷. In the main section of the code we declare two variables ❸, to store the value from the temperature library and to pass to the `displayNumber()` function. We start the ADC for the TMP36 temperature sensor ❹, then set the pins on PORTB to outputs for the LED display ❺.

Finally, we retrieve the temperature from the sensor via the `readTMP36()` function from our thermometer library ❻, convert it to an integer, and show it on the LED display ❼.

For another challenge, see if you can modify `readTMP36()` so that it can return temperatures in either Celsius or Fahrenheit, or make your own ADC initialization or PWM library, or simplify the numerical display code in `displayNumber(uint8_t value)`. Whichever you choose, I hope you see how easy it is to rework your own custom functions into a convenient library. This is a key tool in your box of programming tricks.

In the next chapter, you'll learn how to use many more interesting and useful parts via the SPI data bus, including LED display drivers, shift registers, and analog-to-digital converters.

# AVR AND THE SPI BUS

We can loosely define a *bus* as a connection between two devices that allows us to send data from one device to the other. For example, there are several types of data buses that can connect your AVR microcontroller to a sensor or display device. This chapter introduces the *serial peripheral interface (SPI) bus*, which we use to send bytes of data directly between a primary device and one or more secondary devices.

In this chapter, you will learn how to:

- Implement the SPI bus with AVR microcontrollers.
- Read SPI device data sheets in order to write matching code.
- Add a reset button to your projects.
- Use two different SPI-based devices in the same project.

Along the way, you'll also learn how to use 74HC595 shift register ICs to increase the number of available digital output pins, display eight-digit numbers with a MAX7219 LED display driver IC, and measure voltages with the MCP3008 ADC IC.

## How Buses Work

The SPI bus enables communication between AVR microcontrollers and many popular parts and sensors. It works similarly to the USART that we used in Chapters 4 and 9, in that it transmits data in serial fashion from the microcontroller using one wire and to the microcontroller using another. However, the SPI bus also uses a third connection: a *clock line*, which carries an electrical signal that turns on and off at a constant frequency. Every time the clock changes state from high to low or low to high, a bit of data (an *on* or *off*) is sent along the data line either from or to the microcontroller. The clock signal synchronizes with the data signal, allowing for fast and accurate data transmission.

We can demonstrate the changing states of the data and clock lines using a DSO. For example, consider Figure 11-1, which shows a byte of data traveling along the SPI bus.

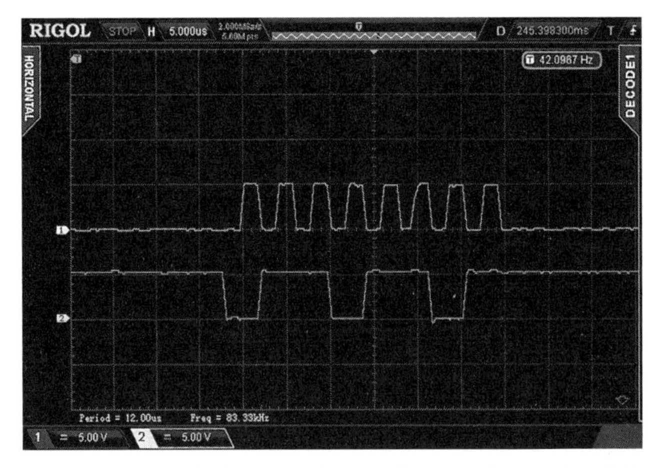


*Figure 11-1: A DSO showing a byte of data traveling on the SPI bus*

In Figure 11-1, the upper waveform (marked 1 in the left margin) is the clock signal, which is activated when we use the SPI bus. The signal starts at 0 V, moves up to 5 V, then returns to 0 V, repeating this pattern if data is being transferred. The lower waveform (marked 2) represents the data, with a 1 being a 5 V signal and a 1 being a 0 V signal. From right to left, the data being sent is 10110110.

The SPI bus can send and receive data simultaneously and at different speeds, depending on the microcontroller or SPI-based device used. Communication with the SPI bus uses a *main–secondary* configuration: the AVR acts as the *main* and determines which device (the *secondary*) it will communicate with at a given time.

In this book we'll use the ATmega328P-PU microcontroller for projects that use the SPI bus, since the ATtiny85 doesn't have enough memory or output pins to run those projects.

## Pin Connections and Voltages

Each SPI device uses four pins to communicate with a main:

- MOSI (main out, secondary in)
- MISO (main in, secondary out)
- SCK (clock)
- SS (secondary select, also known as "latch")

These SPI pins connect to your microcontroller as shown in Figure 11-2.

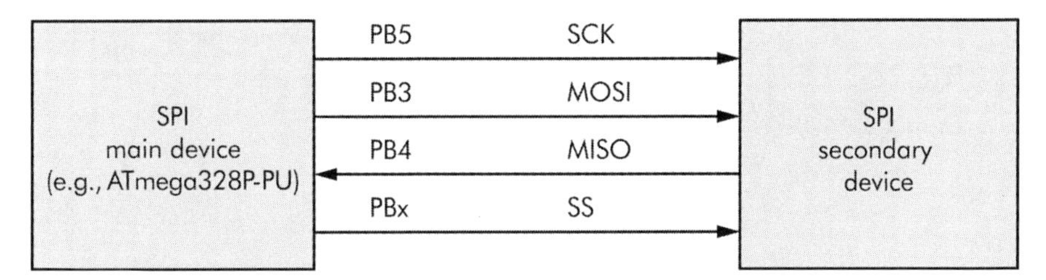

*Figure 11-2: Typical AVR-to-SPI device connection*

The SS pin on the SPI main device is labeled PBx in Figure 11-2; you can use any free GPIO pin, but to keep things simple it's good to use a free pin on PORTB as the connection will be close to the SPI pins. Different manufacturers often use their own terminology for the SPI bus connections, but this should be easy to interpret after a quick examination.

Since our AVR runs on 5 V in the following projects, your SPI device must also operate at or tolerate operating at 5 V, so be sure to check this with the seller or manufacturer before use. If you simply must use an SPI device that operates at a reduced voltage, such as 3.3 V, you can use a *level converter* like PMD Way part number 441079, shown in Figure 11-3. A level converter can convert a 5 V digital signal to a 3.3 V signal, and vice versa.

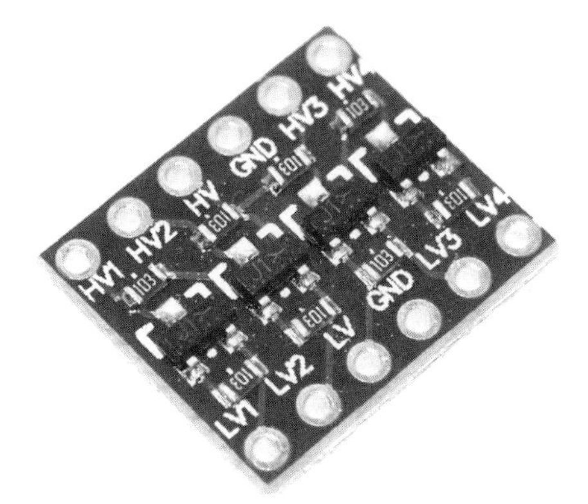


*Figure 11-3: PMD Way part number 441079*

This is a four-channel level converter board, meaning it can convert four independent electrical signals on the one board. To use a level converter, wire it between the four wires on the SPI bus: place the 5 V wires on the HV pads and the matching lower voltage wires on the LV pads and connect GND of both sides to the board. Remember to disconnect your USBasp programmer from your projects once you've uploaded the code, as the programmer's pins share the SPI pins and can sometimes cause interruptions to the data flow between the microcontroller and the SPI-based device.

## Implementing the SPI Bus

Next, let's examine how to implement the SPI bus in our code and how to make the hardware connections. I'll show you some parameters of an example SPI part being used, then how to adjust the SPI Control Register (SPCR) shown in Figure 11-4 to activate the SPI bus to our required parameters. In the following projects, I'll also show you how this is done for various other SPI parts.

| Bit | 7 | 6 | 5 | 4 | 3 | 2 | 1 | 0 | |
|---|---|---|---|---|---|---|---|---|---|
| 0x2C (0x4C) | **SPIE** | **SPE** | **DORD** | **MSTR** | **CPOL** | **CPHA** | **SPR1** | **SPR0** | **SPCR** |
| Read/write | R/W | R/W | R/W | R/W | R/W | R/W | R/W | R/W | |
| Initial value | 0 | 0 | 0 | 0 | 0 | 0 | 0 | 0 | |

*Figure 11-4: SPCR diagram from the ATmega328P-PU data sheet*

Leave SPIE as 0, as we're not working with interrupts, and set SPE to 1 to enable the SPI bus. Next, consider the DORD bit, which determines whether the byte of data is sent with the MSB (most significant bit, bit 7 of the byte) or the LSB (least significant bit, bit 0 of the byte) first. You'll need to determine the direction from the secondary's data sheet, which will include a timing diagram like the one in Figure 11-5, or the supplier.

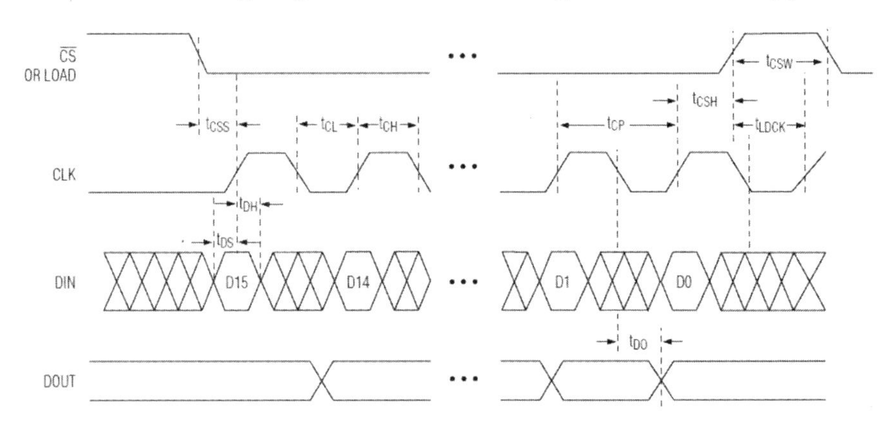


*Figure 11-5: Example timing diagram for an SPI device (the MAX7219)*

Review the timing diagram's DIN line. The data travels down the bus with the MSB first. If D0 were at the start of the line, the LSB would be first. Set DORD to 1 for LSB first, to 0 for MSB first.

Returning to the SPCR, set MSTR to 1 to enable the microcontroller as the main device, and 0 if you need to enable the microcontroller as the secondary device. For all our projects we'll use 1. Next, set CPOL to match the polarity of the clock (SCK) signal in the SPI bus when idle (that is, before and after data is carried on the bus): 0 for a low signal or 1 for a high signal. Again, you can get this information from the timing diagram: in Figure 11-5 the CLK (or clock/SCK) line is low when not in use, then rises when the first bit of data arrives, then alternates repeatedly until the data transmission has finished and it falls low again, so for this device you'd set it to 0.

CPHA, the clock-phase bit, determines whether the data is sampled at the start or the end of the clock bit. For example, review the CLK and DIN lines of Figure 11-5. The data is sampled at the start of the clock bit, as the clock rises at the start time of the data bits. In this case you would set CPHA to 0. If the data bit started as the clock bit ended, CPHA would be 1.

Finally, you'll use the last two bits, SPR1 and SPR0, in conjunction to set the speed of the clock and matching data signals on the SPI bus. Set both to 0 for maximum speed in relation to the microcontroller. Also set the AVR output pin connected to your SPI device's SS pin as an output using a `DDRx` function, then set that pin to high using the typical `PORTx` function.

### Sending Data

Now that you've initialized your SPI bus, it should be ready to receive and send data. We'll practice sending data first, then examine receiving data later in this chapter. To send a byte of data to the SPI device, you need to do four things (all of which we'll do in the next project):

- Set the SS pin low using a `PORTx` command.
- Place the byte of data you wish to send into the SPDR register.
- Wait for the transmission to finish using `while(!(SPSR & (1<<SPIF)));`.
- Set the SS pin high using a `PORTx` command.

This may seem a little complicated, but with practice and the right information from part suppliers, it's easy. I'll explain everything you need to know to build our SPI bus projects, the first of which harnesses a particularly useful shift register IC.

## Project 44: Using the 74HC595 Shift Register

When your AVR-based project doesn't have enough digital output pins, you can connect one or more *shift registers* and still have plenty of output pins for use on the AVR itself. A shift register is an integrated circuit with eight digital output pins that we can control by sending a byte of data to the IC via the SPI bus. The projects in this chapter will use the 74HC595 shift register, as shown in Figure 11-6.

*Figure 11-6: The 74HC595 shift register IC*

The 74HC595 shift register has eight digital outputs that operate in the same way as your AVR's digital outputs, as shown in Figure 11-7.

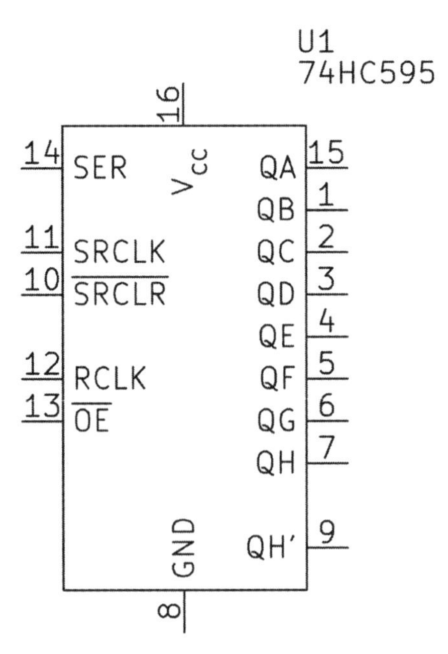


*Figure 11-7: The 74HC595 schematic symbol*

Table 11-1 gives details on how to connect the shift register to the microcontroller.

**Table 11-1:** 74HC595 Connections

| Pin | Connection |
|---|---|
| 16 | 5 V positive power supply |
| 8 | GND |
| QA to QH (15, 1–7) | Digital outputs 0 to 7 |
| 10 | 5 V positive power supply |
| 11 | SPI bus clock |
| 12 | SPI SS |
| 13 | GND |
| 14 | SPI data in |
| 9 | SPI data out |

The principle behind the shift register is simple: we send 1 byte of data (8 bits) to the shift register, which turns the matching eight outputs on or off based on that byte of data. The bits representing the byte match the output pins in order from highest to lowest. Therefore, the MSB of the data represents output pin 7 of the shift register, while the LSB represents output pin 0. For example, if we send the byte `0b10000110` to the shift register via the SPI bus, it will turn on outputs 7, 2, and 1 and will turn off outputs 0 and 3–6 until the shift register receives the next byte of data, or we turn the power off.

Once you send a new byte of data to the shift register, it sends the previous byte of data out via SPI pin 9, the data out pin. Thus, you can harness two or more, sending multiple bytes of data in one operation to control multiple shift registers.

**NOTE** *You can usually draw up to 20 mA of current from an output pin, and the total current drawn from an entire 74HC595 shouldn't exceed 75 mA.*

You get an extra eight digital output pins for every shift register attached to the SPI bus. This makes shift registers very convenient when you want to control lots of LEDs or other devices. In this project, we'll use it to control a seven-segment numeric LED display.

### The Hardware

To build your display circuit, you'll need the following hardware:

- USBasp programmer
- Solderless breadboard
- 5 V breadboard power supply

- ATmega328P-PU microcontroller
- One 74HC595 shift register IC
- One common-cathode seven-segment LED display
- Eight 560 Ω resistors (R1–R8)
- Jumper wires

Assemble your circuit as shown in Figure 11-8.

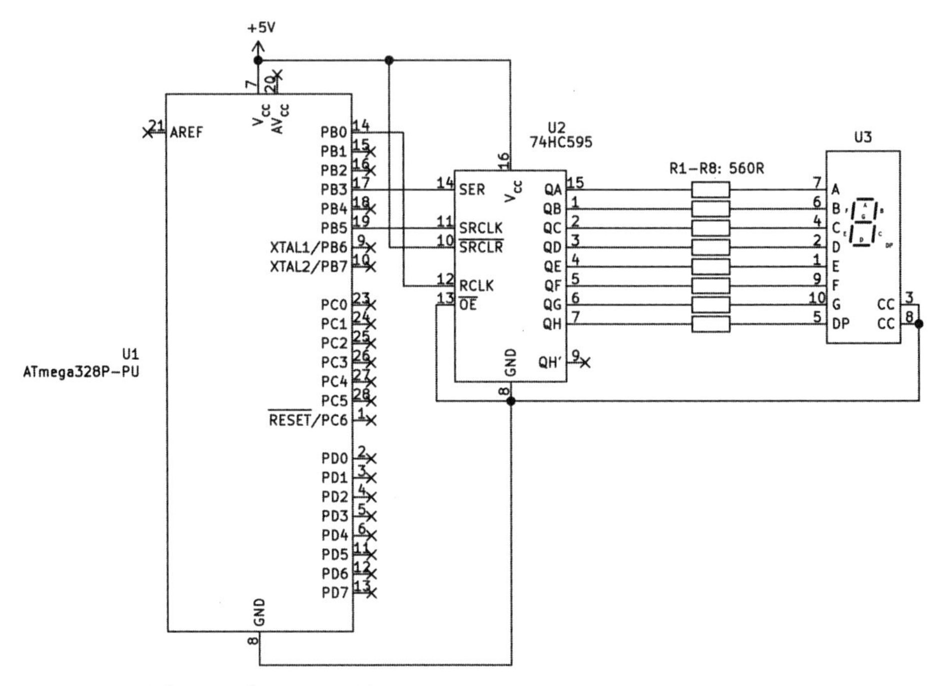


*Figure 11-8: Schematic for Project 44*

## The Code

Open a terminal window, navigate to the *Project 44* subfolder of this book's *Chapter 11* folder, and enter the command **make flash** to compile and upload the data as usual. Don't forget to disconnect the programmer once you've uploaded it to the microcontroller. After a moment, the digits 0 to 9 should show in ascending order on the LED display, and then repeat.

Let's examine the code to see how this works:

```
// Project 44 - Using the 74HC595 Shift Register

#include <avr/io.h>
#include <util/delay.h>

❶ void setupSPI()
{
```

```
    PORTB |= (1 << 0); // SS pin HIGH
    // Set up SPI bus
    SPCR = 0b01110000;
}

❷ void dispNumSR(uint8_t value)
// Displays a number from 0-9 on the seven-segment LED display
{
    // SS pin LOW
  ❸ PORTB &= ~(1 << PORTB0);
    switch(value)
    // Determine which byte of data to send to the 74HC595
    {
      case 0 : SPDR = 0b11111100; break; // 0
      case 1 : SPDR = 0b01100000; break; // 1
      case 2 : SPDR = 0b11011010; break; // 2
      case 3 : SPDR = 0b11110010; break; // 3
      case 4 : SPDR = 0b01100110; break; // 4
      case 5 : SPDR = 0b10110110; break; // 5
      case 6 : SPDR = 0b10111110; break; // 6
      case 7 : SPDR = 0b11100000; break; // 7
      case 8 : SPDR = 0b11111110; break; // 8
      case 9 : SPDR = 0b11100110; break; // 9
    }
  ❹ while(!(SPSR & (1<<SPIF)));          // Wait for SPI transmission to finish
    // SS pin HIGH
    PORTB |= (1 << PORTB0);
}

int main(void)
❺ {
    uint8_t i=0;
    DDRB = 0b11111111;                  // Set PORTB as outputs
    setupSPI();
    while (1)
    {
       for (i=0; i<10; i++)
       {
          dispNumSR(i);
          _delay_ms(250);
       }
    }
}
```

The code contains a couple of custom functions, the first being `setupSPI()` ❶. We use this function to initialize the SPI bus and set the SS pin to high, then set the SPCR register as explained earlier in this chapter. We set the DORD bit in the SPCR register to 1, as we need to send data to the 74HC595 with the LSB first. We can see the requirement to use the 56LSB first from the timing diagram in the 74HC595's data sheet, shown in Figure 11-9: $Q_A$ (the first output pin) is the first to be set high.

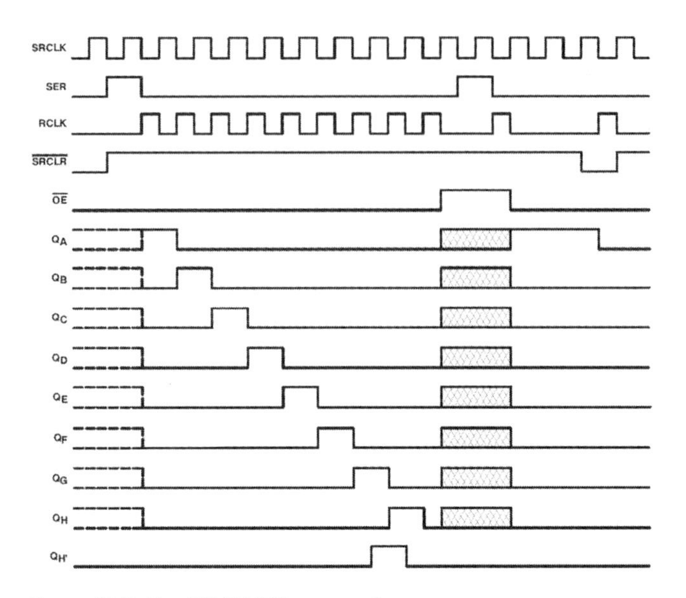


*Figure 11-9: The 74HC595's timing diagram*

We also need to determine what value to use for the CPOL bit. We set this to 0, since the polarity of the clock signal is low, or off, when idle.

The final bit to consider is CPHA, the clock-phase bit mentioned earlier. If you refer to the diagram again and compare the RCLK timing and signals to any of the Q outputs, you'll see that they both change from low to high at the same time. Therefore, the data is sampled at the start, so we set CPHA to 0. The final bits (1 and 0) we leave as 0 to set the SPI bus for maximum possible speed.

Our second custom function, `dispNumSR()` ❷, accepts an integer between 0 and 9 and shows this on the LED display. It first sets the SS pin low ❸, then determines which matching byte of data for the digit to display using a `switch...case` statement.

The microcontroller sends each byte of data with the LSB first. The bits match the eight outputs on the 74HC595, which we wire to the LED display segments A–G and the decimal point, respectively, as shown in Figure 11-8. This byte of data is then placed in the SPDR register in each matching `case` statement. The code waits for the transfer to finish ❹, then sets the SS pin high to complete the data transmission.

The main loop of code ❺ simply sets up PORTB as outputs, as it contains the pins we need for all four SPI bus connections, then calls the `setupSPI()` function to set up the SPI bus as described earlier. It then sends the numbers 0 to 9 to the LED display.

The decimal point in our project is connected, but not in use. You can turn it on and off with the last bit of the byte sent to the shift register. For a challenge, try modifying the `dispNumSR()` function to accept a second variable to turn the decimal point on or off.

Now that you know how to control a single device with the SPI bus, let's try using it to control two devices at the same time.

# Project 45: Using Two 74HC595 Shift Registers

Using two or more shift registers is an inexpensive and simple way to control many more digital outputs with your AVR. As an example, this project uses two 74HC595 shift registers to show double-digit numbers via two LED displays.

## The Hardware

To build your display circuit, you'll need the following hardware:

- USBasp programmer
- Solderless breadboard
- 5 V breadboard power supply
- ATmega328P-PU microcontroller
- Two 74HC595 shift registers
- Two common-cathode seven-segment LED displays
- 16 560 Ω resistors (R1–R16)
- Jumper wires

Assemble your circuit as shown in Figure 11-10.

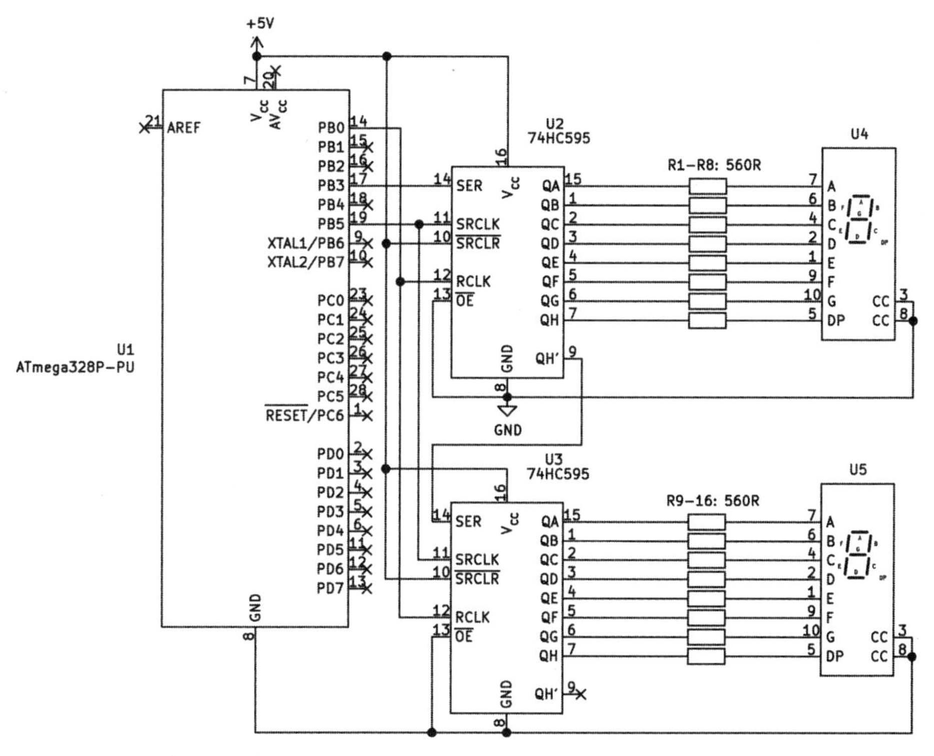


Figure 11-10: Schematic for Project 45

## The Code

Open a terminal window, navigate to the *Project 45* subfolder of this book's *Chapter 11* folder, and enter the command **make flash** to compile and upload the data as usual. After a moment or two the numbers 0 through 99 should appear in ascending order on the LED displays, then repeat.

Let's see how this works:

```
// Project 45 - Using Two 74HC595 Shift Registers

#include <avr/io.h>
#include <util/delay.h>

void setupSPI()
{
   PORTB |= (1 << 0);          // SS pin HIGH
   // Set up SPI bus
   SPCR = 0b01110000;
}

❶ void dispNumSR(uint8_t value)
// Displays a number from 00-99 on the seven-segment LED displays
{
   uint8_t leftDigit;
   uint8_t rightDigit;
 ❷ uint8_t digitData[] = {0b11111100, 0b01100000, 0b11011010, 0b11110010,
0b01100110,
                            0b10110110, 0b10111110, 0b11100000, 0b11111110,
0b11100110};

 ❸ leftDigit = value/10;
   rightDigit = value%10;

 ❹ PORTB &= ~(1 << PORTB0);     // SS pin LOW
 ❺ SPDR = digitData[rightDigit];
 ❻ while(!(SPSR & (1<<SPIF))); // Wait for SPI transmission to finish
   SPDR = digitData[leftDigit];
   while(!(SPSR & (1<<SPIF))); // Wait for SPI transmission to finish
 ❼ PORTB |= (1 << PORTB0);     // SS pin HIGH
}

int main(void)
{
   uint8_t i=0;
   DDRB = 0b11111111;          // Set PORTB as outputs
   setupSPI();
   while (1)
   {
      for (i=0; i<100; i++)
      {
         dispNumSR(i);
         _delay_ms(250);
      }
   }
}
```

The code for this dual-digit version is similar to that of Project 44, except that since we have two shift registers to control, it sends two bytes of data at once. This time, the `dispNumSR()` function ❶ accepts a number between 0 and 99, then divides the number using division and modulo ❸ in order to treat each digit separately and store them in the `leftDigit` and `rightDigit` variables.

Next, the required operation for sending SPI data begins. We set the SS pin low ❹, then send out the byte of data that represents the digit 0 to 9 as specified in the `digitData` array ❷ for the right-hand digit to the shift register ❺. After waiting ❻ for the byte to be transferred, we send the byte of data for the left-hand digit in the same manner. After we've finished waiting for the transmission to complete, the code sets the SS pin high ❼ to complete the data transmission. The SS pins of both shift registers are connected, so we only need one digital output to control them.

The byte for the second (ones) digit is sent first, as it is the second shift register—the byte sits in the first shift register, then the byte for the tens digit is sent, which pushes the first byte into the second shift register. That means the first shift register contains the data for the tens digit, and the second shift register contains the data for the ones digit. Once the SS pin is set high, the shift registers' outputs activate, and the LED displays begin showing the numbers.

Although this project controlled LED displays, you now have the necessary skills to use multiple shift registers to expand your AVR's outputs in other cases. If you need to control larger numerical displays—up to eight digits—the next project is for you.

## Project 46: Using the MAX7219 LED Driver IC

When you need to use more than two seven-segment numerical displays for a project, the wiring and related controls can become quite complex. Thankfully, there's a solution for this: the Maxim MAX7219 LED driver IC, a popular IC that can control up to 64 LEDs at once. In turn, we can use these LEDs to simultaneously display eight numerical digits with only four control wires via the SPI bus. This project shows how to use this display module.

### *The Hardware*

The MAX7219 is available in both through-hole (Figure 11-11) and surface-mount (Figure 11-12) package types.

*Figure 11-11: The MAX7219 in a through-hole package type*

*Figure 11-12: The MAX7219 in a surface-mount package type*

The through-hole version is most useful when working with a solderless breadboard or making your own hand-assembled printed circuit board (PCB). If you have trouble finding a MAX7219, the Allegro AS1107 is a drop-in replacement.

If you're looking to control large numerical displays, you can easily find them preassembled, usually with four to eight digits fitted to a module with the MAX7219. For this project we'll use an eight-digit module, as shown in Figure 11-13.

*Figure 11-13: An eight-digit LED module*

These modules use the surface-mount version of the MAX7219, which is soldered onto the rear of the module's PCB. The modules usually include some inline header pins to allow for attaching control wires. If you haven't already done so, solder these pins to your module as shown in Figure 11-14.

*Figure 11-14: Connecting inline header pins*

To build your circuit, you'll need the following hardware:

- USBasp programmer
- Solderless breadboard
- 5 V breadboard power supply
- ATmega328P-PU microcontroller
- MAX7219 eight-digit module
- 470 μF 16 V electrolytic capacitor
- Jumper wires

Assemble your circuit as shown in Figure 11-15.

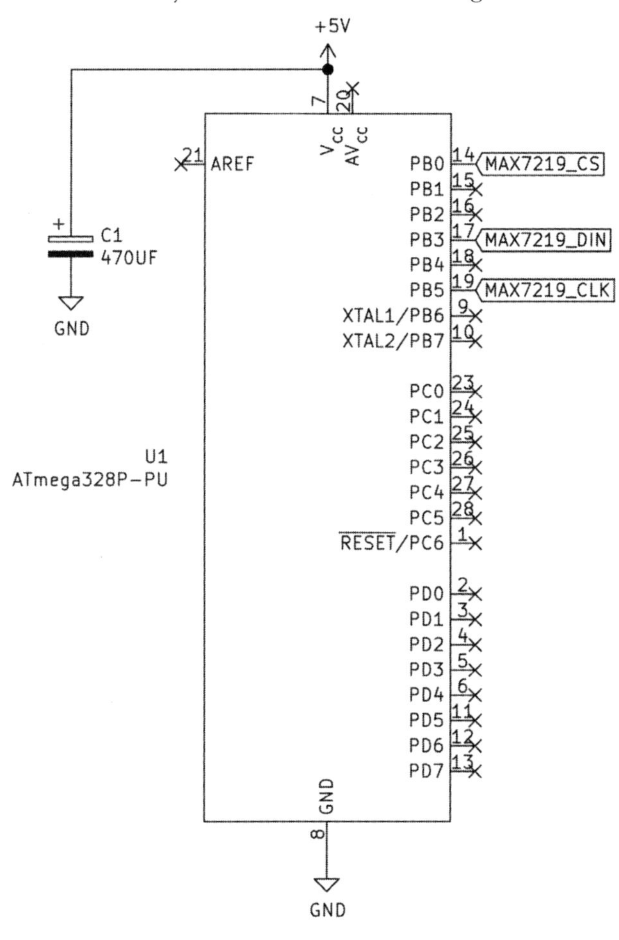


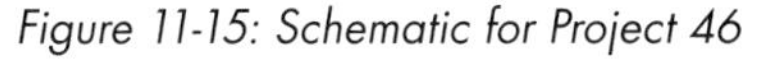
*Figure 11-15: Schematic for Project 46*

As you construct your circuit to follow the schematic, connect your display module as shown in Table 11-2.

**Table 11-2**: ATmega328P-PU to MAX7219 Connections

| ATmega328P-PU | MAX7219 module |
|---|---|
| 7 | $V_{cc}$ |
| 8 | GND |
| 17 | DIN |
| 14 | SS |
| 19 | CLK |

Note that the display may quickly draw and stop drawing current, which sometimes will affect the power supply voltage. Therefore, we use a 470 µF electrolytic capacitor to keep the 5 V power smooth. You can review capacitor types in Chapter 2.

Before we dive into the code, let's consider the parameters required for the SPI setup via the SPCR register. From the timing diagram for the MAX7219 in Figure 11-5, we can see that we should set the DORD bit to 0, as the MAX7219 requires data sent LSB first. We'll set the CPOL bit to 0, as the clock signal is low at the start of data transmission, and we'll set the CPHA bit to 0 too, as the clock signal's polarity is low at idle.

Now we need to explore how to control the MAX7219. Every time we want the IC to do something, we must send two bytes of data. The first byte is the address of a control register (other ICs have registers just as the microcontroller does), and the second byte is the value to be stored in that register. This could be setup configuration such as display brightness, or the value that represents a number to display on a certain digit.

The possible values for each register are described in the MAX7219 using hexadecimal numbers (base-16), so we'll use them to save effort. You can store hexadecimal numbers in `char` variable types. For your own reference and research, you may wish to download and review the MAX7219's data sheet from *https://www.maximintegrated.com/en/products/power/display-power-control/MAX7219.html.*

## The Code

Open a terminal window, navigate to the *Project 46* subfolder of this book's *Chapter 11* folder, and enter the command **make flash** to compile and upload the data as usual. After a moment or two the display should show eight zeros, then count upward until it gets to 9,999,999, then repeat.

Let's see how this is done:

```
// Project 46 - Using the MAX7219 LED Driver IC

#include <avr/io.h>
#include <util/delay.h>

❶ void writeMAX7219(char hexdata1, char hexdata2)
// Sends two bytes in hexadecimal to the MAX7219
{
   PORTB &= ~(1 << PORTB0);     // SS pin LOW
   SPDR = hexdata1;             // Send value of hexdata1
   while(!(SPSR & (1<<SPIF))); // Wait for SPI transmission to finish
   SPDR = hexdata2;             // Send value of hexdata2
   while(!(SPSR & (1<<SPIF))); // Wait for SPI transmission to finish
   PORTB |= (1 << PORTB0);      // SS pin HIGH
}

❷ void blankMAX7219()
// Blanks all digits
{
   uint8_t i;
   for (i=1; i<9; i++)          // Blank all digits
   {
      writeMAX7219(i,15);       // Send blank (15) to digit register (i)
   }
}
```

```
❸ void initMAX7219()
  // Set up MAX7219 for use
  {
     PORTB |= (1 << 0);           // SS pin HIGH
     SPCR = 0b01010000;           // Set up SPI bus for MAX7219
     writeMAX7219(0x09,0xFF);     // Mode decode for digits
     writeMAX7219(0x0B,0x07);     // Set scan limit to 8 digits: 0x09 + 0xFF)
     writeMAX7219(0x0A,0x01);     // Set intensity to 8 - 0x0A + 0x08)
     writeMAX7219(0x0C,0x01);     // Mode display on
     blankMAX7219();
  }

❹ void dispMAX7219(uint8_t digit, uint8_t number, uint8_t dp)
  // Displays "number" in location "digit" with decimal point on/off
  // Digit: 1~8 for location 1~8
  // Number: 0~15 for 0~9, - E, H, L, P, blank
  // dp: 1 on, 0 off
  {
   ❺ if (dp==1)                   // Add decimal point
     {
        number = number + 128;
     }
     writeMAX7219(digit, number);
  }

❻ void numberMAX7219(uint32_t value)
  // Displays a number between 0-99999999

     uint8_t digits[9];
     uint8_t i = 1;

     for (i=1; i<9; i++)
     {
      ❼ digits[i]=15;             // Sending 15 blanks the digit
     }
     i = 1;
     while (value > 0)            // Continue until value > 0
     {
        digits[i] = value % 10;   // Determine and store last digit
        value = value / 10;       // Divide value by 10
        i++;
     }
     for (i=1; i<9; i++)
     {
        dispMAX7219(i, digits[i],0);
     }
  }

  int main(void)
❽ {
     uint32_t i;
     DDRB = 0b11111111;           // Set PORTB as outputs
     initMAX7219();
     while (1)
     {
```

```
        for (i = 0; i<100000000; i++)
        {
           numberMAX7219(i);
           _delay_ms(100);
        }
    }
}
```

To save time, we use a custom function, `writeMAX7219(`*`char hexdata1, char hexdata2`*`)` ❶, to send two hexadecimal bytes of data to the IC via the SPI bus. This function sets the SS pin low, assigns the first byte of data to the SPDR register, waits for transmission to finish, then repeats the process for the second byte and the sets the SS pin high again. After setting up the SPI bus, we initialize the MAX7219 by placing values in four configuration registers using another custom function, `initMAX7219()` ❸.

Before writing any digits to the display, we introduce the `blankMAX7219()` function ❷ to be sure we clear the display between writes. Without this function, if we were to, say, write 32,785 to the display and then write 45, the display would show 32,745.

To show a digit on the display, we use `writeMAX7219()` to send two bytes of data. The first byte of data is the digit location, from right to left (locations 0 to 7). The address for each digit location is conveniently the same as the location plus 1; for example, the address for digit 5 is `0x06` in hexadecimal. The second byte of data is the actual number to display. To display the number nine on the leftmost digit of our module, for instance, we'd send `0x08` then `0x09`, as follows:

```
writeMAX7219(0x08, 0x09);
```

You can also use decimal numbers or integer variables if convenient:

```
writeMAX7219(8, 9);
```

You can view the address map for the digit locations in Table 2 of the MAX7219 data sheet and the characters you can display in Table 5.

Now that we've set up out `writeMAX7219()` function to easily write data to the MAX7219, we harness this function within another function: `dispMAX7219(uint8_t digit, uint8_t number, uint8_t dp)` ❹. We use this to display digits in locations with or without the decimal point. Set *dp* to 1 to show the decimal point, or 0 to not show it. For example, to display the number 3 without the decimal point on the rightmost digit of the module, we would use:

```
dispMAX7219(1, 3, 0)
```

The decimal point is activated by adding 128 (which is `0xF0` in hexadecimal) to the byte representing the number to display ❺.

All the custom functions mentioned so far build up to our final function, `numberMAX7219(`*`uint32_t value`*`)` ❻, which accepts an integer between 0 and 99,999,999 and shows it on our display module. This function uses modulo and division to break down the whole number into separate digits and places them in an array. It then runs through the array and sends each digit to the display.

At the start of the function, we fill the array with the number 15 ❼. This is because sending a 15 as the number value to the MAX7219 causes the IC to blank the digit being addressed, allowing us to avoid displaying leading zeros for unused digits. Finally, the main loop of code ❽ sets up the interface pins used for SPI as outputs, and successively displays the numbers 0 through 99,999,999 on the LED display.

This may seem like a lot of work, but now that you have the tools to easily drive these larger numerical displays, you can reuse the functions in your own projects. If you enjoy a challenge, why not write your own MAX7219 library? In the meantime, to set the stage for later projects, I'd like to introduce a new addition to your AVR circuitry.

## Project 47: Adding a Reset Button

In future projects in this book and in your own creations, there will come a time when you need to literally reset a project so it starts operating again in the same manner as when first turned on. To enable this, your projects will need a reset button, which we'll construct now. Reset buttons save time and are much more convenient than disconnecting then reconnecting the power supply.

To add reset buttons to your AVR projects, you will need the following:

- Pushbutton
- 10 kΩ resistor
- Jumper wires

The reset button circuit for the ATtiny85 is shown in Figure 11-16.

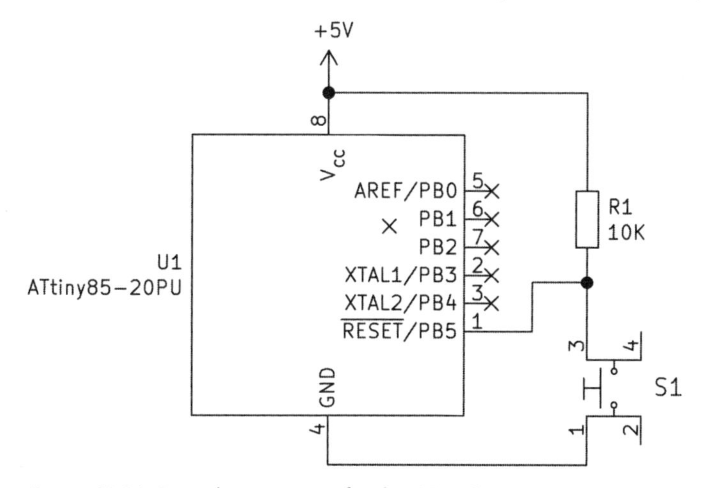


*Figure 11-16: Reset button circuit for the ATtiny85*

Figure 11-17 shows the reset button circuit for the ATmega328P-PU.

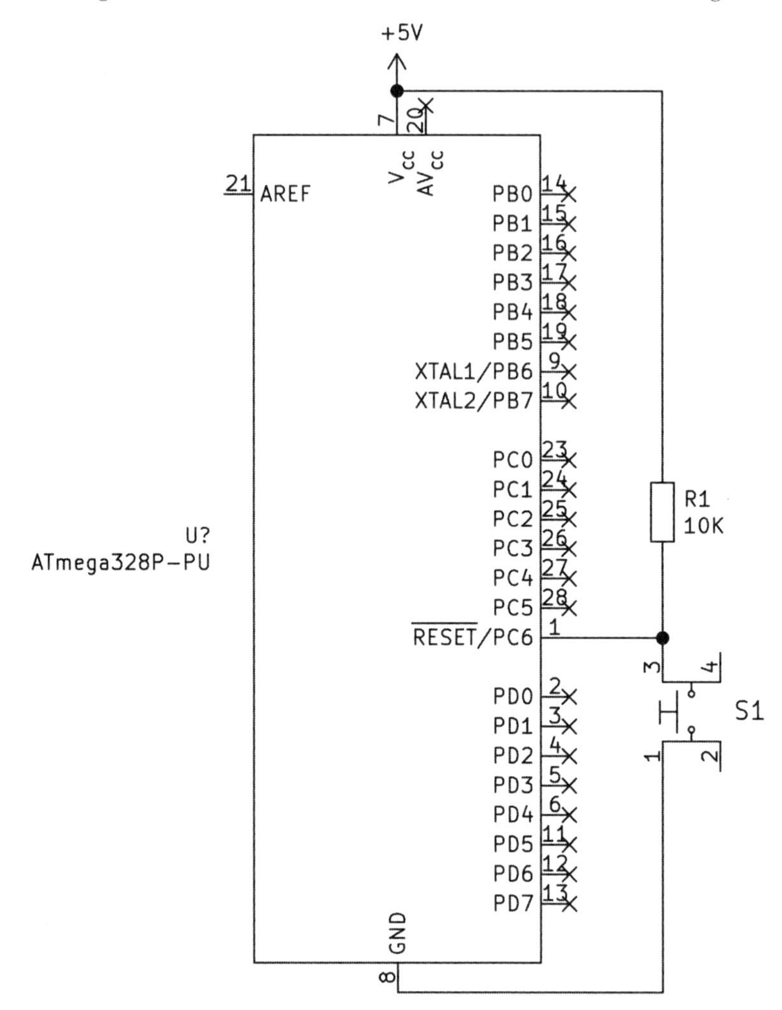


*Figure 11-17: Reset button circuit for the ATmega328P-PU*

If you are comparing this to the buttons discussed in Chapter 3, notice the difference in the wiring. This button configuration, in which the connection of the resistor between the 5 V and RESET pins keeps the pin in a high state for normal operation, is called a *pullup* configuration. When the user presses the button, the RESET pin is set to a low state as the button directly connects the pin to GND.

We know to use this pullup configuration thanks to the part's schematic symbol in the data sheet: there's a solid bar over the RESET pin's label, unlike the labels for the other pins. This bar means that the default input for this pin for normal operation is high, and whatever function the pin is used for will be activated when it is set to low.

You don't need to add any code to your project to allow for the reset button—it's a simple hardware addition. With the button set up, let's move on to controlling two different SPI devices and learning how to receive data from the SPI bus. These sections will prepare us for the final project in this chapter, where we'll create a simple digital voltmeter.

## Multiple SPI Devices on the Same Bus

You can use two or more different SPI-based devices on the same SPI bus, and doing so only requires one extra digital output pin per device. Simply connect all the SCK, MOSI, and MISO pins together, and then connect the SS pins to their own digital output pin on the AVR. For example, Figure 11-18 shows two SPI devices on the one bus, each with its own SS line connected to a unique PORTB pin.

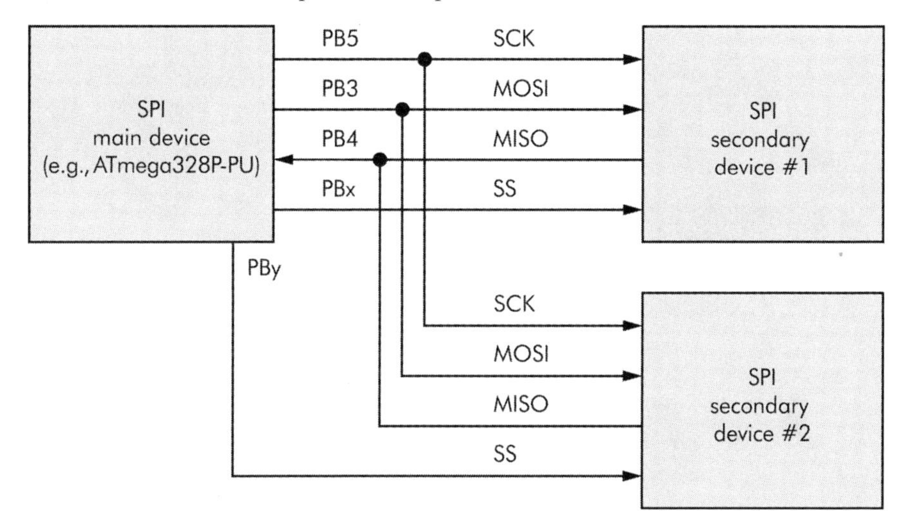


*Figure 11-18: Two SPI devices connected to one microcontroller*

When it comes time to communicate with a particular SPI device, just use the appropriate SS connection and proceed as normal. We'll do that in the next project, which uses the MAX7219 LED display from Project 46 along with a new device.

## Receiving Data from the SPI Bus

As explained previously, we send a byte of data from the microcontroller to the SPI device by placing it in the SPDR register. Receiving a byte of data from an SPI device requires two operations: first the SPI device we're communicating with sends a byte of data, then the AVR places this byte in the SPDR register for our use.

Thus, you can think of the SPI bus as a continuous circle of data, as shown in Figure 11-19.

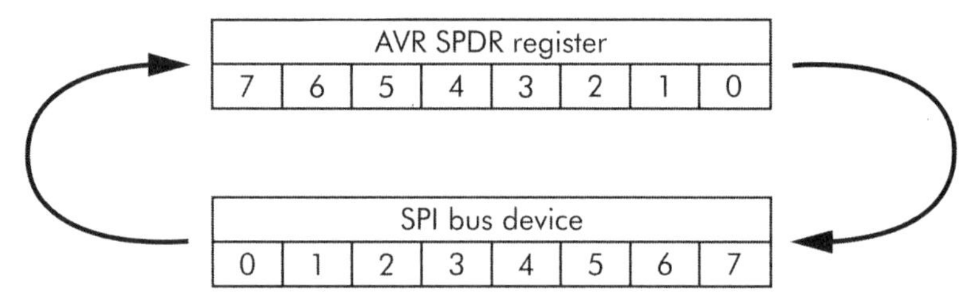


*Figure 11-19: SPI bus data transfer*

As a bit of data leaves the microcontroller on its way to the SPI device, a bit of data leaves the SPI device and heads back into the SPDR register. When you place a full byte of data in the SPDR register, it travels to the SPI device and pushes the data out of the SPI device and into the SPDR register.

This means that when you need a byte of data from the SPI device, you need to send a byte of data to that device to "push" the data from the SPI device back to the SPDR register. You'll see how this works in the following project.

## Project 48: Using the MCP3008 ADC IC

Back in Chapter 3, you started to learn how to use the AVRs' built-in ADC pins to measure connected voltages from external devices such as potentiometers and temperature sensors. However, if you want to use more ADCs, you may have a conflict with the usage of the ADC pins and other uses—that is, you may have already planned to use the pins on the microcontroller that can be used for ADCs for another purpose. An alternative is to use an external ADC IC like the Microchip MCP3008 8-channel ADC IC shown in Figure 11-20, which has eight ADC pins.

*Figure 11-20: Microchip MCP3008 8-channel ADC IC*

Each of these eight pins can measure between 0 V and 5 V DC, and each returns a 10-bit value for measurement. The MCP3008 is easy to use, since it's connected to the SPI bus, and you don't need to worry about any other non-SPI bus AVR registers for setup or control. See the schematic in Figure 11-21.

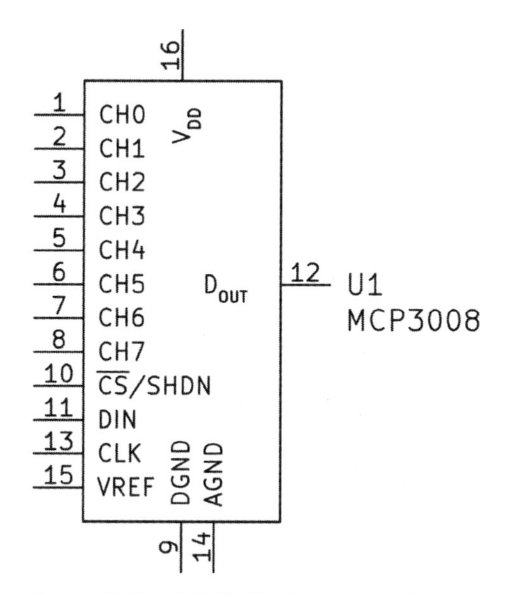


*Figure 11-21: MCP3008 schematic symbol*

To connect the ATmega328P-PU to the MCP3008, follow the guidelines in Table 11-3.

**Table 11-3**: ATmega328P-PU to MCP3008 Connections

| ATmega328P-PU | MCP3008 |
|---|---|
| 5 V | $V_{DD}$ |
| 5 V | $V_{REF}$ |
| GND | DGND, AGND |
| 15 | SS |
| 17 | $D_{IN}$ |
| 18 | $D_{OUT}$ |
| 19 | CLK |

There are also some extra pins on the MCP3008 to consider. The first is the $V_{REF}$, for voltage reference. Our ADC measures analog signals with a 10-bit resolution that represents the signal with a number between 0 and 1,023. In our projects we'll connect the $V_{REF}$ pin to the 5 V power supply, which gives our ADC a *reference voltage*—the upper limit (where the lower is zero) to the signals being measured.

Later, you may wish to measure signals between (for example) 0 and 3 V DC. You can then connect the $V_{REF}$ pin to a 3 V signal. In that case, the reading would be more accurate, since the 1,023 possible values would cover between 0 V and 3 V, rather than spreading farther out over 0 V and 5 V.

The other eight pins are for ADC channels 0 to 7 and can be connected to signals up to 5 V DC. Do not exceed 5 V, as doing so will damage the IC. The negative or GND connections for the signals being measured connect to the AGND pin on the IC.

In this project, you'll use the MCP3008 to measure signals with one ADC, then display the value on our MAX7219 module from Project 48. In addition to familiarizing you with the MCP3008, this project serves as a great example of using two SPI bus devices with the same microcontroller.

## The Hardware

To build your circuit, you'll need the following hardware:

- USBasp programmer
- Solderless breadboard
- Microchip MCP3008 IC
- 5 V breadboard power supply
- ATmega328P-PU microcontroller
- MAX7219 eight-digit module
- 470 μF 16 V electrolytic capacitor
- Jumper wires

Assemble your circuit as shown in Figure 11-22. Along with following the schematic, connect your display module as shown in Table 11-2.

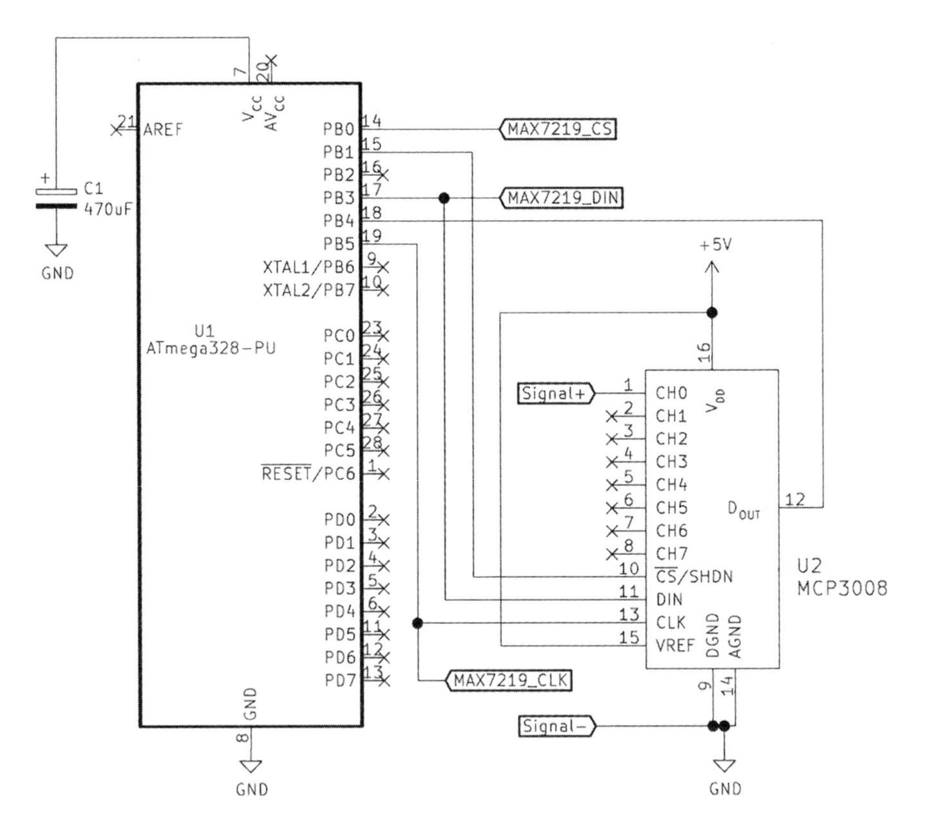


Figure 11-22: Schematic for Project 48

## *The Code*

Open a terminal window, navigate to the *Project 48* subfolder of this book's *Chapter 11* folder, and enter the command `make flash` to compile and upload the data as usual. Since nothing is currently connected to input, we say the ADC has a *floating input.* This means the value returned is somewhat random, and the display should show random numbers.

Now connect something with an output between 0 V and 5 V DC, such as a AA battery or your TMP36 temperature sensor from previous projects, to the Signal+/–connections. Be sure to connect the positive of the signal or battery to the Signal+ pin on the ADC (pin 1) and the negative to GND (also marked as Signal– in the schematic). The display should now show the number of millivolts measured by the ADC (1 volt equals 1,000 millivolts). If you don't have a battery or sensor or anything else to measure, simply connect the ADC input to the 5 V or GND line and see how close it is to 5 V or 0 V, respectively.

As with all SPI devices, we determine the parameters for SPI bus setup from the SPCR register. Figure 11-23 shows the timing diagram from the MCP3008's data sheet (available at *https://www.microchip.com/wwwproducts/en/MCP3008/*).

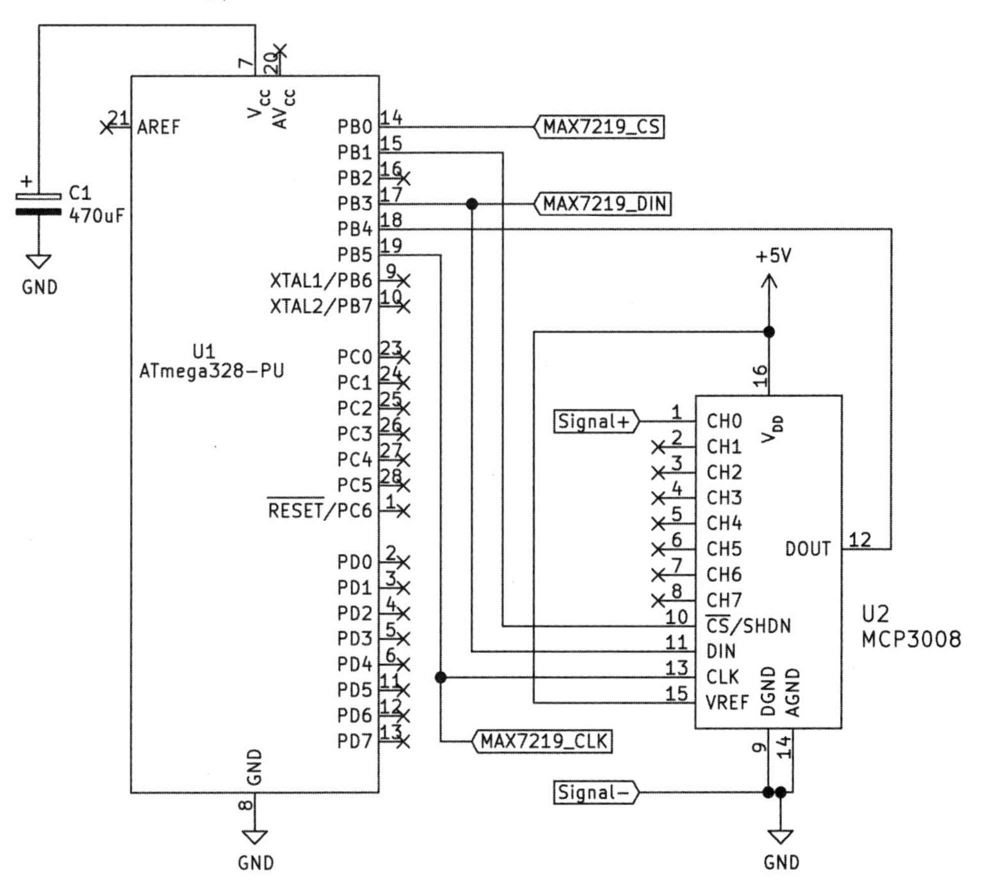


*Figure 11-23: Timing diagram for MCP3008*

We can see that we should set the DORD bit to 0, as the MCP3008 requires data to be sent LSB first. We'll set the CPOL and CPHA bits to 0 as well, since the clock signal is low at the start of data transmission and the polarity of the clock signal is low at idle.

Now we need to learn how the MCP3008 is controlled. We'll be using it in its simplest form, that of a single-ended ADC (so simply measuring between 0 V and $V_{REF}$, in our case 5 V). Every time we want to use the MCP3008, we set the SPCR register to `0b01010010`. (If you're using multiple SPI devices, you will need to set SPCR before communicating with each device.)

Next, we send three bytes of data to the MCP3008 so it will return the required ADC value to the microcontroller over two bytes. We first place `0b00000001` into SPDR as the "start bit" to activate the MCP3008. Next, we place a configuration data byte into SPDR. The first bit is 1 for a single-ended ADC, then the next three bits represent, in binary, which ADC to use (0 to 7). As we're using ADC 0, we set these three bits to 0. The last four bits are unused, so we leave them as 0.

Once we've sent the configuration byte to the MCP3008, it returns a byte of data representing the most significant two bits of the ADC result (bits 0 and 1 of the byte). As mentioned previously, data moves circularly through the SPI bus, so the byte from the MCP3008 is found in the SPDR register. To capture this byte of data, we "push" it out by placing an integer variable in SPDR. The remaining six bits of the byte will contain random data, so to set them to 0 we use the bitwise operation `&` on the captured byte.

Finally, we need the last eight bits of the ADC result, so we push a random byte of data (all 0s is fine) down the SPI bus to receive the byte from the ADC by placing 0 in SPDR. After waiting for the transmission to complete, we then equate another integer variable to SPDR, which now contains the rest of the ADC data.

Now that we have 2 bytes of data, one containing the top 2 bits (the MSB) and the other containing the other 8 bits of data (the LSB), we need to convert them into a single value: a 16-bit integer, which we'll call the *result*. For this, we bit-shift the MSB variable 8 bits to the left into the result integer, then use bitwise operation | to drop the LSB variable into the result. At last, we have the 10-bit value of the ADC in one integer variable, which will be between 0 and 1,023.

Let's examine the code to see how this works:

```
// Project 48 - Using the MCP3008 ADC IC

#include <avr/io.h>
#include <util/delay.h>

void writeMAX7219(char hexdata1, char hexdata2)
// Sends two bytes in hexadecimal to MAX7219
{
   SPCR = 0b01010000;             // Set up SPI bus for MAX7219
   PORTB &= ~(1 << PORTB0);       // SS pin LOW
   SPDR = hexdata1;               // Send value of hexdata1
   while(!(SPSR & (1<<SPIF)));    // Wait for SPI transmission to finish
```

```
    SPDR = hexdata2;                    // Send value of hexdata2
    while(!(SPSR & (1<<SPIF)));         // Wait for SPI transmission to finish
    PORTB |= (1 << PORTB0);             // SS pin HIGH
}

void blankMAX7219()
{
    uint8_t i;
    for (i=1; i<9; i++)
    {
       writeMAX7219(i,15);
    }
}

void initMAX7219()
// Set up MAX7219 for use
{
    PORTB |= (1 << 0);
    SPCR = 0b01010000;
    writeMAX7219(0x09,0xFF);
    writeMAX7219(0x0B,0x07);
    writeMAX7219(0x0A,0x01);
    writeMAX7219(0x0C,0x01);
    blankMAX7219();
}

void dispMAX7219(uint8_t digit, uint8_t number, uint8_t dp)
{
    if (dp==1)                          // Add decimal point
    {
       number = number + 128;
    }
    writeMAX7219(digit, number);
}

void numberMAX7219(uint32_t value)
// Displays a number between 0-99999999

    uint8_t digits[9];
    uint8_t i = 1;

    for (i=1; i<9; i++)
    {
       digits[i]=15;                   // Sending 15 blanks the digit
    }
    i = 1;
    while (value > 0)                   // Continue until value > 0
    {
       digits[i] = value % 10;         // Determine and store last digit
       value = value / 10;             // Divide value by 10
       i++;
    }
    for (i=1; i<9; i++)
    {
       dispMAX7219(i, digits[i],0);
```

```
    }
  }

❶ uint16_t readMCP3008()
  // Read channel 0 and return value
  {
  ❷ uint8_t LSB;
    uint8_t MSB;
    uint16_t ADCvalue;              // Holds data to return to main code
    SPCR = 0b01010010;              // Set up SPI bus for MCP3008

    // SS on PB1 (15)
    PORTB &= ~(1 << PORTB1);        // SS pin LOW

  ❸ SPDR = 0b00000001;              // Send start bit
    while(!(SPSR & (1<<SPIF)));     // Wait for SPI transmission to finish

    SPDR = 0b10000000;              // Select ADC0
    while(!(SPSR & (1<<SPIF)));

    // Place top 2 bits of ADC value in MSB, ignore unwanted bits
  ❹ MSB = SPDR & 0b00000011;

  ❺ SPDR = 0b00000000;              // Request next 8 bits of data
    while(!(SPSR & (1<<SPIF)));
    // Place lower 8 bits of data in LSB
  ❻ LSB = SPDR;
  ❼ PORTB |= (1 << PORTB1);         // SS pin HIGH
  ❽ ADCvalue = MSB << 8 | LSB;      // Construct final 10-bit ADC value
    return ADCvalue;
  }

  int main(void)
  {
    uint16_t ADCoutput;
    DDRB = 0b11111111;              // Set PORTB as outputs
    initMAX7219();
    while (1)
    {
    ❾ ADCoutput = readMCP3008();
      // Convert ADC value to millivolts
    ❿ ADCoutput = ADCoutput * 4.8828;
      numberMAX7219(ADCoutput);
      _delay_ms(100);
    }
  }
```

In this code, we have reused all the MAX7219 functions from Project 48 to display the ADC value in millivolts. Our main code receives an ADC value, converts it to millivolts, then shows it on the MAX7219 display. You should be familiar with the basic structure by now.

We also declare a new function, `readMCP3008()` ❶, that returns a 16-bit integer containing the value measured by the MCP3008's first ADC (0). Inside the function we define three variables—two 8-bit integers to hold the MSB and LSB of the data that the ADC returns and a 16-bit integer to return the full value of the ADC measurement—and set the SPCR register for the MCP3008 ❷. We then set the SS pin low to start the SPI bus as usual.

After sending the start bit to activate the MCP3008 ❸, we wait for the SPI bus transmission to be completed, as always. Following this, we send the configuration byte, telling the MCP3008 we want a single-ended ADC result from channel 0. Once we've done so, the MCP3008 returns the MSB of the result.

We store the MSB in the variable `MSB` ❹ and perform the bitwise `&` to remove the random unnecessary bits. Then we request the LSB of the data by sending a random byte (here, all 0s) ❺, wait for the transmission to end, and store the data in `LSB` ❻. After this, we're finished with the MCP3008, so we remove it from the SPI bus by setting the SS pin high ❼.

We now have the two bytes of data that need to be converted to a single integer for return as the value of this function. We do this by shifting the MSB into the top eight bits of the return variable `ADCvalue` and dropping in the LSB with the | function ❽.

Now that we have a value from the ADC, we move on to the main loop of code, assigning the reading from the ADC to a 16-bit integer ❾. However, this value falls between 0 and 1,023, so we need to convert it to millivolts (mV). Our $V_{REF}$ is 5 V, or 5,000 mV. Therefore, we divide 5,000 by 1,024 to determine the multiplier to convert this ADC value to mV: 4.8828. The program then converts the ADC value to millivolts ❿ and sends it to the display.

At this point you should understand how to implement the SPI bus, including how to examine SPI device data sheets to locate the information required to use them with your AVR. You've also learned how to take advantage of the useful shift register IC, the MAX7219 display driver, and the MCP3008 ADC. This knowledge should prepare you to use other SPI-based parts for your own projects.

In the next chapter, you'll learn how to use many more interesting and useful parts with another type of data bus: the $I^2C$.

# 12

## AVR AND THE $I^2C$ BUS

The *Inter-Integrated Circuit* bus, or *$I^2C$*, is another popular type of data bus. Originally devised by Philips (now NXP), this bus is designed to enable one or many devices to transmit and receive data to and from a host device such as a microcontroller over short distances. This chapter shows you how to set up the $I^2C$ and use it to communicate with external ICs by learning the required functions and hardware and exploring some examples of popular $I^2C$ devices.

You'll learn how to increase the operating speed of AVR microcontrollers, enabling you to implement the $I^2C$ data bus with an AVR. After learning how to control the MCP23017 $I^2C$ 16-bit I/O expander IC and store and retrieve data from an external $I^2C$ EEPROM, you'll use the DS3231 $I^2C$ real-time clock IC for the first time.

There are thousands of devices that use the $I^2C$ bus, from display and motor controllers to sensors and more. After working through this chapter, you'll be prepared to harness these devices to make more complex projects like weather monitoring and display solutions, multiple-servo robots, and projects that require adding more I/O ports to your microcontroller.

## Increasing AVR Speed

Up until now, your AVR projects have been running with a CPU speed of 1 MHz, using the AVR's internal oscillator to generate the required clock signal for timing and other operations. This minimized complications and reduced the number of required parts for the circuits. At times, however, you may need to work with parts that require a faster data bus. These parts include those that communicate with microcontrollers using the $I^2C$ bus. In order to use this bus, you'll need to know how to run your projects with a higher CPU speed to generate faster clock signals.

To increase the ATmega328P-PU's CPU speed, you'll need to make two modifications: one to the hardware and one to the project's Makefile. To alter the hardware, you'll need three more components: two 22 pF ceramic capacitors and one 16 MHz *crystal oscillator*. More commonly known as *crystals*, these oscillators create an electrical signal with an exactly accurate frequency, in this case 16 MHz. Figure 12-1 shows the crystal we'll use.

*Figure 12-1: A 16 MHz crystal oscillator*

Crystals are not polarized. The schematic symbol for our 16 MHz crystal is shown in Figure 12-2.

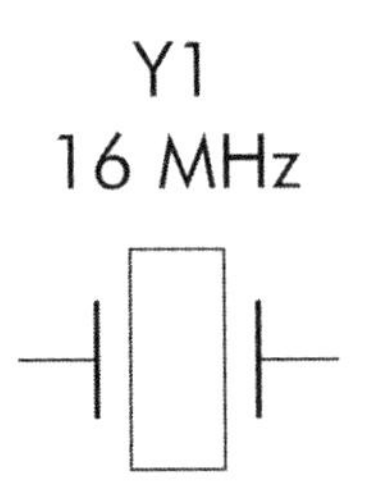


*Figure 12-2: Crystal oscillator schematic symbol*

The crystal determines the microcontroller's speed of operation. For example, the circuit we will be assembling runs at 16 MHz, which means it can execute 16 million processor instructions per second. That doesn't mean it can execute a line of code or function that rapidly, of course, since it takes many processor instructions to execute a single line of code.

The schematic in Figure 12-3 shows the additional circuitry required to connect a crystal to an ATmega328P-PU microcontroller.

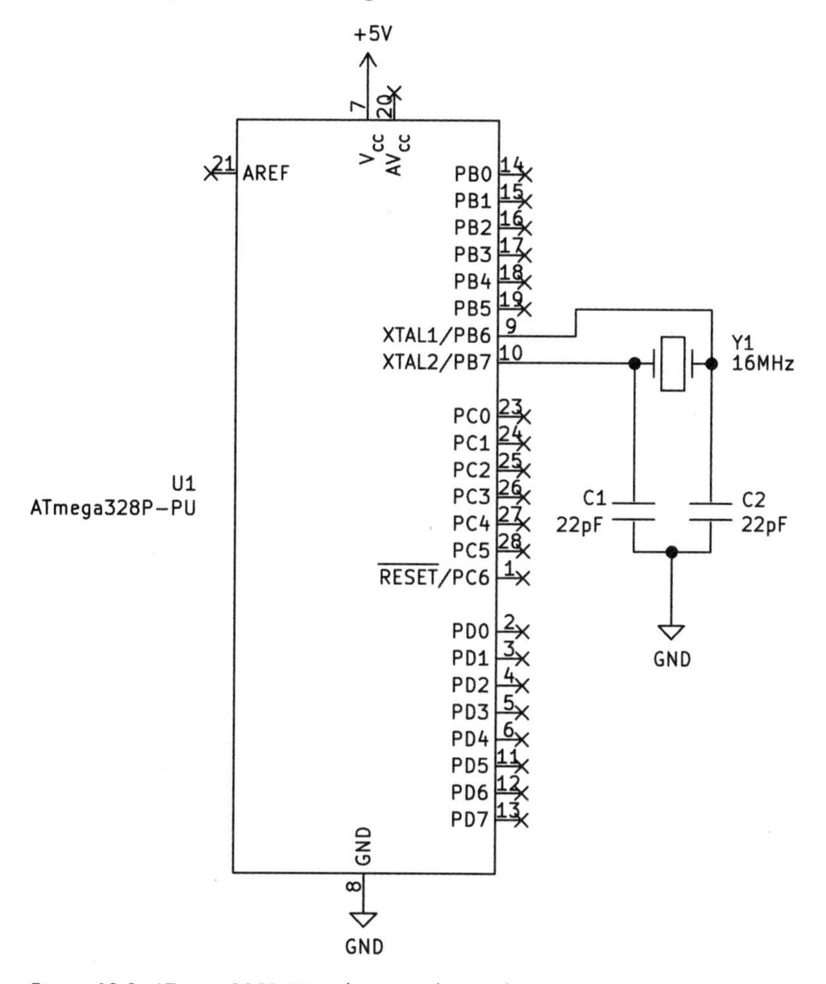


*Figure 12-3: ATmega328P-PU with external crystal circuit*

Along with the hardware changes mentioned above, you'll also need to edit the project's Makefile to tell the toolchain to set the microcontroller to operate at 16 MHz. This will be necessary for all projects using an external crystal. To do this for this chapter's first project, open the Makefile found in the *Project 49* subfolder of the book's *Chapter 11* folder. Scroll down to line 21, labeled `FUSES`, and update it to match the following line of code (don't forget to save the file when you're done):

```
FUSES      = -U lfuse:w:0xff:m -U hfuse:w:0xde:m -U efuse:w:0x05:m
```

**NOTE** *It's a good idea to store different Makefiles separately so you can easily copy them into project folders of your own when required. If you have downloaded the code for this book from the No Starch Press website, the Makefiles are already set up in this way for each project.*

## Introducing the I²C Bus

The I²C bus works similarly to the SPI bus, in that data is again transmitted and received in serial fashion along one wire to or from the microcontroller (the *serial data line*, usually called *SDA*), while another wire carries the clock signal (the *serial clock line*, usually called *SCL* or *SCK*). That signal is synchronized with the data signal for accurate data transmission. The clock signal frequency for our projects is 100 kHz.

**NOTE** *The I²C is a bidirectional bus, in that data is transmitted or received along one wire. Because the clock signal is carried on another wire, some suppliers, such as Microchip, call the I²C the* two-wire serial interface (TWI).

On the I²C bus, the microcontroller acts as the *primary* device, and each IC on the bus is a *secondary*. Each secondary has its own address, a 7-bit number that allows the microcontroller to communicate with that device. Each device usually has a range of I²C bus addresses to choose from, detailed in the manufacturer's data sheet. If an IC has two or more potential I²C bus addresses, we determine which address to use by wiring the IC pins in a certain way.

Figure 12-4 shows an example of the I²C bus in action as captured using a DSO.

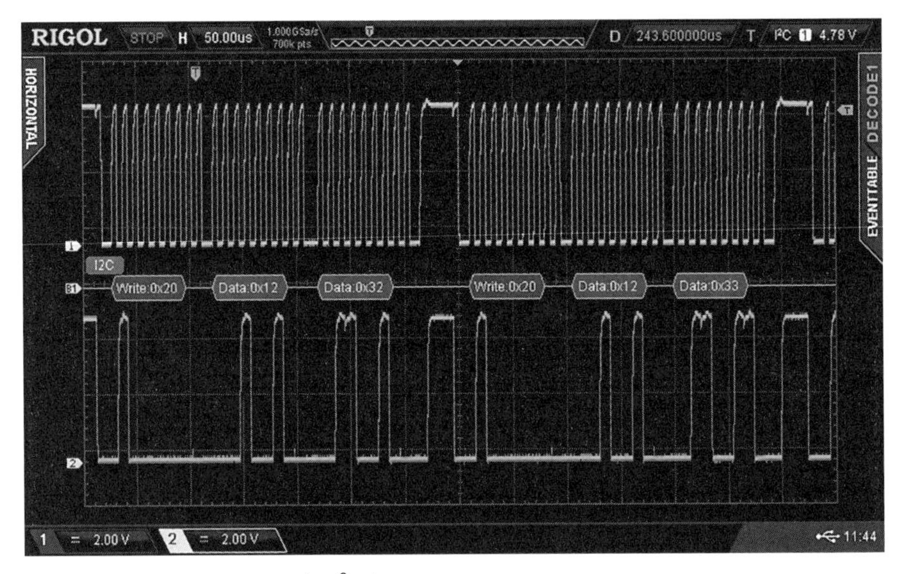


*Figure 12-4: Data traversing the I²C bus*

Once activated, using the $I^2C$ bus follows a simple logical pattern, as shown in Figure 12-4. Both lines of the bus are held at high; we connect them to 5 V via pullup resistors. We send a start signal through the primary device (the microcontroller) by setting the data line low and then starting the clock signal on the clock line. Next, we send the 7 bits of the address of the device with which we want to communicate, followed by a 0 (which tells the device we want to write to it) or a 1 (which tells the device we want it to send us data).

The secondary device will then either acknowledge that it has received the byte from the primary device by sending an *ACK bit* (a 0) or, if there was an error, send a *NACK bit* (a 1) for "not acknowledged," indicating that the primary device should stop sending. The secondary may also send a NACK if it has finished sending data back to the primary.

You can see this in action in Figure 12-4, as the primary starts the $I^2C$ bus, then sends a `0x20+0` to tell the secondary device it is writing data, then sends another two bytes of data to the secondary device. Hexadecimal is the preferred base system used with the $I^2C$ bus in code, but you can use binary or decimal if it's easier or makes more sense to you. This may seem complex, but after examining the projects in this chapter, you'll be well on your way to $I^2C$ expertise.

## *Pin Connections and Voltages*

Each $I^2C$ device uses two pins—usually labeled SCL and SDA, as mentioned earlier—to communicate. These pins connect to matching pins on the microcontroller. If you have multiple $I^2C$ devices, they all make the same connections back to the microcontroller. Finally, a pullup resistor is placed between 5 V (the supply voltage) and each of the $I^2C$ lines. Figure 12-5 shows a simplified example of this.

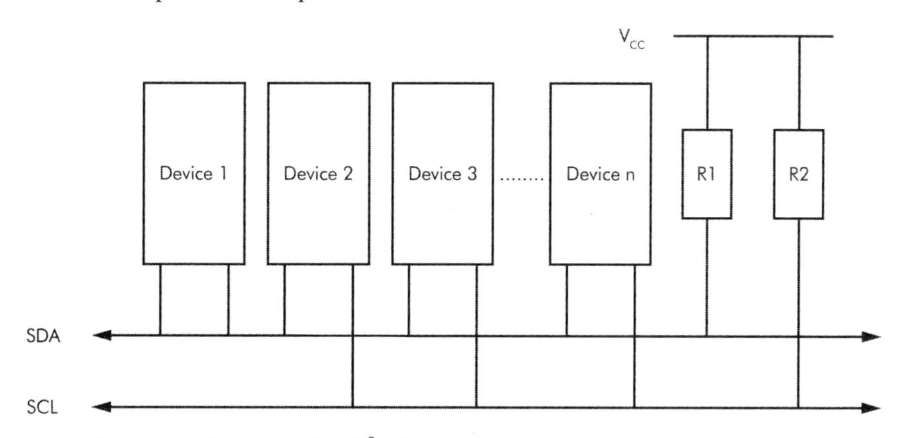


*Figure 12-5: Simplified example of $I^2C$ bus wiring*

As with the SPI bus, your $I^2C$ bus device must operate or tolerate operating at 5 V, since the microcontroller runs on 5 V in our projects. Be sure to check with the seller or manufacturer before use. If you simply must use an $I^2C$ bus device that operates at a reduced voltage, such as 3.3 V, use an $I^2C$-compatible level converter like the unit mentioned in Chapter 11, PMD Way part number 441079.

Again, using these level converters is simple. Use one pair of channels for the $I^2C$ bus, with the 5 V wires on the HV pads and the matching lower-voltage wires on the LV pads, and connect GND on both sides to the board.

## Writing to $I^2C$ Devices

To show you how to implement the $I^2C$ bus, I'll first explain how to write data to $I^2C$ devices, then follow with reading data later in the chapter. Writing data to $I^2C$ bus devices requires five functions that don't exist in our AVR toolchain, so we use the following custom functions to complete the necessary operations. These functions are included in this chapter's $I^2C$ bus projects.

### Enable the $I^2C$ Bus

The `I2Cenable()` function converts the two GPIO pins (PC5 and PC4 on the ATmega328P-PU) from normal use to $I^2C$ bus pins (SCL and SDA, respectively):

```
void I2Cenable()
{
   TWBR = 72;             // 100 kHz I2C bus
   TWCR |= (1 << TWEN); // Enable I2C on PORTC4 and 5
}
```

We first set the *TWBR (TWI Bit Rate Register)*, which is used in the formula provided by Microchip to determine the $I^2C$ bus clock speed. Here we use the value 72, which results in the microcontroller dividing the CPU speed down to 100 kHz for our clock. We also set the *TWEN (Two-Wire Enable)* bit in the *TWCR (Two-Wire Control Register)* to 1, which turns the GPIO pins into $I^2C$ bus pins SCL and SDA.

### Wait for the $I^2C$ Bus to Complete

Bus operations are not instantaneous, so we use the `I2Cwait()` function after other $I^2C$ bus commands to allow for data transmission to complete before executing another operation on the bus:

```
void I2Cwait()
{
    while (!(TWCR & (1<<TWINT)));
}
```

We can check if the bus is busy by looking at the *TWINT (Two-Wire Interrupt)* bit of the TWCR register. This code sets the TWINT bit to 1 when the bus is free for another operation, so the code does nothing while TWINT is 0.

### Start the I²C Bus

The `I2CstartWait()` function starts the process of sending data to a device on the I²C bus. It sends the secondary address to enable the required device on the bus and awaits that device's acknowledgment that it is ready for use:

```
void I2CstartWait(unsigned char address)
{
   uint8_t status;
   while (1)
   {
   ❶ TWCR = (1<<TWINT) | (1<<TWSTA) | (1<<TWEN);
   ❷ I2Cwait();

   ❸ status = TWSR & 0b11111000;
     if ((status != 0b00001000) && (status != 0b00010000)) continue;

   ❹ TWDR = address;
     TWCR = (1<<TWINT) | (1<<TWEN);
     I2Cwait();

   ❺ status = TWSR & 0b11111000;
     if ((status == 0b00100000)||(status == 0b01011000))
     {
     ❻ TWCR = (1<<TWINT) | (1<<TWEN) | (1<<TWSTO);
       while(TWCR & (1<<TWSTO));
       continue;
     }
   break;
   }
}
```

This function first sets a start condition on the bus by setting TWINT in the TWCR register to 1, setting the start condition on with TWSTA (two-wire interface start condition), and activating the bus with TWEN ❶. We wait for those operations to complete ❷, then load the value of the TWSR register ❸ and continue only if the status bits TWS3 and TWS4 are not set to 1. If set to 1, these bits indicate that a start condition or repeated start condition was not successfully sent, so we can't continue.

At this point the bus initialization has been successful, so we now send the secondary address to enable the required device on the bus. We do so by loading the address into the TWDR register ❹, then sending it off by setting TWINT and TWEN in the TWCR register to 1. This is followed by a short wait to give time for the transmission to complete.

Once again we check the status of the I²C bus by loading the value of the TWSR register ❺, and we continue only if the secondary device is not busy or if the secondary device didn't acknowledge the write. If it is busy or didn't acknowledge, we send an I²C stop command ❻ and wait for that instruction to finish.

### Write to the I²C Bus

This function simply sends a byte of data along the I²C bus:

```
void I2Cwrite(uint8_t data)
{
   TWDR = data;
   TWCR |= (1 << TWINT)|(1 << TWEN);
   I2Cwait();
}
```

Once I²C bus communication has been initialized and we've started the I²C bus, we use this function to write a byte of data to the device being addressed. We load the data into the TWDR register, and then send it out using the TWCR register. We then wait a moment for the process to complete. You can call this function two or more times in a row to send multiple bytes of data.

### Stop the I²C Bus

The `I2Cstop()` function releases the GPIO pins from I²C bus duty and sets them back to normal duty:

```
void I2Cstop()
{
   TWCR |= (1 << TWINT)|(1 << TWEN)|(1 << TWSTO);
}
```

When your code is finished with the I²C bus, use this function to stop the bus and release the GPIO pins used for SDA and SCL back to normal duty. The bus is stopped by three bit changes:

- Setting the TWINT bit to 1 tells the microcontroller you're finished using the bus.
- Setting the TWSTO bit to 1 puts a stop condition on the bus, telling the devices on the bus that the bus is being deactivated.
- Setting the TWEN bit to 1 disables bus operation.

Now let's turn theory into practice by using some interesting devices over the I²C bus with the ATmega328P-PU. Our first such device will be the Microchip MCP23017, which adds 16 more I/O pins to our microcontroller.

## Project 49: Using the MCP23017 16-Bit I/O Expander

When your AVR-based project doesn't have enough digital GPIO pins, you can add 16 more at a time with the Microchip MCP23017, shown in Figure 12-6. We'll use the MCP23017 for the first time in this project, which doubles as a demonstration of writing data to the I²C bus.

*Figure 12-6: Microchip MCP23017 16-bit I/O expander*

The MCP23017 has eight possible bus addresses, so you can connect a maximum of eight on the same bus, giving you up to 128 GPIO pins. The I/O pins are arranged in two banks of eight, as shown in the schematic symbol in Figure 12-7. Each pin can handle up to 20 mA of current, though the entire MCP23017 has a continuous maximum of 125 mA.

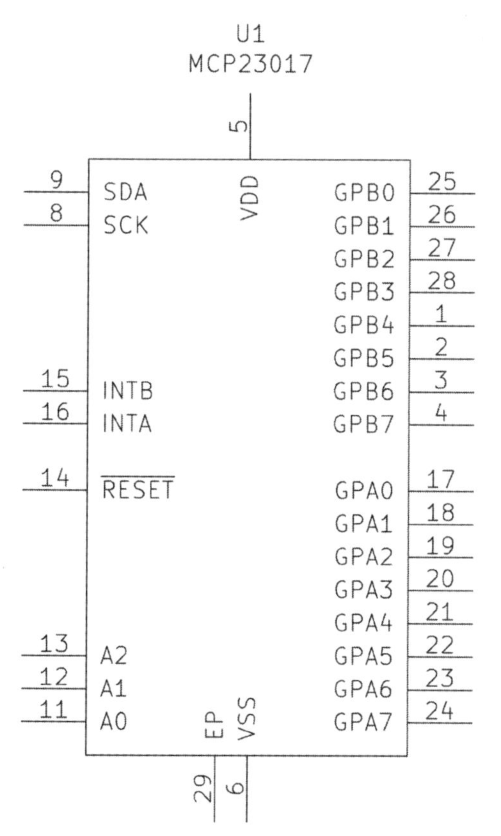

*Figure 12-7: Microchip MCP23017 schematic symbol*

To set the $I^2C$ bus address, you need to connect the pins labeled A0 to A2 to combinations of 5 V or GND. If you're using one MCP23017, you can set the bus address to `0x20` by connecting the three pins to GND. If you're using two or more MCP23017s, or need to let another device use the `0x20` address, refer to Table 12-1 for your configuration.

**Table 12-1**: MCP23017 I$^2$C Bus Address Configuration

| Bus address | Pin A2 | Pin A1 | Pin A0 |
|---|---|---|---|
| `0x20` | GND | GND | GND |
| `0x21` | GND | GND | 5 V |
| `0x22` | GND | 5 V | GND |
| `0x23` | GND | 5 V | 5 V |
| `0x24` | 5 V | GND | GND |
| `0x25` | 5 V | GND | 5 V |
| `0x26` | 5 V | 5 V | GND |
| `0x27` | 5 V | 5 V | 5 V |

As mentioned earlier, the bus address is a 7-bit number, which we complete to an 8-bit number by adding a 0 or 1 at the end for writing to the bus or reading the bus, respectively. You can create this 8-bit number by bit-shifting a 1 from the right, using a `<<1` to create the address for writing—`0x20<<1`, for example—or converting the address to the result, which in this example would be `0x40`.

Controlling the MCP23017 also involves writing to its configuration registers, each of which has its own address. To use the pins as outputs, we set the I/O direction registers for both banks of eight pins. These are known as the GPIOA and GPIOB registers for banks A and B, whose addresses are `0x12` and `0x13`.

Once we've addressed these registers, we send a 0 to each to set the pins as outputs. For example, to set the GPIOA register to outputs, we'd send the following sequence of data to the I$^2$C bus: `0x20<<1` or `0x40` (the MCP23017 I$^2$C bus address for writing), then `0x12` (the GPIOA register address), then `0x00` (0). Let's do this now in our project.

## *The Hardware*

To build your circuit, you'll need the following hardware:

- USBasp programmer
- Solderless breadboard
- 5 V breadboard power supply
- ATmega328P-PU microcontroller
- MCP23017 16-bit I/O IC
- Up to 16 LEDs (D1–D16)
- Up to 16 560 Ω resistors (R1–R16)
- Two 4.7 kΩ resistors (R17–R18)
- Two 22 pF ceramic capacitors
- 16 MHz crystal oscillator
- Jumper wires

Assemble your circuit as shown in Figure 12-8.

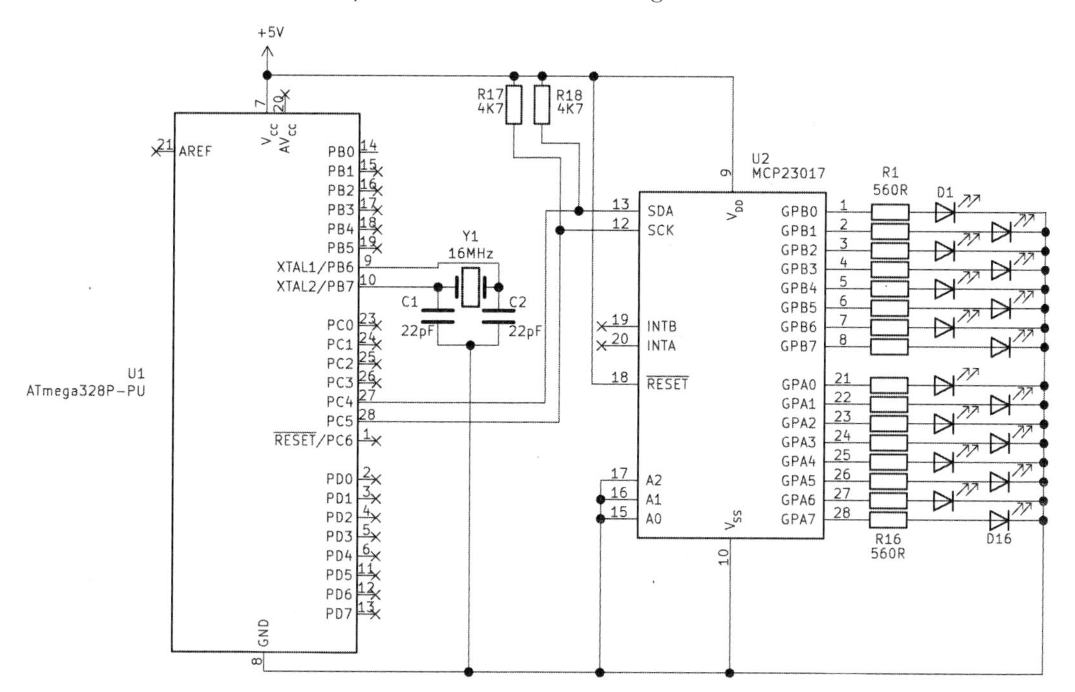


*Figure 12-8: Schematic for Project 49*

## The Code

Open a terminal window, navigate to the *Project 49* subfolder of this book's *Chapter 12* folder, and enter the command **make flash**. After a few moments, each bank of LEDs should repeatedly display the count between 0 and 255 in binary.

Let's examine the code to see how this is done:

```
// Project 49 - Using the MCP23017 16-Bit I/O Expander

#include <avr/io.h>
#include <util/delay.h>

void I2Cenable()
// Enable the I2C bus
{
   TWBR = 72;            // 100 kHz I2C bus
   TWCR |= (1 << TWEN); // Enable I2C on PORTC4 and 5
}

void I2Cwait()
// Wait until I2C finishes an operation
{
   // Wait until bit TWINT in TWCR is set to 1
   while (!(TWCR & (1<<TWINT)));
}
```

```
void I2CstartWait(unsigned char address)
{
   uint8_t status;
   while (1)
   {
      // Send START condition
      TWCR = (1<<TWINT) | (1<<TWSTA) | (1<<TWEN);

      // Wait until transmission completes
      I2Cwait();

      // Check value of TWSR, and mask out status bits
      status = TWSR & 0b11111000;
      if ((status != 0b00001000) && (status != 0b00010000)) continue;

      // Send device address
      TWDR = address;
      TWCR = (1<<TWINT) | (1<<TWEN);

      // Wait until transmission completes
      I2Cwait();

      // Check value of TWSR, and mask out status bits
      status = TWSR & 0b11111000;
      if ((status == 0b00100000 )||(status == 0b01011000))
      {
      // Secondary device is busy, send stop to terminate write operation
         TWCR = (1<<TWINT) | (1<<TWEN) | (1<<TWSTO);
         // Wait until stop condition is executed and I2C bus is released
         while(TWCR & (1<<TWSTO));
         continue;
      }
   break;
   }
}

void I2Cstop()
// Stop I2C bus and release GPIO pins
{
   // Clear interrupt, enable I2C, generate stop condition
   TWCR |= (1 << TWINT)|(1 << TWEN)|(1 << TWSTO);
}

void I2Cwrite(uint8_t data)
// Send 'data' to I2C bus
{
   TWDR = data;
   TWCR |= (1 << TWINT)|(1 << TWEN);
   I2Cwait();
}

❶ void initMCP23017()
// Configure MCP23017 ports for all outputs
{
   I2CstartWait(0x20<<1); // 0x20 write mode
```

```
    I2Cwrite(0x00);        // IODIRA register
    I2Cwrite(0x00);        // Set all register A to outputs
    I2Cstop();

    I2CstartWait(0x20<<1); // 0x20 write mode
    I2Cwrite(0x01);        // IODIRB register
    I2Cwrite(0x00);        // Set all register B to outputs
    I2Cstop();
}

int main(void)
{
   uint8_t i;
❷ I2Cenable();
   initMCP23017();
   while (1)
   {
      for (i = 0; i< 256; i++)
      {
       ❸ I2CstartWait(0x20<<1);
       ❹ I2Cwrite(0x12);  // Control register A 0x12
       ❺ I2Cwrite(i);     // Value to send
       ❻ I2Cstop();

         I2CstartWait(0x20<<1);
       ❼ I2Cwrite(0x13);  // Control register B 0x13
       ❽ I2Cwrite(i);     // Value to send
       ❾ I2Cstop();
         _delay_ms(100);
      }
   }
}
```

This code uses the five $I^2C$ functions described in the previous section to simplify data transmission. The initialization function ❶ facilitates using the MCP23017. This sets up the GPIOA and GPIOB registers to make all the I/O pins outputs.

Next, we initialize the $I^2C$ bus ❷ and address each GPIO bank to control the outputs. We start the $I^2C$ bus ❸, then address the GPIO registers by sending their addresses ❹. Then we send the data to control the registers ❺ and stop the $I^2C$ bus ❻. Following this, after restarting the $I^2C$ bus, the code selects the second bank of the MCP23017 ❼, sends the data ❽, and again stops the bus ❾.

As this project is intended to demonstrate all possible output combinations, it sends a decimal number generated by the `for` loop. However, you can also use a binary number if it helps with visualizing which of the output pins to control. For example, if you want to turn on pins 7, 4, and 0 in a bank, you can send `0b10010001` rather than `145` in decimal, and the corresponding outputs will go high. Notice how the physical pin numbers match the ones in the binary number.

Now that you know how to harness the MCP23017, let's move on to reading data from an $I^2C$ device.

## Reading Data from I²C Devices

Now that you can write to an I²C device, it's time to learn how to read data—such as sensor data, external memory, and other types of output—from these devices. To read data, after initializing the I²C bus as normal, we use this function:

```
I2CstartWait(address);
```

This time we add a 1 at the end of the 7-bit bus address (instead of a 0, for writing). Once the secondary device receives this address byte, it knows to send back one or more bytes of data to the bus for the primary device to receive.

To determine the correct bus address to use to read from a device, take the device's I²C address and convert it to binary. For example, the address `0x50` converts to `1010000` in binary. Add a 1 at the end so you end up with `10100001`, then convert this back to hexadecimal, which results in `0xA1`.

**NOTE** *If you don't have a calculator capable of converting from binary to decimal to hexadecimal and vice versa, a useful website for various mathematical conversions is* https://www.rapidtables.com/convert/number/.

Next, we use one of two new functions, `I2Cread()` or `I2CreadACK()`. `I2Cread()` waits for a byte of data to come back from the secondary device (without an acknowledge bit) and places it in a byte variable:

```
uint8_t I2Cread()
{
   TWCR |= (1 << TWINT)|(1 << TWEN);
   I2Cwait();
   // Incoming byte is placed in TWDR register
   return TWDR;
}
```

This function enables the primary device to receive a byte of data by first setting TWINT and TWEN to enable the bus and free it for use. After waiting for the operation to complete, the byte of data received from the secondary device is available in the TWDR register, and it's then passed as the result of the function with `return TWDR;`.

Like `I2Cread()`, `I2CreadACK()` waits for a byte from the secondary device and places it in a byte variable, but it also considers the acknowledge bit from the secondary device:

```
uint8_t I2CreadACK()
{
   TWCR |= (1 << TWINT)|(1 << TWEN)|(1 << TWEA);
   I2Cwait();
   // Incoming byte is placed in TWDR register
   return TWDR;
}
```

This time, in addition to setting TWINT and TWEN, we set TWEA in TWCR to 1. This generates the ACK (acknowledge bit) on the bus when the primary device has successfully received the byte of data.

We pick which read function to use based on the parameters of the secondary device. Some devices require the ACK bit before sending more data, and some do not.

Now that we have the full complement of functions for using the I²C bus, let's put them to use in the following projects.

## Project 50: Using an External IC EEPROM

Project 39 in Chapter 9 showed how to use the ATmega328P-PU's internal EEPROM to store data that you don't want to delete when you remove the power source. Taking this idea further, you can also use external EEPROM ICs that have more storage space and allow you to build projects with microcontrollers that don't have their own EEPROMs.

For this project, we'll use the Microchip 24LC512-E/P EEPROM IC, an example of which is shown in Figure 12-9.

*Figure 12-9: Microchip 24LC512-E/P external EEPROM IC*

Figure 12-10 shows the Microchip 24LC512-E/P's schematic symbol.

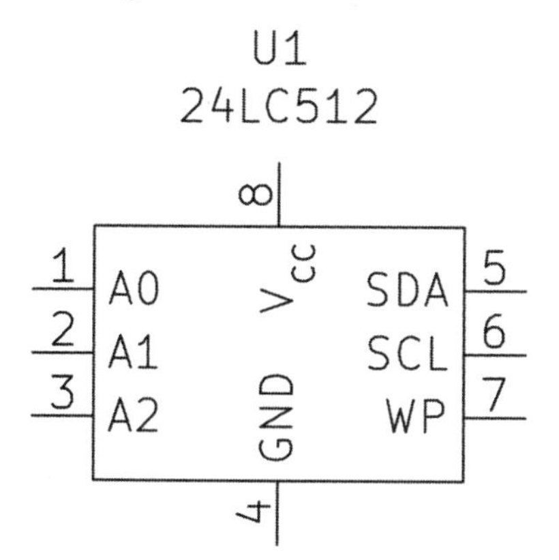


*Figure 12-10: Microchip 24LC512-E/P schematic symbol*

As with most other $I^2C$ devices, you can set this IC's 7-bit address by connecting the combination of pins A0, A1, and A2 to power or GND. If you're using one 24LC512-E/P, you can set the bus address to `0x50` by connecting the three A pins to GND. If you're using two or more, or need to let another secondary device use the `0x50` address; refer to Table 12-2 for your configuration.

**Table 12-2**: 24LC512-E/P $I^2C$ Bus Address Configuration

| Bus address | Pin A2 | Pin A1 | Pin A0 |
|---|---|---|---|
| `0x50` | GND | GND | GND |
| `0x51` | GND | GND | 5 V |
| `0x52` | GND | 5 V | GND |
| `0x53` | GND | 5 V | 5 V |
| `0x54` | 5 V | GND | GND |
| `0x55` | 5 V | GND | 5 V |
| `0x56` | 5 V | 5 V | GND |
| `0x57` | 5 V | 5 V | 5 V |

The 24LC512-E/P can store up to 512KB of data (or, divided by 8, 64,000 bytes). This project demonstrates how to write and read bytes of data to the EEPROM for integration into other projects.

### *The Hardware*

To build your circuit, you'll need the following hardware:

- USBasp programmer
- Solderless breadboard
- 5 V breadboard power supply
- ATmega328P-PU microcontroller
- Microchip 24LC512-E/P EEPROM IC
- Two 4.7 kΩ resistors (R1–R2)
- Two 22 pF ceramic capacitors (C1–C2)
- 470 μF 16 V electrolytic capacitor (C3)
- 16 MHz crystal oscillator
- MAX7219 eight-digit module
- Jumper wires

Assemble your circuit as shown in Figure 12-11.

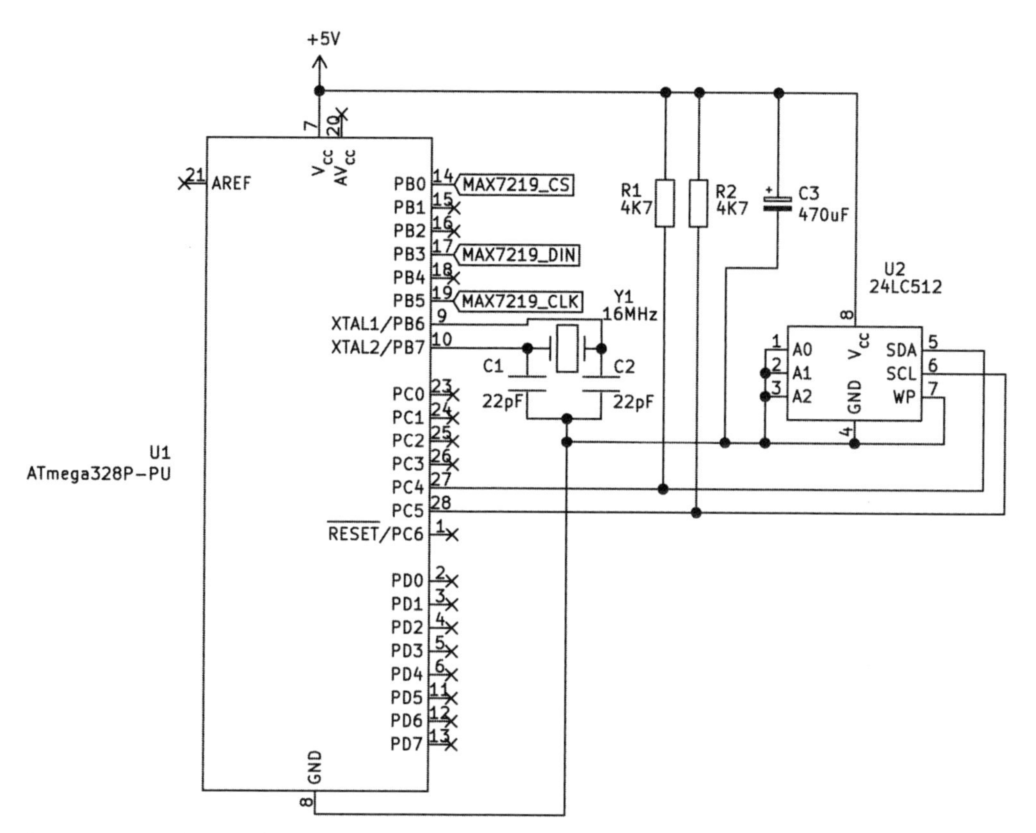


*Figure 12-11: Schematic for Project 50*

## The Code

Open a terminal window, navigate to the *Project 50* subfolder of this book's *Chapter 12* folder, and enter the command **make flash**. After a few moments, the MAX7219 display should rapidly display the numbers between 0 and 255, writing these values to the EEPROM. Then it should display these numbers again at a slower pace as it reads them from the EEPROM.

Let's examine the code to see how this is done:

```
// Project 50 - 24LC512 I2C EEPROM

#include <avr/io.h>
#include <util/delay.h>

void I2Cenable()
// Enable I2C bus
{
   TWBR = 72;            // 100 kHz I2C bus
```

```
   TWCR |= (1 << TWEN); // Enable I2C on PORTC4 and 5
}

void I2Cwait()
// Wait until I2C finishes an operation
{
   // Wait until bit TWINT in TWCR is set to 1
   while (!(TWCR & (1<<TWINT)));
}

void I2CstartWait(unsigned char address)
{
   uint8_t status;
   while (1)
   {
      // Send START condition
      TWCR = (1<<TWINT) | (1<<TWSTA) | (1<<TWEN);

      // Wait until transmission completes
      I2Cwait();

      // Check value of TWSR, and mask out status bits
      status = TWSR & 0b11111000;
      if ((status != 0b00001000) && (status != 0b00010000)) continue;

      // Send device address
      TWDR = address;
      TWCR = (1<<TWINT) | (1<<TWEN);

      // Wait until transmission completes
      I2Cwait();

      // Check value of TWSR, and mask out status bits
      status = TWSR & 0b11111000;
      if ((status == 0b00100000 )||(status == 0b01011000))
      {   // Secondary device is busy, so send stop condition to terminate
          // write operation
          TWCR = (1<<TWINT) | (1<<TWEN) | (1<<TWSTO);
          // Wait until stop condition is executed and I2C bus released
          while(TWCR & (1<<TWSTO));
          continue;
       }
       break;
    }
}

void I2Cstop()
// Stop I2C bus and release GPIO pins
{
   // Clear interrupt, enable I2C, generate stop condition
   TWCR |= (1 << TWINT)|(1 << TWEN)|(1 << TWSTO);
}

void I2Cwrite(uint8_t data)
// Send 'data' to I2C bus
```

```
{
   TWDR = data;
   TWCR |= (1 << TWINT)|(1 << TWEN);
   I2Cwait();
}

uint8_t I2Cread()
// Read incoming byte of data from I2C bus
{
   TWCR |= (1 << TWINT)|(1 << TWEN);
   I2Cwait();
   // Incoming byte is placed in TWDR register
   return TWDR;
}
uint8_t I2CreadACK()
// Read incoming byte of data from I2C bus and ACK signal
{
   TWCR |= (1 << TWINT)|(1 << TWEN)|(1 << TWEA);
   I2Cwait();
   // Incoming byte is placed in TWDR register
   return TWDR;
}

void writeMAX7219(char hexdata1, char hexdata2)
// Sends two bytes in hexadecimal to MAX7219
{
   PORTB &= ~(1 << PORTB0);      // SS pin LOW
   SPDR = hexdata1;              // Send value of hexdata1
   while(!(SPSR & (1<<SPIF))); // Wait for SPI transmission to finish
   SPDR = hexdata2;              // Send value of hexdata2
   while(!(SPSR & (1<<SPIF))); // Wait for SPI transmission to finish
   PORTB |= (1 << PORTB0);      // SS pin HIGH
}

void blankMAX7219()
// Blanks all digits
{
   uint8_t i;
   for (i=1; i<9; i++)          // Blank all digits
   {
      writeMAX7219(i,15);
   }
}

void initMAX7219()
// Set up MAX7219 for use
{
   PORTB |= (1 << 0);            // SS pin HIGH
   SPCR = (1<<SPE)|(1<<MSTR);  // Set up SPI bus for MAX7219
   // Mode decode for digits (table 4 page 7 - 0x09 + 0xFF)
   writeMAX7219(0x09,0xFF);
   writeMAX7219(0x0B,0x07);     // Set scan limit to 8 digits - 0x09 + 0xFF)
   writeMAX7219(0x0A,0x01);     // Set intensity to 8 - 0x0A + 0x08)
   // Mode display on (table 4 page 7 - 0x09 + 0xFF)
   writeMAX7219(0x0C,0x01);
```

```
   blankMAX7219();
}

void dispMAX7219(uint8_t digit, uint8_t number, uint8_t dp)
// Displays "number" in location "digit" with decimal point on/off
// Digit: 1~8 for location 1~8
// Number: 0~15 for 0~9, - E, H, L, P, blank
// dp: 1 on, 0 off
{
   if (dp==1)                          // Add decimal point
   {
      number = number + 128;
   }
   writeMAX7219(digit, number);
}

void numberMAX7219(uint32_t value)
// Displays a number between 0-99999999 on MAX7219-controlled 8-digit display
{
   uint8_t digits[8];
   uint8_t i = 1;

   for (i=1; i<9; i++)
   {
      digits[i]=15;
   }

   i = 1;
   while (value > 0)                   // Continue until value > 0
   { // Determine and store last digit of number
      digits[i] = value % 10;
      value = value / 10;              // Divide value by 10
      i++;
   }
   for (i=1; i<9; i++)
   {
      dispMAX7219(i, digits[i],0);
   }
}

int main(void)
{
   uint16_t i;
   uint16_t j;
   DDRB = 0b11111111;                  // Set PORTB as outputs
   I2Cenable();
   initMAX7219();
   while (1)
   {
      dispMAX7219(0,10,0);
      for (i = 0; i<256; i++)          // Write loop
      {
      ❶ I2CstartWait(0x50<<1);         // 0x50 << 1 - 0b10100000
      ❷ I2Cwrite(i >> 8);
      ❸ I2Cwrite(i);
```

```
       ❹ I2Cwrite(i);
         I2Cstop();
         numberMAX7219(i);
         _delay_ms(1);
       }

       for (i = 0; i<256; i++)       // Read loop
       {
       ❺ I2CstartWait(0x50<<1);      // Write address
       ❻ I2Cwrite(i >> 8);
       ❼ I2Cwrite(i);
       ❽ I2CstartWait((0x50<<1)+1); // Read address - 0b10100001
       ❾ j = I2Cread();
       ❿ I2Cstop();
         numberMAX7219(j);
         _delay_ms(5);
       }
       _delay_ms(100);
    }
}
```

This code reuses functions from previous projects, such as the MAX7219 functions from Chapter 11 and the I²C bus functions from this chapter's Project 49; the new code is contained in the main loop. In summary, it writes and retrieves the values 0 through 255 to and from EEPROM locations 0 through 255. This is accomplished in the write and read loops in the main section of the code.

To write values to the EEPROM, we first start the bus and wait for acknowledgment ❶ in our write loop, using the "write" form of the bus address. The EEPROM is now expecting 2 bytes of data, which represent the address (or location) in the EEPROM's memory to deal with (in this case, write to). It expects 2 bytes of data because there are more than 256 possible locations. Therefore, the code creates the variable `i` in the loop as a 16-bit integer.

We send the "high byte" of the address ❷, which details the part of the address above 255, and follow it with the "low byte," which details the part of the address equal to or less than 255 ❸. Then we send the value to store to the EEPROM ❹, and the MAX7219 display shows that value for our reference.

To read the values from the EEPROM, we first start the bus and wait for acknowledgment ❺ in our read loop, again using the "write" form of the bus address, then send the high ❻ and low ❼ bytes of the address as we did previously.

Next, to retrieve data from the EEPROM, we restart the I²C bus ❽ by using the read form of the bus address. We then use our new `I2Cread()` function to take the byte sent from the EEPROM back to the microcontroller and store it in the variable `j` ❾. Now that we have the data from the EEPROM, we stop using the I²C bus ❿ and show the values on the MAX7219 display module for reference.

> **MORE ON THE HIGH BYTE AND LOW BYTE**
>
> We need to split 16-bit integers into high and low bytes to send them along the $I^2C$ (or SPI) data bus. This involves bit-shifting the whole 16-bit number 8 bits to the right to determine the high byte, then sending the low byte by simply using the 16-bit number in an 8-bit operation, as doing this effectively removes the high byte.
>
> For example, consider the number 41,217. That's greater than 255, so we need 2 bytes of data to represent it in AVR operations. If you convert 41,217 to binary, you'll see that it uses 16 bits:
>
> ```
> 1010000100000001
> ```
>
> We create the 8-bit high byte by bit-shifting the entire number 8 bits to the right. For example:
>
> ```
> 1010000100000001 >> 8 = 10100001 // Our high byte
> ```
>
> We then create the low byte by simply using it in an 8-bit operation. For example:
>
> ```
> I2Cwrite(1010000100000001)
> ```
>
> This has the same effect as `I2Cwrite(00000001)`.

This project provided a neat demonstration not only of writing and reading bytes of data to and from an $I^2C$-based device but also of the framework for storing data in external EEPROM IC. In the next project, we'll move on to our final $I^2C$ bus device, the DS3231—a *real-time clock (RTC)* IC that allows you to add time and date information to your projects.

## Project 51: Using the DS3231 Real-Time Clock

Once set with the current time and date, an RTC provides accurate time and date data on request. RTCs allow you to build a variety of interesting projects, from simple clocks to data-logging devices, alarms, and more. In this project, you'll create a clock that displays the current date and time in 24-hour format using an RTC and the MAX7219 display module.

You'll find many different RTC ICs on the market, some more accurate than others. In this chapter, we'll use the Maxim DS3231; it doesn't require any external circuitry other than a backup battery, it's incredibly accurate, and it's quite robust in module form.

The DS3231 is available as a breakout board from various retailers, including the version from PMD Way (part number 883422) shown in Figure 12-12.

*Figure 12-12: A DS3231 real-time clock IC module*

Using a breakout board means you don't need to worry about support circuitry like pullup resistors for the DS3231, nor connecting a backup battery, as the board takes care of all this for you. All you need to do is insert a CR2032 coin-cell battery for the backup and connect jumper wires to your project.

Connecting the breakout board to your project is easy: just use $V_{CC}$ (for 5 V), GND, SCL, and SDA connections. The DS3231 has a fixed $I^2C$ bus address of `0x68`, which converts to a write address of `0xD0` and a read address of `0xD1`.

It has a group of registers it uses to store time and date information, starting at `0x00` and increasing sequentially, as shown in Table 12-3.

**Table 12-3**: DS3231 Data Registers

| Address | Function |
|---|---|
| `0x00` | Seconds |
| `0x01` | Minutes |
| `0x02` | Hours |
| `0x03` | Day of week (1 = Sunday, 7 = Saturday) |
| `0x04` | Day of month |
| `0x05` | Month |
| `0x06` | Year (two-digit) |

In this project, we'll use only the registers shown in Table 12-3. However, the data sheet details more that you can investigate; it's available at *https://www.maximintegrated.com/en/products/analog/real-time-clocks/DS3231.html.*

Data is stored in the DS3231 registers using *binary-coded decimal (BCD)* format, which assigns a four-digit binary code to each digit in a decimal number. Therefore, we'll use simple BCD-to-decimal conversions in our code.

To set the time and date, we'll write the bytes of data in order from `0x00` using our $I^2C$ bus write function. To retrieve the data, we can either read from a particular address, as we did with the EEPROM in Project 50, or start a read at `0x00` with ACK, causing the DS3231 to send the rest of the data one byte at a time from each sequential register. We'll use this latter method in our project's code. But first, let's assemble the hardware.

### *The Hardware*

To build your circuit, you'll need the following hardware:

- USBasp programmer
- Solderless breadboard
- 5 V breadboard power supply
- ATmega328P-PU microcontroller
- DS3231 RTC module with backup battery
- Two 22 pF ceramic capacitors (C1–C2)
- 470 μF 16 V electrolytic capacitor (C3)
- 16 MHz crystal oscillator
- MAX7219 eight-digit module
- Jumper wires

Assemble your circuit as shown in Figure 12-13. Don't forget to connect the MAX7219 and DS3231 boards to 5 V and GND as well.

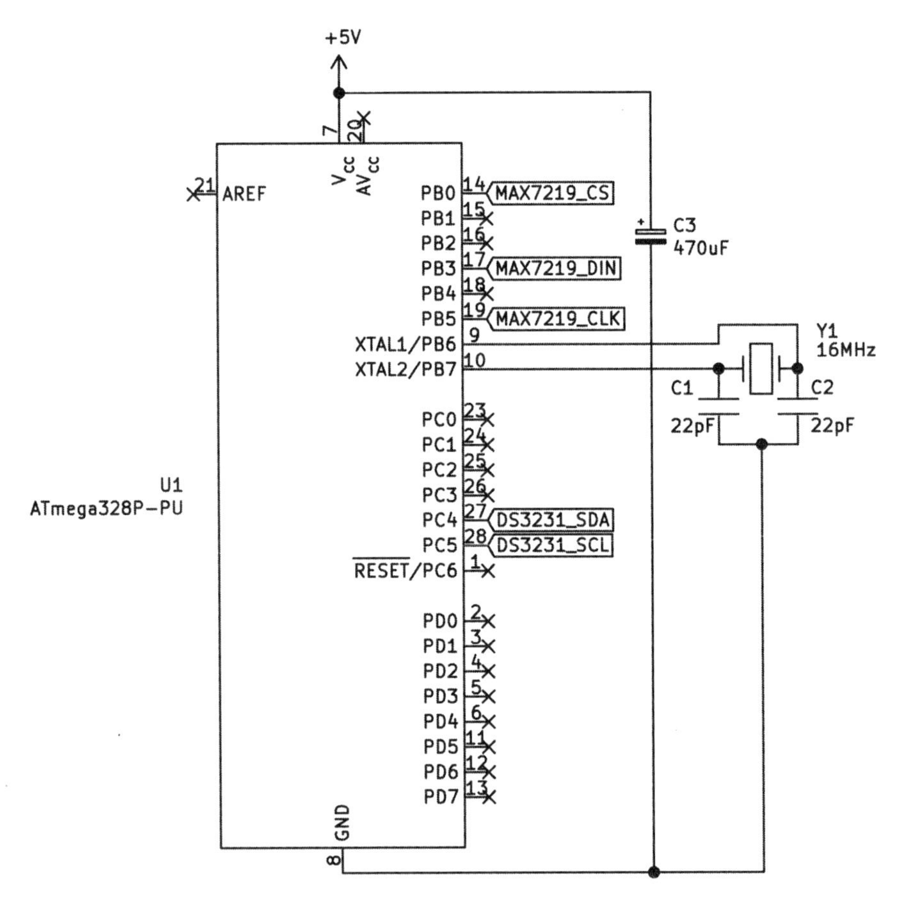


*Figure 12-13: Schematic for Project 51*

## The Code

Before uploading the code to the microcontroller as normal, open the *main.c* file found in the *Project 51* subfolder of this book's *Chapter 12* folder in your text editor and scroll down to line 309. Remove the comment slashes in front of the function `setTimeDS3231()`. Next, update the parameters in that function to match your current date and time. The parameters, in order, are: hours (in 24-hour format), minutes, seconds, day of the week (1 to 7), date (1 to 31), month, year (00 to 99). For example, assuming you consider Sunday the first day of the week—in some areas Monday is considered the first day, so Monday would be 1—if the time is 2:32 PM (that is, 14:32) on Tuesday, November 1, 2022, you would change the line to:

```
setTimeDS3231(14,32,0,3,6,11,22);
```

Now save the file, and execute the **`make flash`** command as usual from the terminal window. Then reopen the *main.c* file, replace the comment slashes in front of the `setTimeDS3231()` function, save the file, and reflash the code. The first flash sets the time and date, and the second deactivates the setting every time the microcontroller is reset or power-cycled. If you skipped this second flash, the project would set the same time and date after every reset.

Once you've finished, you should see the current time then date alternately displayed on the MAX7219 module. Congratulations—you've made your own digital clock!

Now let's examine the code to see how this works:

```
// Project 51 - Using the DS3231 I2C Real-Time Clock

#include <avr/io.h>
#include <util/delay.h>

// Variables to store time and date
❶ uint8_t hours, minutes, seconds, dow, dom, mo, years;

void I2Cenable()
// Enable I2C bus
{
   TWBR = 72;            // 100 kHz I2C bus
   TWCR |= (1 << TWEN); // Enable I2C on PORTC4 and 5
}

void I2Cwait()
// Wait until I2C finishes an operation
{
   // Wait until bit TWINT in TWCR is set to 1
   while (!(TWCR & (1<<TWINT)));
}

void I2CstartWait(unsigned char address)
{
   uint8_t status;
   while (1)
```

```
   {
      // Send START condition
      TWCR = (1<<TWINT) | (1<<TWSTA) | (1<<TWEN);

      // Wait until transmission completes
      I2Cwait();

      // Check value of TWSR, and mask out status bits
      status = TWSR & 0b11111000;
      if ((status != 0b00001000) && (status != 0b00010000)) continue;

      // Send device address
      TWDR = address;
      TWCR = (1<<TWINT) | (1<<TWEN);

      // Wait until transmission completes
      I2Cwait();

      // Check value of TWSR, and mask out status bits
      status = TWSR & 0b11111000;
      if ((status == 0b00100000 )||(status == 0b01011000))
      {
         TWCR = (1<<TWINT) | (1<<TWEN) | (1<<TWSTO);

         // Wait until stop condition is executed and I2C bus is released
         while(TWCR & (1<<TWSTO));
         continue;
      }
      break;
   }
}

void I2Cstop()
// Stop I2C bus and release GPIO pins
{
   // Clear interrupt, enable I2C, generate stop condition
   TWCR |= (1 << TWINT)|(1 << TWEN)|(1 << TWSTO);
}

void I2Cwrite(uint8_t data)
// Send 'data' to I2C bus
{
   TWDR = data;
   TWCR |= (1 << TWINT)|(1 << TWEN);
   I2Cwait();
}

uint8_t I2Cread()
// Read incoming byte of data from I2C bus
{
   TWCR |= (1 << TWINT)|(1 << TWEN);
   I2Cwait();
   // Incoming byte is placed in TWDR register
   return TWDR;
}
```

```
uint8_t I2CreadACK()
// Read incoming byte of data from I2C bus and ACK signal
{
   TWCR |= (1 << TWINT)|(1 << TWEN)|(1 << TWEA);
   I2Cwait();
   // Incoming byte is placed in TWDR register
   return TWDR;
}
void writeMAX7219(char hexdata1, char hexdata2)
// Sends two bytes in hexadecimal to MAX7219
{
   PORTB &= ~(1 << PORTB0);       // SS pin LOW
   SPDR = hexdata1;               // Send value of hexdata1
   while(!(SPSR & (1<<SPIF)));    // Wait for SPI transmission to finish
   SPDR = hexdata2;               // Send value of hexdata2
   while(!(SPSR & (1<<SPIF)));    // Wait for SPI transmission to finish
   PORTB |= (1 << PORTB0);        // SS pin HIGH
}

void blankMAX7219()
// Blanks all digits
{
   uint8_t i;
   for (i=1; i<9; i++)            // Blank all digits
   {
      writeMAX7219(i,15);
   }
}

void initMAX7219()
// Set up MAX7219 for use
{
   PORTB |= (1 << 0);             // SS pin HIGH (SS)
   SPCR = 0b01010000;             // Set up SPI bus for MAX7219
   // Mode decode for digits (table 4 page 7 - 0x09 + 0xFF
   writeMAX7219(0x09,0xFF);
   writeMAX7219(0x0B,0x07);       // Set scan limit to 8 digits - 0x09 + 0xFF)
   writeMAX7219(0x0A,0x01);       // Set intensity to 8 - 0x0A + 0x08)
   // Mode display on (table 4 page 7 - 0x09 + 0xFF)
   writeMAX7219(0x0C,0x01);
   blankMAX7219();
}

void dispMAX7219(uint8_t digit, uint8_t number, uint8_t dp)
// Displays "number" in location "digit" with decimal point on/off
// Digit: 1~8 for location 1~8
// Number: 0~15 for 0~9, - E, H, L, P, blank
// dp: 1 on, 0 off
{
   if (dp==1)                    // Add decimal point
   {
      number = number + 128;
   }
   writeMAX7219(digit, number);
}
```

```
  void numberMAX7219(uint32_t value)
  // Displays a number between 0-99999999 on MAX7219-controlled 8-digit display
  {
     uint8_t digits[8];
     uint8_t i = 1;

     for (i=1; i<9; i++)
     {
        digits[i]=15;
     }

     i = 1;
     while (value > 0)              // Continue until value > 0
     {
        // Determine and store last digit of number
        digits[i] = value % 10;
        value = value / 10;         // Divide value by 10
        i++;
     }
     for (i=1; i<9; i++)
     {
        dispMAX7219(i, digits[i],0);
     }
  }

❷ uint8_t decimalToBcd(uint8_t val)
  // Convert integer to BCD
  {
     return((val/10*16)+(val%10));
  }

  uint8_t bcdToDec(uint8_t val)
  // Convert BCD to integer
  {
     return((val/16*10)+(val%16));
  }

❸ void setTimeDS3231(uint8_t hh, uint8_t mm, uint8_t ss, uint8_t dw,
                     uint8_t dd, uint8_t mo, uint8_t yy)
  // Set the time on DS3231
  {
     I2CstartWait(0xD0);            // DS3231 write
     I2Cwrite(0x00);                // Start with seconds register
     I2Cwrite(decimalToBcd(ss));    // Seconds
     I2Cwrite(decimalToBcd(mm));    // Minutes
     I2Cwrite(decimalToBcd(hh));    // Hours
     I2Cwrite(decimalToBcd(dw));    // Day of week
     I2Cwrite(decimalToBcd(dd));    // Date
     I2Cwrite(decimalToBcd(mo));    // Month
     I2Cwrite(decimalToBcd(yy));    // Year
     I2Cstop();
  }

❹ void readTimeDS3231()
  // Retrieve time and date from DS3231
```

```
{
   I2CstartWait(0xD0);            // DS3231 write
   I2Cwrite(0x00);                // Seconds register
   I2CstartWait(0xD1);            // DS3231 read
   seconds = bcdToDec(I2CreadACK());
   minutes = bcdToDec(I2CreadACK());
   hours = bcdToDec(I2CreadACK());
   dow = bcdToDec(I2CreadACK());
   dom = bcdToDec(I2CreadACK());
   mo = bcdToDec(I2CreadACK());
   years = bcdToDec(I2CreadACK());
}

❺ void displayTimeMAX7219()
// Display time then date on MAX7219 module
{
   blankMAX7219();
   readTimeDS3231();
   // Display seconds
   if (seconds == 0)
   {  // Display '00'
      dispMAX7219(1,0,0);
      dispMAX7219(2,0,0);
   } else if (seconds >0 && seconds <10)
   {  // Display leading zero
      dispMAX7219(1,seconds,0);
      dispMAX7219(2,0,0);
   } else
   {  // Seconds > 10
      numberMAX7219(seconds);
   }
   dispMAX7219(3,10,0);           // Display a dash
   // Display minutes
   if (minutes == 0)
   {  // Display '00'
      dispMAX7219(4,0,0);
      dispMAX7219(5,0,0);
   } else if (minutes >0 && minutes <10)
   {  // Display leading zero
      dispMAX7219(4,minutes,0);
      dispMAX7219(5,0,0);
   } else
   {  // Minutes > 10
      dispMAX7219(4,(minutes % 10),0);
      dispMAX7219(5,(minutes / 10),0);
   }
   dispMAX7219(6,10,0);           // Display a dash
   // Display hours
   if (hours == 0)
   {  // Display '00'
      dispMAX7219(7,0,0);
      dispMAX7219(8,0,0);
   } else if (hours >0 && hours <10)
   {  // Display leading zero
      dispMAX7219(7,hours,0);
```

```
      dispMAX7219(8,0,0);
   } else
   {  // Hours > 10
      dispMAX7219(7,(hours % 10),0);
      dispMAX7219(8,(hours / 10),0);
   }

   _delay_ms(1000);
   // Display date
   if (dom >0 && dom <10)
   {  // Display leading zero
      dispMAX7219(7,dom,0);
      dispMAX7219(8,0,0);
   } else
   {  // Seconds > 10
      dispMAX7219(8,(dom / 10), 0);
      dispMAX7219(7,(dom % 10), 0);
   }

   dispMAX7219(6,10,0);                 // Display a dash
   // Display month
   if (mo >0 && mo <10)
   {  // Display leading zero
      dispMAX7219(4,mo,0);
      dispMAX7219(5,0,0);
   } else
   {  // Seconds > 10
      dispMAX7219(5,(mo / 10), 0);
      dispMAX7219(4,(mo % 10), 0);
   }

   dispMAX7219(3,10,0);                 // Display a dash
   // Display year
   if (years == 0)
   {   // Display '00'
     dispMAX7219(1,0,0);
      dispMAX7219(2,0,0);
   } else
   if (years >0 && years <10)
   {  // Display leading zero
      dispMAX7219(1,years,0);
      dispMAX7219(2,0,0);
   } else
   {  // Years > 10
      dispMAX7219(2,(years / 10), 0);
      dispMAX7219(1,(years % 10), 0);
   }
   _delay_ms(1000);
}

int main(void)
{
   DDRB = 0b11111111;                   // Set PORTB as outputs
   I2Cenable();
   initMAX7219();
```

```
    // Uncomment to set time and date, then comment and reflash code
    // setTimeDS3231(9,13,0,5,29,4,21); // h,m,s,dow,dom,m,y
    while (1)
    {
    ❻ displayTimeMAX7219();
       _delay_ms(250);
    }
}
```

Once again, this code reuses some functions you've seen in previous projects (namely, the functions for the MAX7219 display in Project 46 from Chapter 11 and the I²C bus functions from this chapter's Project 49).

First, we declare the variables to deal with the time and date information ❶. These variables will hold the data to write to the DS3231 and receive data from the DS3231. As mentioned earlier, the DS3231 works with data in binary-coded decimal format, so the code includes functions to convert integers to and from BCD ❷.

The `setTimeDS3231()` function accepts the time, day of week, and date and sends them to the DS3231 ❸. It first writes to the DS3231 to set the register to address (`0x00`), then sequentially writes each byte of data in the order described in Table 12-3. Note that each I²C write function uses the decimal-to-BCD function.

The `readTimeDS3231()` function ❹ retrieves the time and date information. It requests a byte of data from the DS3231 register `0x00`, and as the function uses ACK in the read process, the DS3231 will sequentially send the following bytes of data from the registers. This means we can simply use `I2CreadACK()` seven times to retrieve all the required data. As we're retrieving data from the DS3231, we use the BCD-to-decimal function within the I²C read functions.

Next comes the `displayTimeMAX7219()` function ❺, which organizes the time and date data into digits and shows them on the MAX7219 display. It shows the time first, then the date after a short delay. You can remove the date display and just let your clock run continuously, if you prefer.

The entire project is wrapped up in the main loop, where we initialize the GPIO, I²C, and SPI bus, then simply call the display function ❻ and delay until it is called again. For a challenge, why not write your own I²C and DS3231 libraries for future reference, or make an alarm clock as well?

There's still plenty more to learn, including new information on how to display data on popular character liquid crystal display modules, which we'll explore in the next chapter.

# AVR AND CHARACTER LIQUID CRYSTAL DISPLAYS

In previous chapters you've used LEDs, numerical LED displays, and the larger MAX7219 to display numerical values. However, a common *liquid crystal display (LCD)* module can allow your AVR projects to show a more versatile range of output, including text, numerical data, and your own custom characters.

In this chapter, you'll use character LCD modules to display all three types of data. To do so, you'll learn to convert integers into string variables and display floating-point numbers on the LCD. Along the way, you'll build your own digital clock and a digital thermometer that can display the minimum and maximum temperature over time.

## Introducing LCDs

Our LCD-based projects will use inexpensive LCDs that can display 2 rows of 16 characters. Any LCD with an HD44780- or KS0066-compatible interface and a 5 V backlight, such as the one in Figure 13-1, should work with these projects.

*Figure 13-1: A 16×2-character LCD module*

Some rare LCDs have a 4.5 V rather than a 5 V backlight. If this is true of your LCD, place a 1N4004 diode in series between the 5 V power supply and the LCD's LED+ or A pin.

LCDs like the one in Figure 13-1 usually come without any wiring or connectors. To use an LCD with a solderless breadboard, you'll need to solder in some 0.1 inch / 2.54 mm pitch inline header pins (such as PMD Way part number 1242070A) like those shown in Figure 13-2. These are usually supplied in 40-pin lengths; however, you can easily trim them down to the required 16-pin length.

*Figure 13-2: Inline header pins*

Once assembled, your LCD will fit easily into the solderless breadboard, as shown in Figure 13-3. Note the labels on pins 1 through 16.

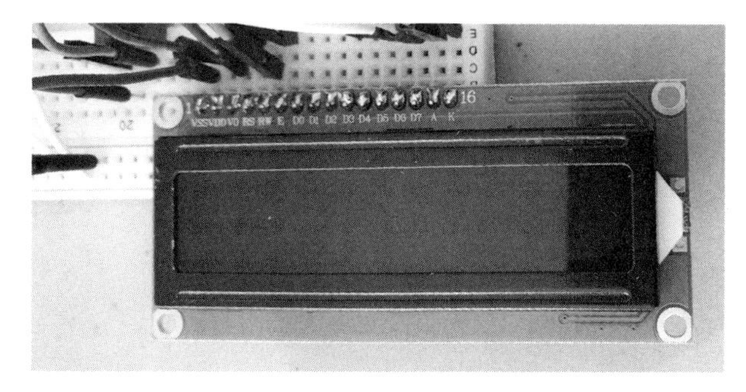

*Figure 13-3: An LCD in a solderless breadboard*

The schematic symbol for our LCD is shown in Figure 13-4.

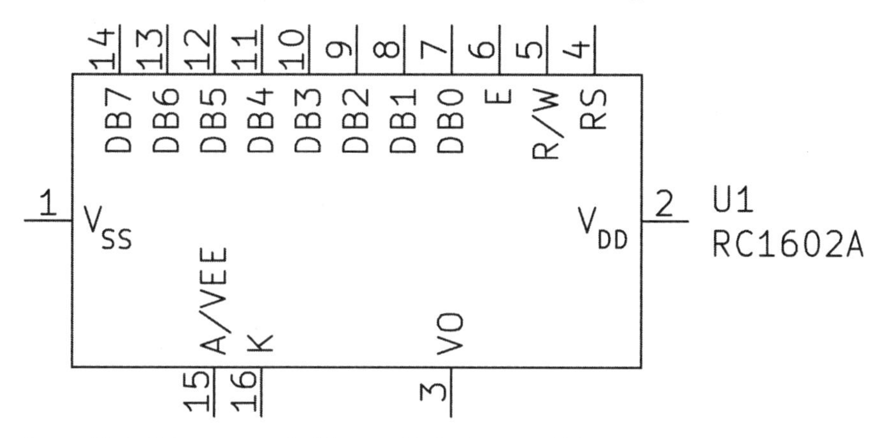


*Figure 13-4: Schematic symbol for our 16×2-character LCD*

Pins DB0 through DB7 constitute the 8-bit data interface of the LCD, which communicates with our ATmega328P-PU microcontroller. If you need to save wiring, you can also use the LCD in a 4-bit mode, which only requires DB4 through B7. We'll use this method in our projects.

Finally, you'll also need a small 10 kΩ trimpot to control the contrast of the display. You can get breadboard-compatible trimpots that don't require any extra soldering, like the one shown in Figure 13-5.

*Figure 13-5: An example of a breadboard trimpot*

The schematic symbol for the trimpot in Figure 13-5 is shown in Figure 13-6.

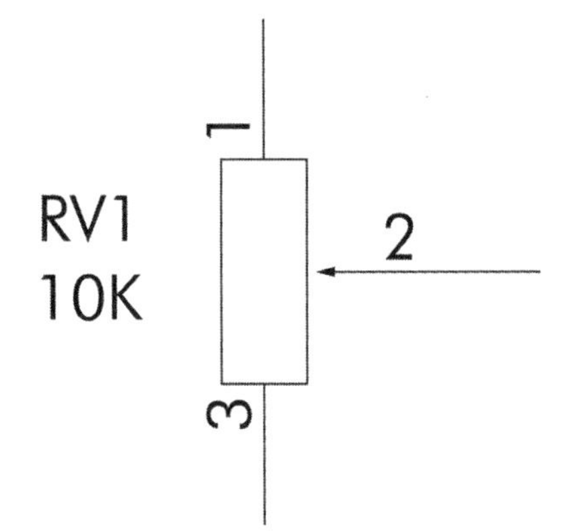


*Figure 13-6: Schematic symbol for our breadboard trimpot*

Once you've gotten your LCD ready to use with a solderless breadboard, it's time to understand how to display various kinds of data. To use the LCD in your projects, you will need functions for the following tasks:

- Converting instructions into suitable control signals in order to send commands to the LCD
- Initializing the LCD for use
- Clearing all data from the LCD
- Moving the cursor to the required position on the LCD
- Displaying text on the LCD

As there are no functions to complete these tasks in our AVR toolchain, we'll use the custom functions described in the following sections.

You'll notice that each of these functions sends values to the LCD to cause a given effect. For example, sending `0x01` to the LCD clears the screen. To determine which values we should use to accomplish certain tasks, we refer to the LCD's instruction table, which is Table 6 in the HD44780's data sheet (widely available and included with the book's code download at *https://nostarch.com/avr-workshop/*). This table shows the status of the RS and R/_W pins required for a particular command, along with the binary representation of the command. Figure 13-7 shows the clear display command, `0x01`.

| | Code | | | | | | | | | | |
|---|---|---|---|---|---|---|---|---|---|---|---|
| **Instruction** | **RS** | **R/$\overline{W}$** | **DB7** | **DB6** | **DB5** | **DB4** | **DB3** | **DB2** | **DB1** | **DB0** | **Description** |
| Clear display | 0 | 0 | 0 | 0 | 0 | 0 | 0 | 0 | 0 | 1 | Clears entire display and sets DDRAM address 0 in address counter. |

*Figure 13-7: A numerical description of the LCD's "clear screen" command*

As the figure shows, to clear the display we need to set the LCD pins RS and R/_W to low, then send 0b00000001 (or 0x01) to the LCD. We'll do this with the commandLCD() function (introduced in the following section), which is then called from the clearLCD() function (described shortly in "Clear the LCD").

In the following sections, refer to the table in the HD44780's data sheet to understand which values I use to construct the other LCD commands. Later, you can use the table to create commands to suit your own needs.

### *Send Commands to the LCD*

All information sent to the LCD, be it setup commands or data to display, is sent in bytes. However, as we're using the LCD in 4-bit mode to save on hardware connections, we'll need to use the following function to split the bytes of data into nibbles (4 bits) and send them to the LCD in the correct order:

```
void commandLCD(uint8_t _command)
{
❶ PORTD = (PORTD & 0x0F)|(_command & 0xF0);
❷ PORTD &= ~(1<<PD0);
❸ PORTD |= (1<<PD1);
  _delay_us(1);
❹ PORTD &= ~(1<<PD1);
  _delay_us(200);
❺ PORTD = (PORTD & 0x0F)|(_command << 4);
❻ PORTD |= (1<<PD1);
  _delay_us(1);
❼ PORTD &= ~(1<<PD1);
  _delay_ms(2);
}
```

To understand what's happening in this code, recall that a byte of data consists of 8 bits, or 2 nibbles: the higher nibble, which consists of bits 7 to 4, and the lower nibble, which consists of bits 3 to 0. For example:

```
11110000    // Ones are the higher nibble
00001111    // Ones are the lower nibble
```

The commandLCD() function first takes the upper nibble of the command byte _command ❶ and uses bitwise arithmetic (see Chapter 2) to clear the GPIO pins back to low. It then ensures the GPIO pins are set to match the upper nibble, the first half of the command byte.

Next, it sets the RS pin on the LCD to low ❷, which tells the LCD we need to send data to its instruction register, and quickly sets the LCD's E pin on ❸ and off ❹, which tells the LCD more data is coming.

The function then uses bitwise arithmetic to shift the 4 bits of the lower nibble up into the higher nibble ❺, which will match the GPIO pins used for sending data to the LCD. Finally, it again sets the LCD's E pin on ❻ and off ❼ to finalize the data transmission. We use the _delay_us() (delay in microseconds, not milliseconds) function to give the LCD time to process the changes.

### Initialize the LCD for Use

Like many other devices, the LCD needs to be initialized with various parameters before we first use it in our code. We'll use the `initLCD()` function to do this:

```
void initLCD()
{
❶ DDRD = 0b11111111;
   _delay_ms(100);
❷ commandLCD(0x02);
❸ commandLCD(0x28);
❹ commandLCD(0x0C);
❺ commandLCD(0x06);
❻ commandLCD(0x01);
   _delay_ms(2);
}
```

This function first sets the required GPIO pins to digital outputs ❶. After a short delay to give the LCD time to wake up, it then sends the command to set the cursor (the position at which data is first displayed) back to the top left of the screen ❷. The next command configures the LCD's controller IC to use it as a 16×2-character unit with a 4-bit data interface and to select a default font with characters that consist of 5×8 pixels ❸.

The following command ❹ tells the LCD to not use a block cursor, to not blink the cursor, and to turn the display on. We then tell the LCD controller we need the cursor to move in incremental stages ❺ so that if we wish to display more than one character in turn, we don't need to explicitly set the cursor position after each character. Finally, we clear the LCD of all characters ❻ and give it a little time to process the change.

### Clear the LCD

The convenient `clearLCD()` function simply clears the LCD of all data:

```
void clearLCD()
{
❶ commandLCD(0x01);
   _delay_ms(2);
❷ commandLCD(0x80);
   _delay_ms(2);
}
```

We send the command to clear the screen ❶, then the command to return the cursor to the top left of the LCD ❷.

### Set the Cursor

The `cursorLCD()` function sets the cursor to a given location on the LCD, following which you can display data starting from that position:

```
void cursorLCD(uint8_t column, uint8_t row)
{
```

```
    if (row == 0 && column<16)
    {
  ❶ commandLCD((column & 0x0F)|0x80);
    }
    else if (row == 1 && column<16)
    {
  ❷ commandLCD((column & 0x0F)|0xC0);
    }
}
```

With our LCD, we have 2 rows of 16 characters: rows 0 and 1, with 16 columns numbered 0 to 15. This function creates the required LCD command based on the position data received for a row 0 location ❶ and for a row 1 location ❷.

### Print to the LCD

The `printLCD()` function is used to display data on the LCD, such as text or numbers:

```
void printLCD(char *_string)
{
    uint8_t i;
❶ for (i=0; _string[i]!=0; i++)
    {
    ❷ PORTD = (PORTD & 0x0F)|(_string[i] & 0xF0);
    ❸ PORTD |= (1<<PD0);
    ❹ PORTD |= (1<<PD1);
      _delay_us(1);
    ❺ PORTD &= ~(1<<PD1);
      _delay_us(200);
    ❻ PORTD = (PORTD & 0x0F)|(_string[i] << 4);
    ❼ PORTD |= (1<<PD1);
      _delay_us(1);
    ❽ PORTD &= ~(1<<PD1);
      _delay_ms(2);
    }
}
```

This function can accept text in quotes, like this:

```
printLCD("AVR Workshop!");
printLCD("3.141592654");
```

or an array of characters, like this:

```
char resultsArray[9];
printLCD(resultsArray);
```

The function sends each character in turn from the array using its `for` loop ❶, representing characters as numerical values from a standard ASCII table (discussed in Chapter 4). All LCD displays on the market should support the values 33 to 125, which includes the lower- and uppercase alphabet, numbers, and standard popular symbols and punctuation marks. We set the location of the first (or only) character to display using the `cursorLCD()` or `clearLCD()` functions.

The `printLCD()` function is very similar to the `commandLCD()` function. It first takes the upper nibble of the character byte `_string[i]` ❷ and uses bitwise arithmetic to clear the GPIO pins back to low. It then ensures the GPIO pins are set to match the upper nibble, the first half of the command byte.

Next, it sets the RS pin on the LCD to high ❸, which tells the LCD we need to send data to its instruction register, and quickly sets the LCD's E pin on ❹ and off ❺, which tells the LCD more data is coming.

The function then uses bitwise arithmetic to shift the 4 bits of the lower nibble up into the higher nibble ❻, which will match the GPIO pins used for sending data to the LCD. Finally, it again sets the LCD's E pin on ❼ and off ❽ to finalize the data transmission. We use the `_delay_us()` (delay in microseconds) function to give the LCD time to process the changes.

**NOTE** *To use* `printLCD()` *to display the contents of an integer variable, first convert the variable to an array of characters with* `itoa(a,b,c)`. *This takes the integer* `a` *and places it in an array of characters* `b` *with a maximum length of* `c` *characters. You'll need to include the* stdlib.h *library along with the other* `include` *statements in your code, as it contains the* `itoa()` *function.*

In the following projects, you'll put the LCD to use.

## Project 52: Using a Character LCD with Your AVR

In this project you'll consolidate the information presented so far about controlling the LCD by building your own LCD circuit and displaying various information. This will introduce you to using LCDs in your own projects.

### The Hardware

To build your circuit, you'll need the following hardware:

- USBasp programmer
- Solderless breadboard
- 5 V breadboard power supply
- ATmega328P-PU microcontroller
- 16×2-character LCD with fitted inline header pins
- 10 kΩ breadboard-compatible trimpot (variable resistor)
- Two 22 pF ceramic capacitors (C1–C2)
- 470 μF 16 V electrolytic capacitor (C3)
- 16 MHz crystal oscillator
- Jumper wires

Assemble your circuit as shown in Figure 13-8.

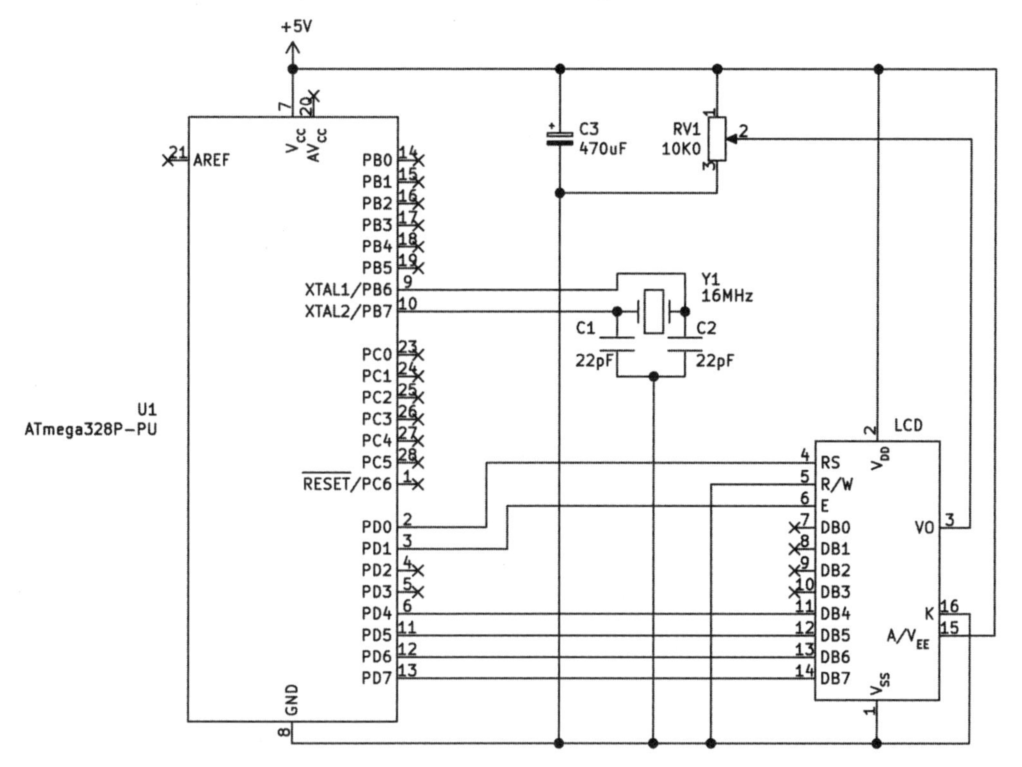


*Figure 13-8: Schematic for Project 52*

Once you've finished with this circuit, keep it assembled, as you'll use it again for Project 55.

## The Code

Open a terminal window, navigate to the *Project 52* subfolder of this book's *Chapter 13* folder, and enter the command `make flash` as usual. After a few moments, the LCD should show the text in Figure 13-9.

*Figure 13-9: First example of text displayed using Project 52*

That text should soon be replaced with an incrementing digit, as shown in Figure 13-10.

*Figure 13-10: Example of the counting display routine from Project 52*

Let's examine the code and review the functions that make this possible:

```
// Project 52 - Using a Character LCD with Your AVR

#include <avr/io.h>
#include <util/delay.h>
#include <stdlib.h>

void initLCD()
{
   DDRD = 0b11111111;
   _delay_ms(100);

   commandLCD(0x02);
   commandLCD(0x28);
   commandLCD(0x0C);
   commandLCD(0x06);
   commandLCD(0x01);
   _delay_ms(2);
}

void commandLCD(uint8_t _command)
{
   PORTD = (PORTD & 0x0F) | (_command & 0xF0
   PORTD &= ~(1<<PD0);
   PORTD |= (1<<PD1);
   _delay_us(1);
   PORTD &= ~(1<<PD1)
   _delay_us(200);
   PORTD = (PORTD & 0x0F) | (_command << 4);
   PORTD |= (1<<PD1);
   _delay_us(1);
   PORTD &= ~(1<<PD1);
   _delay_ms(2);
}

void clearLCD()
{
   commandLCD (0x01);
   _delay_ms(2);
   commandLCD (0x80);
   _delay_ms(2);
}

void printLCD(char *_string)
{
```

```
    uint8_t i;
    for(i=0; _string[i]!=0; i++)
    {
        PORTD = (PORTD & 0x0F) | (_string[i] & 0xF0);
        PORTD |= (1<<PD0);
        PORTD |= (1<<PD1);
        _delay_us(1);
        PORTD &= ~(1<<PD1);
        _delay_us(200);
        PORTD = (PORTD & 0x0F) | (_string[i] << 4);
        PORTD |= (1<<PD1);
        _delay_us(1);
        PORTD &= ~(1<<PD1);
        _delay_ms(2);
    }
}

void cursorLCD(uint8_t column, uint8_t row)
// Move cursor to desired column (0-15), row (0-1)
{
    if (row == 0 && column<16)
    {
        commandLCD((column & 0x0F)|0x80);
    }
    else if (row == 1 && column<16)
    {
        commandLCD((column & 0x0F)|0xC0);
    }
}

int main()
{
 ❶ initLCD();
 ❷ char numbers[9];
 ❸ int i;
    while(1)
    {
     ❹ cursorLCD(1,0);
        printLCD("AVR Workshop!");
        cursorLCD(0,1);
        printLCD("Learning LCD use");
        _delay_ms(1000);
     ❺ clearLCD();
        cursorLCD(1,0);
        printLCD("Counting up:");
     ❻ for (i = 0; i<10; i++)
        {
         ❼ itoa(i,numbers,10);
            cursorLCD(i,1);
         ❽ printLCD(numbers);
            _delay_ms(1000);
        }
        clearLCD();
    }
}
```

This code puts the LCD functions described earlier to use. In the main section of the code, we first initialize the LCD ❶, then declare a character array for displaying numbers ❷ and the required variable for counting ❸.

Next, we set up the display operation. We position and display text using the `cursorLCD()` and `printLCD()` functions ❹, then clear the display with `clearLCD()` ❺. The `for` loop ❻ displays the numbers from zero to nine along the second row of the LCD (as shown in Figure 13-10). We use `itoa()` ❼ to convert the integer variable `i` into a character array of numbers, then display that array using `printLCD()` ❽.

Now that you know how to set up and use a character LCD, let's put this skill to good use by creating a digital clock.

## Project 53: Building an AVR-Based LCD Digital Clock

In this project you'll combine the DS3231 real-time clock module and an LCD to build your own digital clock.

### *The Hardware*

To build your circuit, you'll need the following hardware:

- USBasp programmer
- Solderless breadboard
- 5 V breadboard power supply
- ATmega328P-PU microcontroller
- 16×2-character LCD with fitted inline header pins
- 10 kΩ breadboard-compatible trimpot (variable resistor)
- DS3231 RTC module with backup battery
- Two 22 pF ceramic capacitors (C1–C2)
- 470 μF 16 V electrolytic capacitor (C3)
- 16 MHz crystal oscillator
- Jumper wires

Assemble your circuit as shown in Figure 13-11. Don't forget to connect the DS3231 board to 5 V and GND as well.

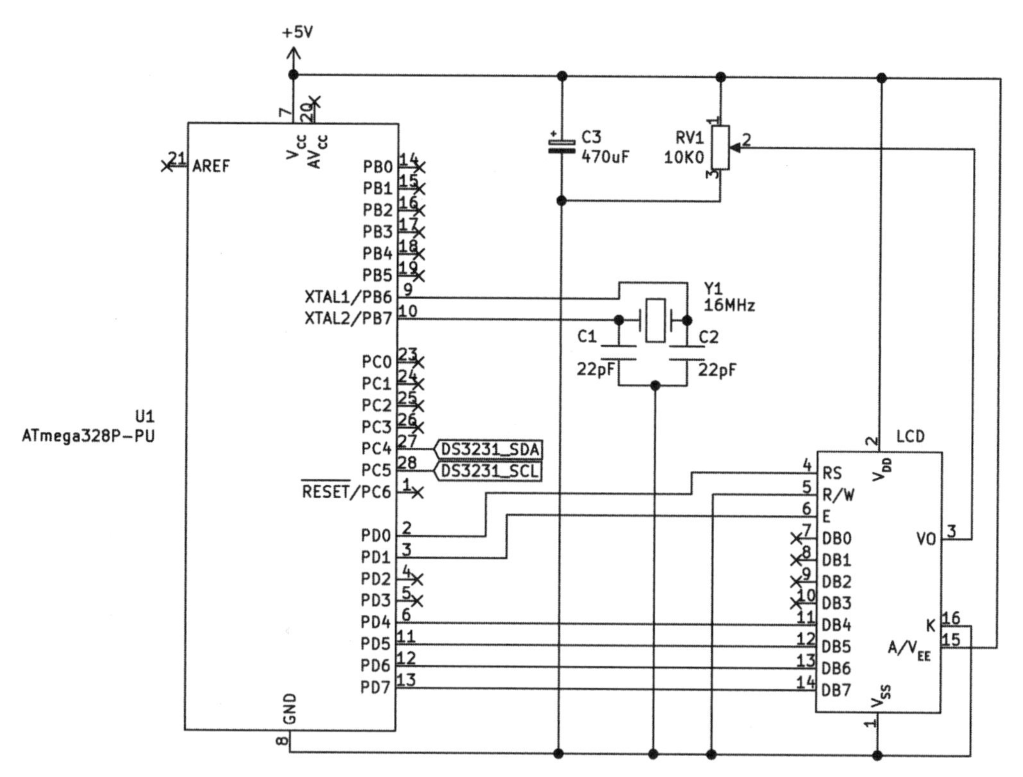


*Figure 13-11: Schematic for Project 53*

## The Code

As with Project 51 from Chapter 12, you'll first need to set the time and date in the DS3231 module. In your text editor, open the *main.c* file in the *Project 53* subfolder of this book's *Chapter 13* folder and remove the comment slashes in front of the function `setTimeDS3231()`. Update the parameters in that function to match your current date and time.

Now save the file, then use the **`make flash`** command as usual from the terminal window. Reopen the *main.c* file and place the comment slashes in front of the same function, save the file, and reflash the code. Once you've done so, you should see the current time and date displayed on your LCD module. An example of this is shown in Figure 13-12. Congratulations—you've made your own LCD digital clock!

*Figure 13-12: Example operation of Project 53*

Let's examine the code to see how this works:

```
// Project 53 - Building an AVR-Based LCD Digital Clock

#include <avr/io.h>
#include <util/delay.h>
#include <stdlib.h>

// Variables to store time and date
uint8_t hours, minutes, seconds, dow, dom, mo, years;

void I2Cenable()
// Enable I2C bus
{
   TWBR = 72;            // 100 kHz I2C bus
   TWCR |= (1 << TWEN); // Enable I2C on PORTC4 and 5
}

void I2Cwait()
// Wait until I2C finishes an operation
{
   // Wait until bit TWINT in TWCR is set to 1
   while (!(TWCR & (1<<TWINT)));
}

void I2CstartWait(unsigned char address)
{
   uint8_t status;
   while (1)
   {
      // Send START condition
      TWCR = (1<<TWINT) | (1<<TWSTA) | (1<<TWEN);

      // Wait until transmission completes
      I2Cwait();

      // Check value of TWSR, and mask out status bits
      status = TWSR & 0b11111000;
      if ((status != 0b00001000) && (status != 0b00010000)) continue;

      // Send device address
      TWDR = address;
      TWCR = (1<<TWINT) | (1<<TWEN);

      // Wait until transmission completes
      I2Cwait();

      // Check value of TWSR, and mask out status bits
      status = TWSR & 0b11111000;
      if ((status == 0b00100000 )||(status == 0b01011000))
      {
         TWCR = (1<<TWINT) | (1<<TWEN) | (1<<TWSTO);

         // Wait until stop condition is executed and I2C bus released
         while(TWCR & (1<<TWSTO));
```

```
            continue;
        }
        break;
    }
}

void I2Cstop()
// Stop I2C bus and release GPIO pins
{
    // Clear interrupt, enable I2C, generate stop condition
    TWCR |= (1 << TWINT)|(1 << TWEN)|(1 << TWSTO);
}

void I2Cwrite(uint8_t data)
// Send 'data' to I2C bus
{
    TWDR = data;
    TWCR |= (1 << TWINT)|(1 << TWEN);
    I2Cwait();
}

uint8_t I2Cread()
// Read incoming byte of data from I2C bus
{
    TWCR |= (1 << TWINT)|(1 << TWEN);
    I2Cwait();
    // Incoming byte is placed in TWDR register
    return TWDR;
}

uint8_t I2CreadACK()
// Read incoming byte of data from I2C bus and ACK signal
{
    TWCR |= (1 << TWINT)|(1 << TWEN)|(1 << TWEA);
    I2Cwait();
    // Incoming byte is placed in TWDR register
    return TWDR;
}

uint8_t decimalToBcd(uint8_t val)
// Convert integer to BCD
{
    return((val/10*16)+(val%10));
}

uint8_t bcdToDec(uint8_t val)
// Convert BCD to integer
{
    return((val/16*10)+(val%16));
}

void setTimeDS3231(uint8_t hh, uint8_t mm, uint8_t ss, uint8_t dw, uint8_t dd,
                   uint8_t mo, uint8_t yy)
// Set the time on DS3231
{
```

```
    I2CstartWait(0xD0);              // DS3231 write
    I2Cwrite(0x00);                  // Start with hours register
    I2Cwrite(decimalToBcd(ss));      // Seconds
    I2Cwrite(decimalToBcd(mm));      // Minutes
    I2Cwrite(decimalToBcd(hh));      // Hours
    I2Cwrite(decimalToBcd(dw));      // Day of week
    I2Cwrite(decimalToBcd(dd));      // Date
    I2Cwrite(decimalToBcd(mo));      // Month
    I2Cwrite(decimalToBcd(yy));      // Year
    I2Cstop();
}

void readTimeDS3231()
// Retrieve time and date from DS3231
{
    I2CstartWait(0xD0);              // DS3231 write
    I2Cwrite(0x00);                  // Seconds register
    I2CstartWait(0xD1);              // DS3231 read
    seconds = bcdToDec(I2CreadACK());
    minutes = bcdToDec(I2CreadACK());
    hours = bcdToDec(I2CreadACK());
    dow = bcdToDec(I2CreadACK());
    dom = bcdToDec(I2CreadACK());
    mo = bcdToDec(I2CreadACK());
    years = bcdToDec(I2CreadACK());
}

void commandLCD(uint8_t _command)
{
    // Takes command byte and sends upper nibble, lower nibble to LCD
    PORTD = (PORTD & 0x0F) | (_command & 0xF0);
    PORTD &= ~(1<<PD0);
    PORTD |= (1<<PD1);
    _delay_us(1);
    PORTD &= ~(1<<PD1);
    _delay_us(200);
    PORTD = (PORTD & 0x0F) | (_command << 4);
    PORTD |= (1<<PD1);
    _delay_us(1);
    PORTD &= ~(1<<PD1);
    _delay_ms(2);
}

void initLCD()
{
    DDRD = 0b11111111;
    _delay_ms(100);
    commandLCD(0x02);
    commandLCD(0x28);
    commandLCD(0x0C);
    commandLCD(0x06);
    commandLCD(0x01);
    _delay_ms(2);
}
```

```
void clearLCD()
{
   commandLCD (0x01);
   _delay_ms(2);
   commandLCD (0x80);
   _delay_ms(2);
}

void printLCD(char *_string)
{
   uint8_t i;
   for(i=0; _string[i]!=0; i++)
   {
      PORTD = (PORTD & 0x0F) | (_string[i] & 0xF0);
      PORTD |= (1<<PD0);
      PORTD |= (1<<PD1);
      _delay_us(1);
      PORTD &= ~(1<<PD1);
      _delay_us(200);
      PORTD = (PORTD & 0x0F) | (_string[i] << 4);
      PORTD |= (1<<PD1);
      _delay_us(1);
      PORTD &= ~(1<<PD1);
      _delay_ms(2);
   }
}

void cursorLCD(uint8_t column, uint8_t row)
// Move cursor to desired column (0-15), row (0-1)
{
   if (row == 0 && column<16)
   {
      commandLCD((column & 0x0F)|0x80);
   }
   else if (row == 1 && column<16)
   {
      commandLCD((column & 0x0F)|0xC0);
   }
}

int main()
{
   initLCD();
   I2Cenable();
   char numbers[9];
   // Uncomment to set time and date, then comment and reflash code
❶ // setTimeDS3231(8,50,0,3,16,6,21); // h, m, s, dow, dom, m, y

   while(1)
   {
    ❷ readTimeDS3231();
    ❸ itoa(hours,numbers,10);           // Hours
      cursorLCD(4,0);
    ❹ if (hours==0)
      {
```

```
    printLCD("00");
❺ } else if (hours>0 && hours <10)
  {
    printLCD("0");
    printLCD(numbers);
  } else if (hours>=10)
  {
    printLCD(numbers);
  }
  cursorLCD(6,0);
  printLCD(":");

  itoa(minutes,numbers,10);        // Minutes
  cursorLCD(7,0);
  if (minutes==0)
  {
    printLCD("00");
  } else if (minutes>0 && minutes <10)
  {
    printLCD("0");
    printLCD(numbers);
  } else if (minutes>=10)
  {
    printLCD(numbers);
  }

  cursorLCD(0,9);
  printLCD(":");
  itoa(seconds,numbers,10);        // Seconds
  cursorLCD(10,0);
  if (seconds==0)
  {
    printLCD("00");
  } else if (seconds>0 && seconds <10)
  {
    printLCD("0");
    printLCD(numbers);
  } else if (seconds>=10)
  {
    printLCD(numbers);
  }

  cursorLCD(2,1);                  // Day of week
❻ switch(dow)
  {
    case 1 : printLCD("Mon"); break;
    case 2 : printLCD("Tue"); break;
    case 3 : printLCD("Wed"); break;
    case 4 : printLCD("Thu"); break;
    case 5 : printLCD("Fri"); break;
    case 6 : printLCD("Sat"); break;
    case 7 : printLCD("Sun"); break;
  }

  itoa(dom,numbers,10);            // Day of month
```

```
        cursorLCD(6,1);
        if (dom<10)
        {
           printLCD("0");
        }
        printLCD(numbers);

        cursorLCD(8,1);
        printLCD("/");

        itoa(mo,numbers,10);            // Month
        cursorLCD(9,1);
        if (mo<10)
        {
           printLCD("0");
        }
        printLCD(numbers);

        cursorLCD(11,1);
        printLCD("/");

        itoa(years,numbers,10);         // Year
        cursorLCD(12,1);
        printLCD(numbers);

     ❼ _delay_ms(900);
        clearLCD();                     // Refresh LCD
    }
}
```

The first section of the code includes all the I²C functions required to read and write data with our DS3231 RTC module as described in Project 51 in Chapter 12, using the same method of working with the time and date information we employed with the MAX7219 display module. It also includes each LCD function explained previously in this chapter. Then we need to ensure the time and date are set using the `setTimeDS3231()` function ❶, and retrieve that information and display it in a nice format on the LCD.

The code displays time in 24-hour format, using two digits to represent each of the hour, minute, and second parts. It first retrieves the data from the DS3231 ❷ in the same way as Project 51 in Chapter 12. then converts the hour, minute and second information using `itoa()` ❸ and displays each part at the correct place on the LCD with `cursorLCD()`.

To maintain correct spacing and display of information, we must ensure the LCD displays single-digit values with a zero preceding them (representing the sixth day of the month as 06, for example). To do so, the code checks if the value from the time clock is zero ❹ or between one and nine ❺, then writes the required zeros before any single-digit time data. It does this for the hours, minutes, seconds, day of month, and month values.

The `switch...case` statement ❻ then takes the day of week data—a value from 1 to 7 corresponding to Sunday through Saturday or Monday through

Sunday, depending on your region and preference—and displays the day in abbreviated form. After all the information has been displayed, the clock waits for 900 ms ❼ before clearing the display, then starting over.

For a challenge, you might convert this project into a 12-hour clock with an AM/PM display, or perhaps add an alarm that sounds a piezo buzzer at a certain time every day.

## Displaying Floating-Point Numbers on the LCD

Our next project requires us to display a floating-point number on the LCD. As with integers, floating-point numbers first need to be converted from floats to character arrays. To do this we use the `dtostrf()` function, as described in Chapter 4, then display the character array as usual with the `printLCD()` function. Always ensure you declare your character array with enough space to cover the entire number and fraction.

For example, to display the numbers 1.2345678 and 12345.678, replace the `int main()` loop from Project 54 with the following code:

```
int main()
{
❶ float a = 1.2345678;
   float b = 12345.678;
❷ char displayNumber[10];
❸ initLCD();
   while(1)
   {
    ❹ cursorLCD(0,0);
    ❺ dtostrf(a,9,7, displayNumber);
      printLCD(displayNumber);
      cursorLCD(0,1);
    ❻ dtostrf(b,9,3, displayNumber);
      printLCD(displayNumber);
      _delay_ms(1000);
   }
}
```

We declare variables holding two sample numbers to display on the LCD for the purposes of demonstration ❶, and the character array used in the display process ❷. We then initialize the LCD as usual ❸ and move the cursor to the top left of the display ❹.

The code then converts the number 1.2345678 to a string displayed using 10 characters, with 7 of them after the decimal point ❺. Finally, it displays the number 12345.678 using 9 characters, this time with 3 of them after the decimal point ❻.

Flash the code and you should see a display like the one in Figure 13-13.

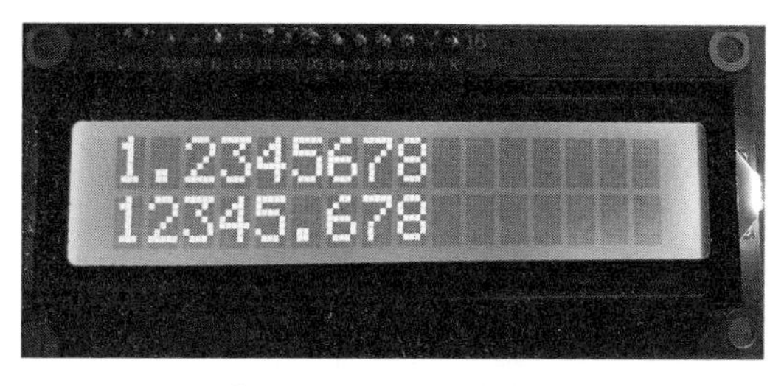


*Figure 13-13: Floating-point numbers on the LCD*

This example displayed two positive numbers. If you'd like to display negative numbers, remember to allow one character space for the negative sign in front of the first digit. For example, to display –123.45, you would need to allocate seven character spaces.

You'll put this new skill to work in the next project.

## Project 54: LCD Digital Thermometer with Min/Max Display

With this project you'll make a digital thermometer that can display the minimum and maximum temperature over time along with the current and average temperature over time. This project is another example of how to incorporate functions from previous chapters into new and more complicated projects.

### *The Hardware*

To build your circuit, you'll need the following hardware:

- USBasp programmer
- Solderless breadboard
- 5 V breadboard power supply
- ATmega328P-PU microcontroller
- 16×2-character LCD with fitted inline header pins
- 10 kΩ breadboard-compatible trimpot (variable resistor)
- One TMP36 temperature sensor
- Two 22 pF ceramic capacitors (C1–C2)
- 470 μF 16 V electrolytic capacitor (C3)
- 0.1 μF ceramic capacitor (C4)
- 16 MHz crystal oscillator
- Jumper wires

Assemble your circuit as shown in Figure 13-14. Don't forget to connect the microcontroller's $AV_{CC}$ pin to 5 V!

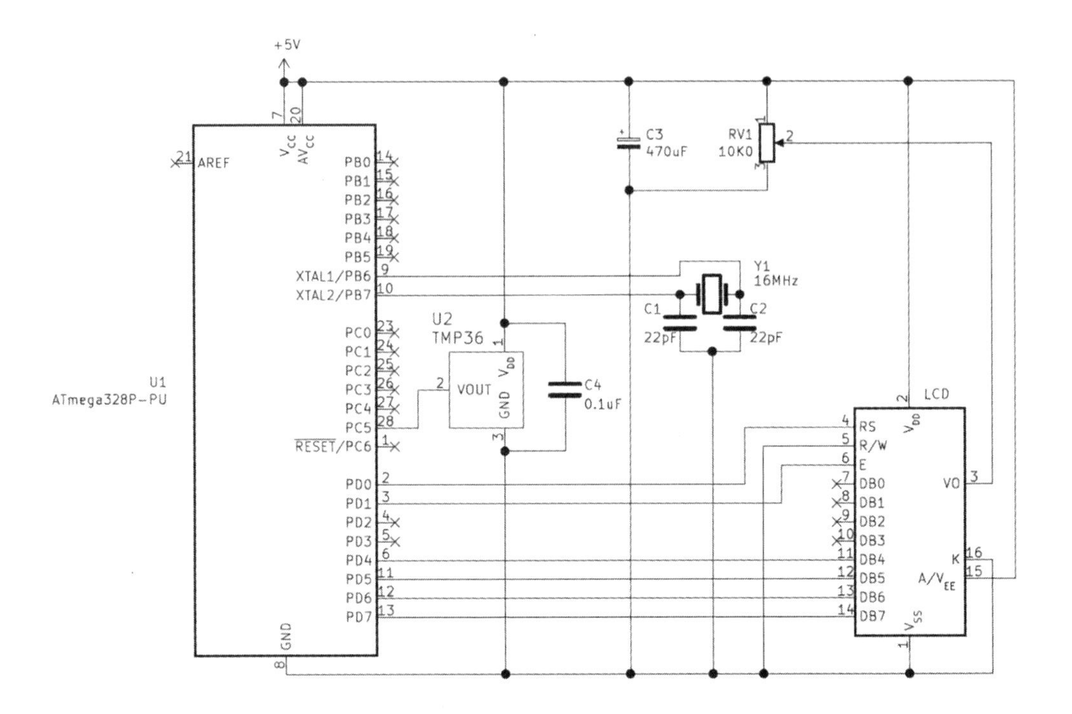


Figure 13-14: Schematic for Project 54

## The Code

Open a terminal window, navigate to the *Project 54* subfolder of this book's *Chapter 13* folder, and enter the command `make flash` as usual. After a moment or two, the LCD should alternate between showing the minimum and maximum temperatures, as shown in Figure 13-15, and the current and average temperatures, as shown in Figure 13-16. The temperature readings are in degrees Celsius and cover the period since the project was last reset or turned on.

Figure 13-15: The LCD displaying the minimum and maximum temperatures

*Figure 13-16: The LCD displaying the current and average temperatures*

Let's take a look at the code to see how this works:

```
// Project 54 - LCD Digital Thermometer with Min/Max Display

#include <avr/io.h>
#include <util/delay.h>
#include <stdlib.h>
#include <math.h>

❶ void startADC()
// Set up the ADC
{
   ADMUX |= (1 << REFS0);              // Use AVcc pin with ADC
   ADMUX |= (1 << MUX2) | (1 << MUX0); // Use ADC5 (pin 28)
   // Prescaler for 16MHz (/128)
 ❷ ADCSRA |= (1 << ADPS2) |(1 << ADPS1) | (1 << ADPS0);
   ADCSRA |= (1 << ADEN);              // Enable ADC
}

void commandLCD(uint8_t _command)
{
   PORTD = (PORTD & 0x0F) | (_command & 0xF0);
   PORTD &= ~(1<<PD0);
   PORTD |= (1<<PD1);
   _delay_us(1);
   PORTD &= ~(1<<PD1);
   _delay_us(200);
   PORTD = (PORTD & 0x0F) | (_command << 4);
   PORTD |= (1<<PD1);
   _delay_us(1);
   PORTD &= ~(1<<PD1);
   _delay_ms(2);
}

void initLCD()
{
   DDRD = 0b11111111;
   _delay_ms(100);
   commandLCD(0x02);
   commandLCD(0x28);
   commandLCD(0x0C);
```

```
   commandLCD(0x06);
   commandLCD(0x01);
   _delay_ms(2);
}

void clearLCD()
{
   commandLCD (0x01);
   _delay_ms(2);
   commandLCD (0x80);
   _delay_ms(2);
}

void printLCD(char *_string)
{
   uint8_t i;
   for(i=0; _string[i]!=0; i++)
   {
      PORTD = (PORTD & 0x0F) | (_string[i] & 0xF0);
      PORTD |= (1<<PD0);
      PORTD |= (1<<PD1);
      _delay_us(1);
      PORTD &= ~(1<<PD1);
      _delay_us(200);
      PORTD = (PORTD & 0x0F) | (_string[i] << 4);
      PORTD |= (1<<PD1);
      _delay_us(1);
      PORTD &= ~(1<<PD1);
      _delay_ms(2);
   }
}

void cursorLCD(uint8_t column, uint8_t row)
// Move cursor to desired column (0-15), row (0-1)
{
   if (row == 0 && column<16)
   {
      commandLCD((column & 0x0F)|0x80);
   }
   else if (row == 1 && column<16)
   {
      commandLCD((column & 0x0F)|0xC0);
   }
}

int main()
{
   DDRC = 0b00000000;                // Set PORTC as inputs
   startADC();
   initLCD();
   char numbers[9];

   float temperature;
   float voltage;
   float average;
```

```
❸ float minimum = -273;              // Needs an initial value
  float maximum;
  uint16_t ADCvalue;

  while(1)
  {
  ❹ // Take reading from TMP36 via ADC
     ADCSRA |= (1 << ADSC);          // Start ADC measurement
     while (ADCSRA & (1 << ADSC) ); // Wait for conversion
     _delay_ms(10);

     // Get value from ADC (which is 10-bit) register
     ADCvalue = ADC;

     // Convert reading to temperature value (Celsius)
     voltage = (ADCvalue * 5);
     voltage = voltage / 1024;
  ❺ temperature = ((voltage - 0.5) * 100);

     // Min/max and average
  ❻ if (temperature < minimum)
     {
        minimum = temperature;
     }
     if (temperature > maximum)
     {
        maximum = temperature;
     }
  ❼ average = ((minimum+maximum)/2);

  ❽ // Display information
     cursorLCD(0,0);
     printLCD("Current:");
     dtostrf(temperature,6,2,numbers);
     printLCD(numbers);
     cursorLCD(15,0);
     printLCD("C");

     cursorLCD(0,1);
     printLCD("Average:");
     dtostrf(average,6,2,numbers);
     printLCD(numbers);
     cursorLCD(15,1);
     printLCD("C");

     _delay_ms(1000);
     clearLCD();

     cursorLCD(0,0);
     printLCD("Minimum:");
     dtostrf(minimum,6,2,numbers);
     printLCD(numbers);
     cursorLCD(15,0);
     printLCD("C");
```

```
        cursorLCD(0,1);
        printLCD("Maximum:");
        dtostrf(maximum,6,2,numbers);
        printLCD(numbers);
        cursorLCD(15,1);
        printLCD("C");

        _delay_ms(1000);
        clearLCD();
    }
}
```

The code for this project breaks down into two concepts: determining the temperature from the TMP36 sensor (as demonstrated in Chapter 3), then displaying the temperature values using the LCD.

We first use a series of functions and commands to activate the ADC on pin 28 and call it into action in the main code ❶. The `startADC()` function is slightly different from its equivalent in previous projects; since we're now operating the microcontroller at 16 MHz rather than 1 MHz, we need a larger prescaler to operate the ADC. Therefore, we set the ADCSRA register to use a prescaler of 128 ❷. We arrive at this value by dividing 16 MHz by 200 kHz (the ideal speed for the ADC), which results in 80; the closest prescaler value is 128, so we use that.

The code reads the raw data from the ADC ❹ and converts it to degrees Celsius ❺. It then determines if the current temperature is a minimum or maximum ❻ and calculates the average temperature measured since the last reset ❼. Note that the variable `minimum` is declared with a value of –273 degrees ❸. If we leave it without an initial value, it will default to 0 and we won't get a true minimum temperature value (unless the sensor is outside and the temperature never goes below freezing!). Finally, we display all this temperature data over two screens, using the LCD functions from earlier in this chapter ❽.

You can of course change the temperature values displayed to Fahrenheit by multiplying them by 1.8 and adding 32. Or, if you feel like a challenge, why not modify this project by combining it with what you learned in Project 53 to build a clock that shows the current temperature?

When you've finished experimenting, let's move on to creating our final type of output: custom characters.

## Displaying Custom Characters on the LCD

In addition to using the standard letters, numbers, and symbols available on most keyboards, you can define up to eight of your own characters in each project. As you know, each character in the LCD module is made up of eight rows of five pixels, as shown in Figure 13-17.

*Figure 13-17: Each LCD character is made up of eight rows of five pixels.*

To display your own custom characters, you must first define each character using an array consisting of eight elements (one element per character line). The value of the element defines the state of the pixels in that line. For example, to create a simple "smiley face," plan out the pixels on a grid as shown in Figure 13-18.

| | | | | | Binary | Decimal |
|---|---|---|---|---|---|---|
| 1 | 1 | | 1 | 1 | 0b11011 | 27 |
| 1 | 1 | | 1 | 1 | 0b11011 | 27 |
| | | | | | 0b00000 | 0 |
| | | 1 | | | 0b00100 | 4 |
| | | | | | 0b00000 | 0 |
| 1 | | | | 1 | 0b10001 | 17 |
| | 1 | | 1 | | 0b01010 | 10 |
| | | 1 | | | 0b00100 | 4 |

*Figure 13-18: Elements of a custom smiley face character*

Convert each horizontal line into a value by converting it from a binary number matching the pixels' on or off state to a decimal number. Then create an array to define your custom character by entering in the eight decimal values, as shown below for the elements in Figure 13-18:

```
uint8_t smiley[] = {27,27,0,4,0,17,10,4};
```

The code for this chapter includes a spreadsheet that simplifies this array creation process.

Once you've created the array, you need to program it into the LCD's *character generator RAM (CGRAM)*. This is a type of RAM used in the LCD's controller chip that stores the design of the characters to display. There are eight possible positions in our LCD's CGRAM. To write this character data and use the custom characters, we'll use the three custom functions defined in the following sections.

### Write Data to CGRAM

The writeLCD() function writes an individual line of data to the LCD's CGRAM:

```
void writeLCD(uint8_t _data)
{
   PORTD |= (1<<PD0); // RS high
   PORTD = (PORTD & 0x0F) | (_data & 0xF0);
   PORTD |= (1<<PD1);
   _delay_us(1);
   PORTD &= ~(1<<PD1);
   _delay_us(200);
   PORTD = (PORTD & 0x0F) | (_data);
   PORTD |= (1<<PD1);
   _delay_us(1);
   PORTD &= ~(1<<PD1);
   _delay_ms(2);
}
```

This function operates in the same way as our commandLCD() function, except that writeLCD() sets the RS pin on the LCD high instead of low, which tells the LCD that the incoming data is for the CGRAM and not a regular command. It's used in conjunction with the following two functions.

### Send Custom Character Data to LCD

The createCC() function directs the custom character data array (ccdata[]) into the specified CGRAM memory position (slot), from 0 to 7:

```
void createCC(uint8_t ccdata[], uint8_t slot)
{
   uint8_t x;
❶ commandLCD(0x40+(slot*8)); // Select character memory (0-7)
   for (x = 0; x<8; x++)
   {
    ❷ writeLCD(ccdata[x]<<4);
   }
}
```

We command the LCD to prepare for character data and to store it in the character position in the variable slot ❶, then send each element of the character array in turn to the LCD's CGRAM with the writeLCD() function ❷.

### Display Custom Characters on LCD

The printCCLCD() function displays one of the LCD's eight custom characters, storing it in position slot:

```
void printCCLCD(uint8_t slot)
{
   PORTD = (PORTD & 0x0F) | (slot & 0xF0);
   PORTD |= (1<<PD0);
   PORTD |= (1<<PD1);
   _delay_us(1);
   PORTD &= ~(1<<PD1);
   _delay_us(200);
   PORTD = (PORTD & 0x0F) | (slot << 4);
   PORTD |= (1<<PD1);
   _delay_us(1);
   PORTD &= ~(1<<PD1);
   _delay_ms(2);
}
```

This function operates similarly to `printLCD()`, but it doesn't need the string decoding and goes straight to showing the character in CGRAM location 0 through 7 (`slot`) at the current cursor position.

The next project demonstrates how to use these functions to display custom characters.

## Project 55: Displaying Custom LCD Characters

In this project, you'll reuse the hardware from Project 52 to practice creating and displaying custom characters on an LCD. Open a terminal window, navigate to the *Project 55* subfolder of this book's *Chapter 13* folder, and enter the command **`make flash`** as usual. After a moment or two, the LCD should display eight custom characters, as shown in Figure 13-19.

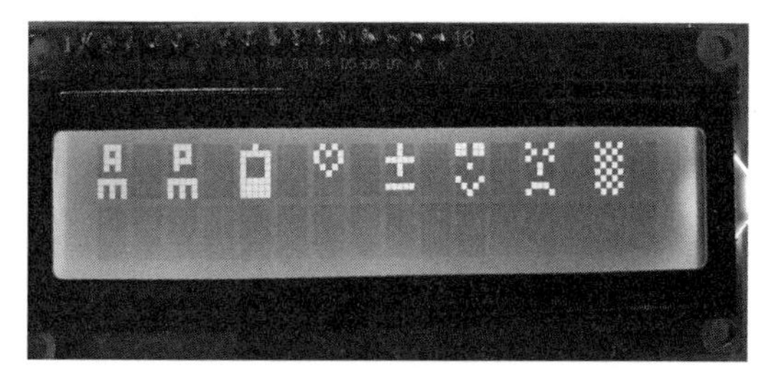

*Figure 13-19: The results of Project 55*

Let's look at the code to see how this works:

```
// Project 55 - Displaying Custom LCD Characters

#include <avr/io.h>
#include <util/delay.h>

uint8_t ch0[] = {14,10,14,10,0,31,21,21};  // "AM"
uint8_t ch1[] = {14,10,14,8,0,31,21,21};   // "PM"
uint8_t ch2[] = {4,31,17,17,17,31,31,31};  // "Battery"
uint8_t ch3[] = {10,21,17,10,4,0,0,0};     // "Heart"
uint8_t ch4[] = {4,4,31,4,4,0,31,0};       // "+ -"
```

```
uint8_t ch5[] = {27,27,0,4,0,17,10,4};     // "Happy face"
uint8_t ch6[] = {17,10,17,4,4,0,14,17};    // "Sad face"
uint8_t ch7[] = {21,10,21,10,21,10,21,10}; // "Pattern"

❶ void writeLCD(uint8_t _data)
// Used for writing to CGRAM
{
   PORTD |= (1<<PD0); // RS high
   PORTD = (PORTD & 0x0F) | (_data & 0xF0);
   PORTD |= (1<<PD1);
   _delay_us(1);
   PORTD &= ~(1<<PD1);
   _delay_us(200);
   PORTD = (PORTD & 0x0F) | (_data);
   PORTD |= (1<<PD1);
   _delay_us(1);
   PORTD &= ~(1<<PD1);
   _delay_ms(2);
}

❷ void commandLCD(uint8_t _command)
{
   PORTD = (PORTD & 0x0F) | (_command & 0xF0);
   PORTD &= ~(1<<PD0);                           // RS low
   PORTD |= (1<<PD1);
   _delay_us(1);
   PORTD &= ~(1<<PD1);
   _delay_us(200);
   PORTD = (PORTD & 0x0F) | (_command << 4);
   PORTD |= (1<<PD1);
   _delay_us(1);
   PORTD &= ~(1<<PD1);
   _delay_ms(2);
}

❸ void createCC(uint8_t ccdata[], uint8_t slot)
// Sends custom character data to LCD
{
   uint8_t x;
   commandLCD(0x40+(slot*8));                    // Select character memory (0-7)
   for (x = 0; x<8; x++)
   {
      writeLCD(ccdata[x]<<4);
   }
}

❹ void printCCLCD(uint8_t slot)
{
   PORTD = (PORTD & 0x0F) | (slot & 0xF0);
   PORTD |= (1<<PD0);
   PORTD |= (1<<PD1);
   _delay_us(1);
   PORTD &= ~(1<<PD1);
   _delay_us(200);
   PORTD = (PORTD & 0x0F) | (slot << 4);
```

```
    PORTD |= (1<<PD1);
    _delay_us(1);
    PORTD &= ~(1<<PD1);
    _delay_ms(2);
}

void initLCD()
{
    DDRD = 0b11111111;
    _delay_ms(100);
    commandLCD(0x02);
    commandLCD(0x28);
    commandLCD(0x0C);
    commandLCD(0x06);
    commandLCD(0x01);
    _delay_ms(2);
}

void clearLCD()
{
    commandLCD (0x01);
    _delay_ms(2);
    commandLCD (0x80);
    _delay_ms(2);
}

void printLCD(char *_string)
{
    uint8_t i;
    for(i=0; _string[i]!=0; i++)
    {
        PORTD = (PORTD & 0x0F) | (_string[i] & 0xF0);
        PORTD |= (1<<PD0);
        PORTD |= (1<<PD1);
        _delay_us(1);
        PORTD &= ~(1<<PD1);
        _delay_us(200);
        PORTD = (PORTD & 0x0F) | (_string[i] << 4);
        PORTD |= (1<<PD1);
        _delay_us(1);
        PORTD &= ~(1<<PD1);
        _delay_ms(2);
    }
}

void cursorLCD(uint8_t column, uint8_t row)
// Move cursor to desired column (0-15), row (0-1)
{
    if (row == 0 && column<16)
    {
        commandLCD((column & 0x0F)|0x80);
    }
    else if (row == 1 && column<16)
    {
        commandLCD((column & 0x0F)|0xC0);
    }
```

```
}

int main()
{
   initLCD();
   while(1)
   {
   ❺ createCC(ch0,0); // "AM"
      createCC(ch1,1); // "PM"
      createCC(ch2,2); // "Battery"
      createCC(ch3,3); // "Heart"
      createCC(ch4,4); // "+ -"
      createCC(ch5,5); // "Happy face"
      createCC(ch6,6); // "Sad face"
      createCC(ch7,7); // "Pattern"

   ❻ cursorLCD(0,0);
      printCCLCD(0);
      cursorLCD(2,0);
      printCCLCD(1);
      cursorLCD(4,0);
      printCCLCD(2);
      cursorLCD(6,0);
      printCCLCD(3);
      cursorLCD(8,0);
      printCCLCD(4);
      cursorLCD(10,0);
      printCCLCD(5);
      cursorLCD(12,0);
      printCCLCD(6);
      cursorLCD(14,0);
      printCCLCD(7);
      _delay_ms(1000);
      clearLCD();
   }
}
```

This project demonstrates how easy it is to create custom characters when we use the three custom functions described previously to do the heavy lifting: respectively, they write the custom character data ❶, send commands to the LCD ❷, and send the custom character data to the LCD ❸. We simply insert the required data for the characters ❹, then feed that data to each location in the LCD's CGRAM in turn with the `createCC()` function ❺. Finally, we position the cursor and display each custom character in turn with the `cursorLCD()` and `printCCLCD()` functions ❻.

After working through this chapter you have the skills required to display all sorts of text and numerical data, as well as your own custom characters, on inexpensive and popular LCD modules. For a challenge, try creating your own AVR LCD library to make this code easier to include in your own future projects; every time you want to use the LCD, this library will save you development time and reduce complexity.

In the next and final chapter, you'll add yet another tool to your growing AVR toolbox: the ability to control servos.

# CONTROLLING SERVOS

Various projects in Chapter 8 used DC motors, which are ideal for rotating devices such as wheels for robots. However, for more precise motor control options, you can use a *servo*, short for *servomechanism*. Servos contain electric motors that you can rotate to a specific angular position using PWM signals.

Servos come in handy for a variety of applications. For example, you might use a servo to steer a remote-controlled car by connecting it to a *horn*, a small arm or bar that the servo rotates. You might also connect a physical pointer to a servo so it can indicate information such as temperature on a scale, or use a servo to raise or lower a rotary drill.

In this chapter, you will:

- Learn how to connect the ATmega328P-PU microcontroller to a servo and use PWM to control it.

- Learn how to independently control two servos at once.
- Build an analog thermometer and an analog clock.

## Setting Up Your Servo

There are a large variety of servos on the market, from tiny units used in portable devices such as digital cameras to large units used in robotic manufacturing assembly devices. When you're selecting a servo, consider several parameters:

**Speed** The time it takes for the servo to rotate, usually measured in seconds per angular degree.

**Rotational range** The angular range through which the servo can rotate—for example, 180 degrees (half of a full rotation) or 360 degrees (one complete rotation).

**Current** How much current the servo draws. When using a servo with an Arduino, you may need to use an external power supply for the servo.

**Torque** The amount of force the servo can exert when rotating. The greater the torque, the heavier the item the servo can control. The torque produced is generally proportional to the amount of current used.

For the examples in this chapter, we'll be using an inexpensive and compact servo like the one in Figure 14-1, commonly known as an SG90-type servo. We'll combine this servo with three types of horns, also shown in the figure.

*Figure 14-1: Servo and various horns*

This servo can rotate up to 180 degrees, as shown in Figure 14-2.

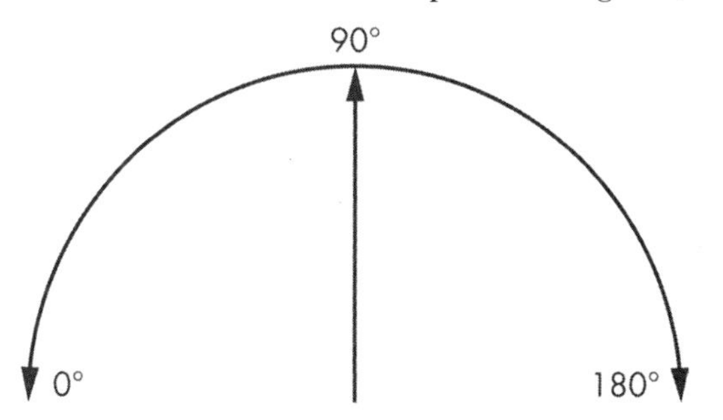

*Figure 14-2: Example of servo rotation range*

Inside the servo is a small DC motor connected to the horn spindle via *reduction gears*, which reduce the rotational speed of the DC motor to a much slower pace for the servo. The servo also contains a *feedback controller*, which measures the rotational position of the DC motor's shaft in order to position it more exactly.

## Connecting a Servo

You need just three wires to connect a servo to your microcontroller. If you're using the SG90, the darkest wire connects to GND, the center wire connects to 5 V, and the lightest wire (the *pulse* or *PWM* wire) connects to a digital pin with PWM capability. If you're using a different servo, check its data sheet for the correct wiring. We'll use the standard schematic symbol for servos shown in Figure 14-3.

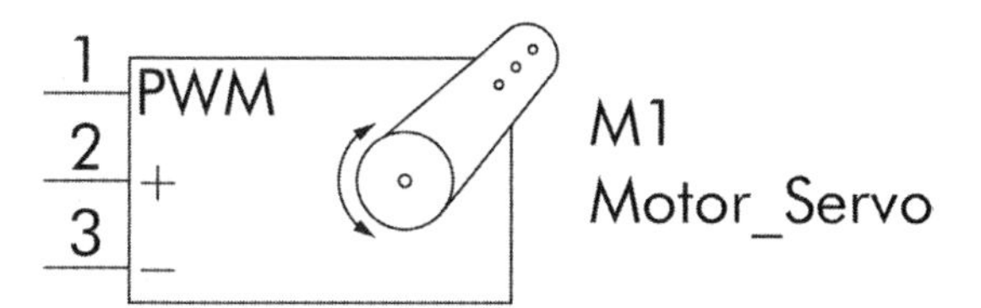

*Figure 14-3: Schematic symbol for a servo*

All the servos you will come across in the hobbyist and experimenting range of products use this same schematic symbol.

## Controlling a Servo

We set a servo's rotational angle by changing the duty cycle of a PWM signal connected to the servo's pulse wire. In general, servos require a PWM signal with a frequency of 50 Hz and a period of 20 ms. Setting the signal's duty cycle to different values causes the servo's internal controller to move the horn to an angle to which the duty cycle is inversely proportionate.

Using our SG90 servo as an example, if we set the duty cycle to 12 percent (or 2.4 ms out of the total period of 20 ms), as shown in Figure 14-4, the horn will rotate to 0 degrees.

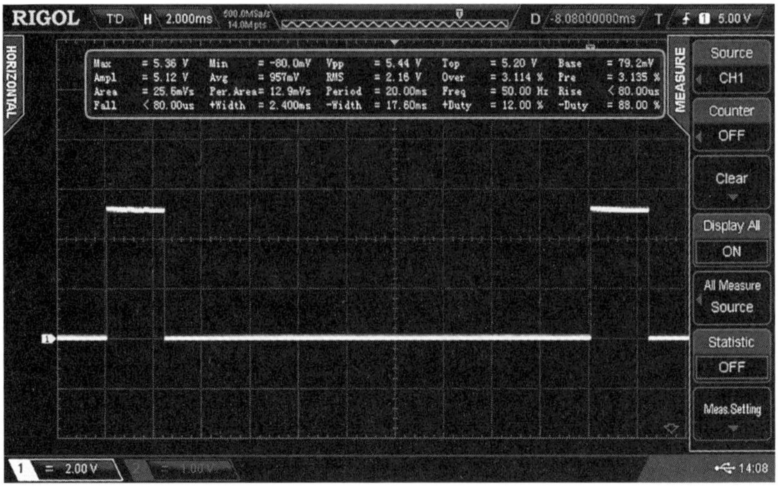

Figure 14-4: PWM signal for 0 degrees

If we set the duty cycle to 3 percent, as shown in Figure 14-5, the horn will rotate to 180 degrees.

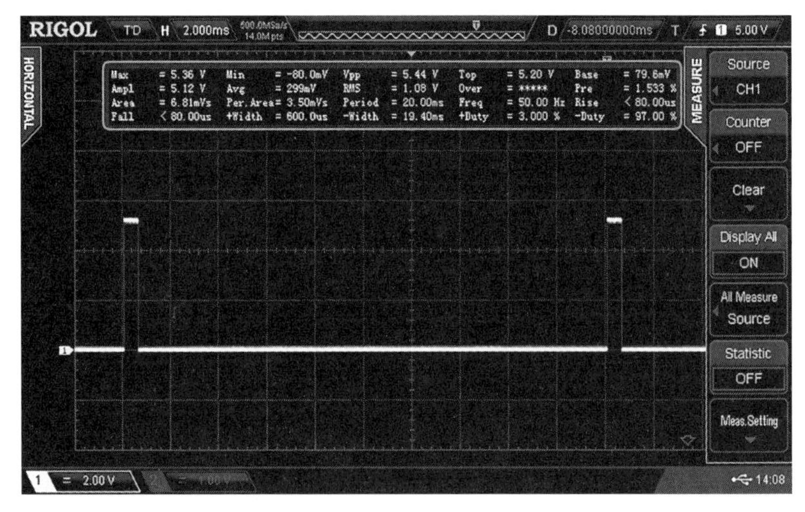

Figure 14-5: PWM signal for 180 degrees

We'll put the code required to enable the PWM output for our SG90 servo in a function called `initPWM()`:

```
void initPWM()
{ // Activate PWM on PB1
❶ TCCR1A |= (1 << WGM11);
  TCCR1B |= (1 << WGM12)|(1 << WGM13)|(1 << CS11);

  // Connect PWM to PB1
❷ TCCR1A |= (1 << COM1A1); // PWM to OCR1A - PB1
❸ ICR1=39999;
}
```

This function sets TIMER1 up for fast PWM. It sets the prescaler to 8 for a timer frequency of 2 MHz ❶, and sends the output to PB1 ❷. (To refresh your memory of how to generate PWM signals, refer to Chapter 7.) The timer will count from 0 to 39,999 then reset ❸, with each period being 0.0000005 seconds in length (*time* = 1/*frequency*). This gives a full pulse period of 20 ms.

We'll then use OCR1A to set the duty cycle and thus position the servo. We know that a 12 percent duty cycle results in a rotation to 0 degrees, so we can calculate the required OCR1A value by multiplying 40,000 (remember that the counter starts at 0 and counts to 39,999) by 0.12, which gives us 4,799. For a full rotation to 180 degrees, we would set OCR1A to 1,199 ($40{,}000 \times 0.12$).

If you're using a servo other than the SG90, determine the duty cycle values required for 0 and 180 degree rotation, then use the calculations described in the previous paragraph to determine your required OCR1A values. You should be able to get the duty cycle information from the servo supplier or retailer.

Now, let's put what you've just learned into practice by rotating a servo in various ways.

## Project 56: Experimenting with Servos

With this project you'll learn the basic of servo control, including the required circuitry and commands for servo movement.

### *The Hardware*

To build your circuit, you'll need the following hardware:

- USBasp programmer
- Solderless breadboard
- 5 V breadboard power supply
- ATmega328P-PU microcontroller
- Two 22 pF ceramic capacitors (C1–C2)
- 470 μF 16 V electrolytic capacitor (C3)
- 16 MHz crystal oscillator
- SG90 servo
- Jumper wires

Assemble your circuit as shown in Figure 14-6.

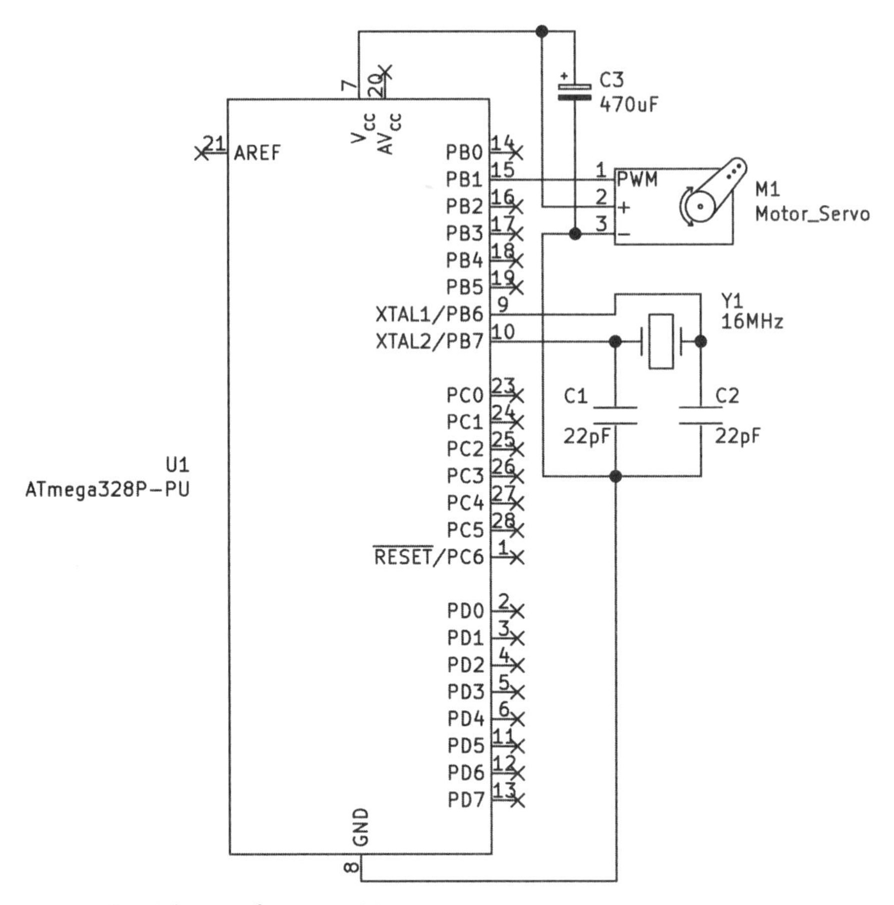


*Figure 14-6: Schematic for Project 56*

Note the use of the large electrolytic capacitor over the 5 V and GND lines. When servos are operating, they can sometimes generate a varying voltage due to the motor turning on and off quickly, so we use the capacitor to smooth out the power to a more consistent 5 V.

## The Code

Open a terminal window, navigate to the *Project 56* subfolder of this book's *Chapter 14* folder, and enter the command `make flash` as usual. After a moment or two, the servo should quickly rotate across its full range from 0 to 180 degrees, then do so again at a slower speed, then return to 0 degrees at an even slower speed.

Let's take a look at the code to see how this works:

```
// Project 56 - Experimenting with Servos
#include <avr/io.h>
#include <util/delay.h>

❶ void initPWM()
{
```

```
    // Activate PWM on PB1
    TCCR1A |= (1 << WGM11);
    TCCR1B |= (1 << WGM12)|(1 << WGM13)|(1 << CS11);
    // Connect PWM to PB1
    TCCR1A |= (1 << COM1A1); // PWM to OCR1A - PB1
    ICR1=39999;
}

❷ void servoRange()
{
    OCR1A=4799;             // 0 degrees
    _delay_ms(1000);
    OCR1A=1199;             // 180 degrees
    _delay_ms(1000);
}

❸ void servoAngle(uint8_t angle)
{
    // Rotate servo to 'angle' position
  ❹ OCR1A = ((angle-239.95)/-0.05);
    // Convert angle to OCR1A (duty cycle) value
}

int main()
{
  ❺ DDRB|=(1<<PB1);
    initPWM();
    uint8_t i;

    while(1)
    {
      ❻ servoRange();
        _delay_ms(1000);

        for (i=0; i<=180; i++)
        {
          ❼ servoAngle(i);
            _delay_ms(25);
        }
        for (i=180; i>0; --i)
        {
          ❽ servoAngle(i);
            _delay_ms(5);
        }
    }
}
```

We start by defining three functions: `initPWM()` ❶, which handles PWM initialization; `servoRange()` ❷, which for demonstrative purposes simply rotates the servo arm between 0 and 180 degrees by setting OCR1A with the duty cycle values for 0 degrees, then 180 degrees; and the useful custom function `servoAngle(uint8_t angle)` ❸, which accepts a number (the rotational angle for our desired servo position) and converts this into the required duty cycle value to be stored in OCR1A ❹. This simplifies the task of commanding the servo, automatically converting the angle we want into the correct duty cycle between 4,799 and 1,199 with the formula *angle* = (*counter* – 239.95) / –0.05. These values are generally used by most common small servos, but check with your supplier if you're unsure.

In the main section of the code, we first set the pin connected to the servo's pulse wire to an output ❺, then call the `initPWM()` function to enable PWM. We call `servoRange()` ❻ to rotate the servo arm quickly from 0 to 180 degrees and back again, then this is repeated in a slower fashion using the `for` loops at ❼ and ❽, respectively. Each introduces a delay between movement of the servo arm one degree in either direction.

**NOTE** *The formula for the* `servoAngle()` *function was created using linear algebra, based on two sets of points: (4799,0) and (1199,180). You can use an online tool such as GeoGebra (*https://www.geogebra.org/m/UyfrABcN*) to determine the equation for your own formula if your servo requires different duty cycle values.*

Now that you have the code framework to control a servo, let's combine it with your prior knowledge about using the TMP36 temperature sensor to build an analog thermometer.

## Project 57: Creating an Analog Thermometer

You can use a servo to display a temperature reading by attaching an arrow to the servo horn and creating a backing sheet with the temperature range on it. This project will display temperatures between 0 and 30 degrees Celsius, but you can modify it to show different ranges.

### The Hardware

To build your circuit, you'll need the following hardware:

- USBasp programmer
- Solderless breadboard
- 5 V breadboard power supply
- ATmega328P-PU microcontroller
- One TMP36 temperature sensor
- Two 22 pF ceramic capacitors (C1–C2)
- 470 μF 16 V electrolytic capacitor (C3)
- 0.1 μF ceramic capacitor (C4)
- 16 MHz crystal oscillator
- SG90-compatible servo
- Jumper wires

Assemble your circuit as shown in Figure 14-7. Don't forget to connect the microcontroller's $AV_{CC}$ pin to 5 V.

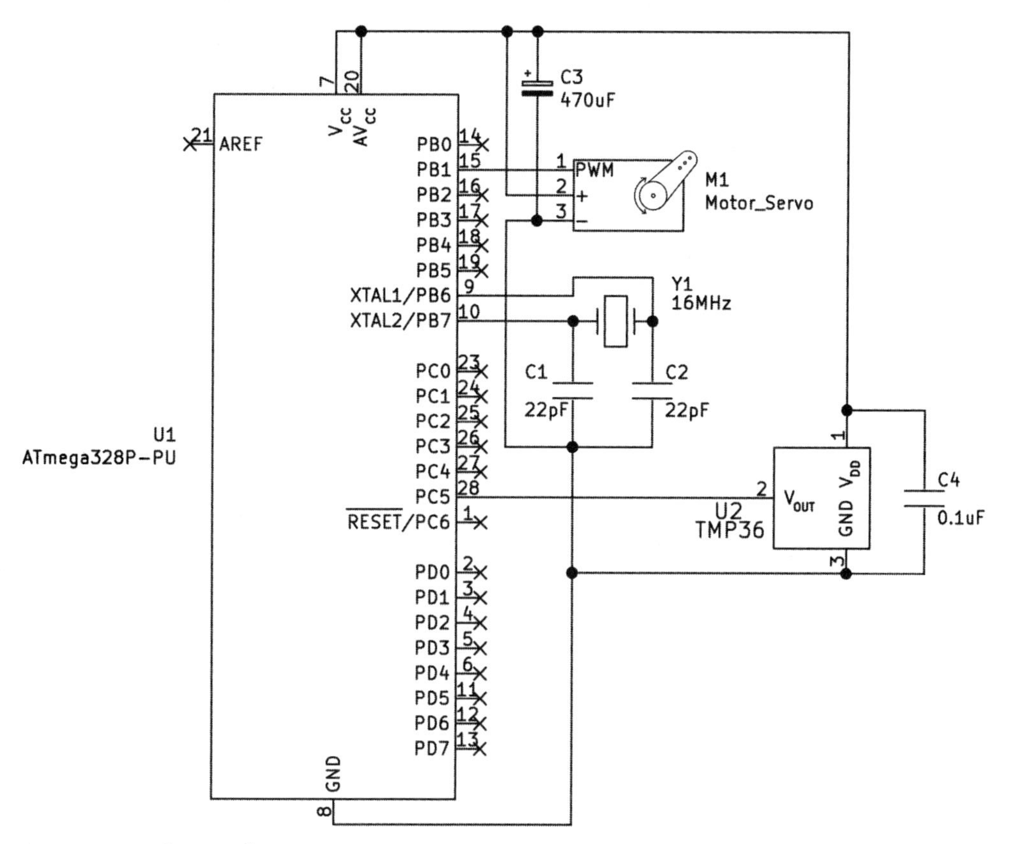


*Figure 14-7: Schematic for Project 57*

Figure 14-8 shows what the backing sheet representing the range of temperatures that the servo will display might look like, with a small arrow attached to the horn as a pointer.

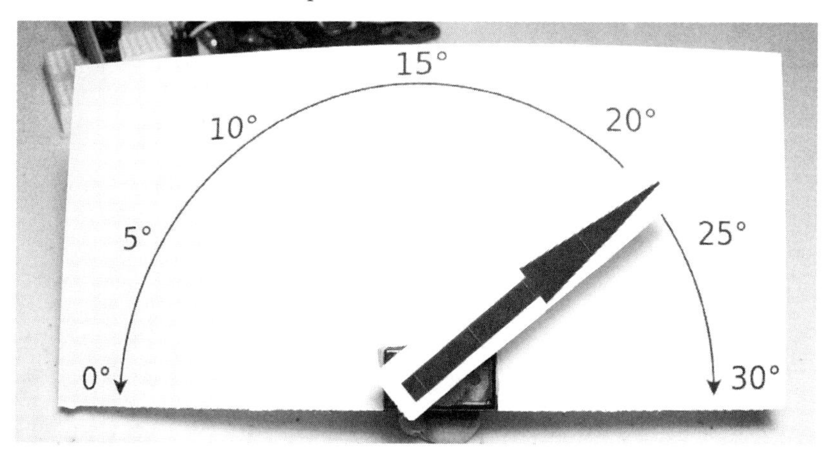


*Figure 14-8: The backing sheet indicating the temperature*

## The Code

Open a terminal window, navigate to the *Project 57* subfolder of this book's *Chapter 14* folder, and enter the command **make flash** as usual. After a moment or two, the servo horn should swing to an angular position that represents the temperature in degrees Celsius.

Let's take a look at the code to see how this works:

```
// Project 57 - Creating an Analog Thermometer

#include <avr/io.h>
#include <util/delay.h>
#include <stdlib.h>
#include <math.h>

void startADC()
// Set up the ADC
{
   ADMUX |= (1 << REFS0);                    // Use AVcc pin with ADC
   ADMUX |= (1 << MUX2) | (1 << MUX0);       // Use ADC5 (pin 28)
   ADCSRA |= (1 << ADPS2) |(1 << ADPS1) | (1 << ADPS0);
   // Prescaler for 16MHz (/128)
   ADCSRA |= (1 << ADEN);                    // Enable ADC
}

void initPWM()
{
   // Activate PWM on PB1
   TCCR1A |= (1 << WGM11);
   TCCR1B |= (1 << WGM12)|(1 << WGM13)|(1 << CS11);

   // Connect PWM to PB1
   TCCR1A |= (1 << COM1A1);
   // PWM to OCR1A - PB1
   ICR1=39999;
}

void servoAngle(uint8_t angle)
{
   // Rotate servo to 'angle' position
   OCR1A = ((angle-239.95)/-0.05);
   // Convert angle to OCR1A (duty cycle) value
}

int main()
{
❶ DDRB|=(1<<PB1);  // Set PORTB1 as output for servo control
❷ DDRC|=(0<<PC5);  // Set PORTC5 as input for TMP36 measurement

❸ float temperature;
   float voltage;
   uint16_t ADCvalue;
   uint8_t finalAngle;
```

```
❹ startADC();
❺ initPWM();

   while(1)
   {
   ❻ ADCSRA |= (1 << ADSC);        // Start ADC measurement
     while (ADCSRA & (1 << ADSC)); // Wait for conversion
     _delay_ms(10);
   ❼ ADCvalue = ADC;

     // Convert reading to temperature value (Celsius)
   ❽ voltage = (ADCvalue * 5);
     voltage = voltage / 1024;
     temperature = ((voltage - 0.5) * 100);

     // Display temperature using servo
   ❾ finalAngle = 6 * temperature;
     servoAngle(finalAngle);
     _delay_ms(500);
   }
}
```

We start by performing the usual steps to set the required pins for the servo as an output ❶ and the TMP36 sensor as an input ❷, then declare the variables needed for storage and conversion of temperature data from the TMP36 sensor ❸. We then call the functions to start the ADC ❹ and initialize PWM ❺. Next, we determine the temperature in Celsius by first reading the ADC ❻ and storing its value into `ADCvalue` ❼, then doing the mathematical conversion to Celsius ❽. Finally, we convert the temperature to an angle for the servo by multiplying it by 6 (since the servo range is 0 to 180 degrees) ❾ and tell the servo to move to the appropriate angle.

At this point, you can use what you've learned in this book so far to make a variety of controllable analog displays with your servo—for example, a low-voltage meter or a countdown timer. But if anything's better than one servo, it's using two servos at once; you'll see how to do that next.

## Project 58: Controlling Two Servos

Since there are multiple PWM-capable output pins on the ATmega328P-PU microcontroller, we can control two servos at once for more involved projects. This project will show you how.

### *The Hardware*

To build your circuit, you'll need the following hardware:

- USBasp programmer
- Solderless breadboard
- 5 V breadboard power supply
- ATmega328P-PU microcontroller

- One TMP36 temperature sensor
- Two 22 pF ceramic capacitors (C1–C2)
- 470 μF 16 V electrolytic capacitor (C3)
- 0.1 μF ceramic capacitor (C4)
- 16 MHz crystal oscillator
- Two SG90-compatible servos
- Jumper wires

Assemble your circuit as shown in Figure 14-9.

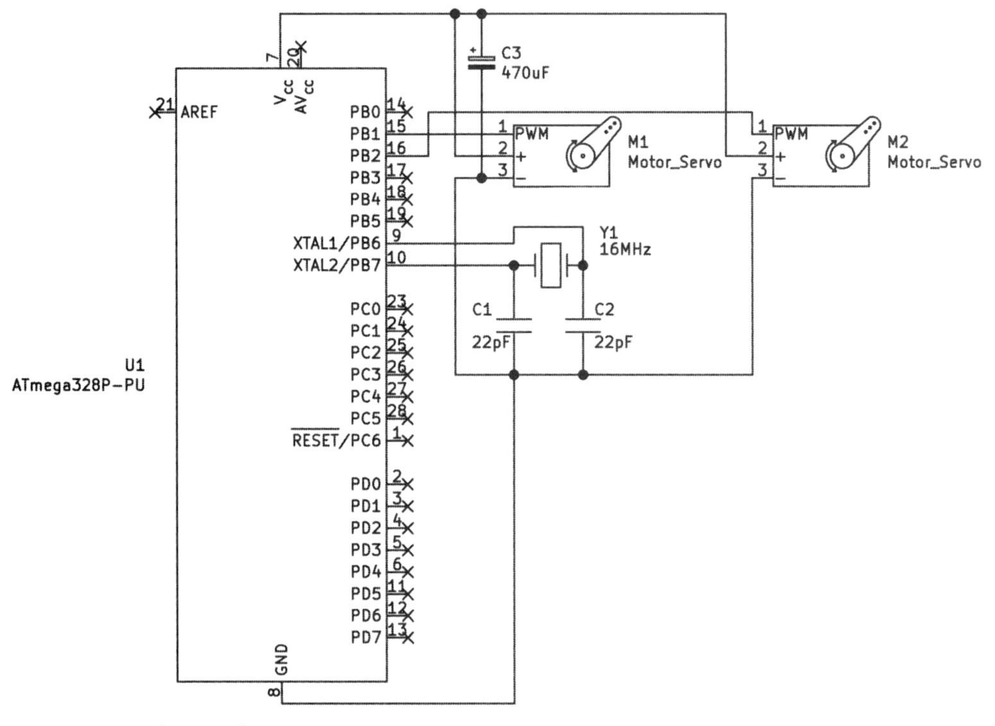

Figure 14-9: Schematic for Project 58

## The Code

Open a terminal window, navigate to the *Project 58* subfolder of this book's *Chapter 14* folder, and enter the command `make flash` as usual. After a moment or two both servos should imitate the motion demonstrated in Project 56, quickly rotating across their full range from 0 to 180 degrees, then repeating this at a slower speed, and then returning to 0 degrees at an even slower speed.

Let's see how this works:

```
// Project 58 - Controlling Two Servos

#include <avr/io.h>
#include <util/delay.h>
```

```
❶ void initPWM()
  {
     // Activate PWM
     TCCR1A |= (1 << WGM11);
     TCCR1B |= (1 << WGM12)|(1 << WGM13)|(1 << CS11);

     // Connect PWM to PB1 and PB2
   ❷ TCCR1A |= (1 << COM1A1)|(1 << COM1B1);
     // PWM to OCR1A - PB1 and OCR1B - PB2
     ICR1=39999;
  }

❸ void servoAngleA(uint8_t angle)
  {
     // Rotate servo on OCR1A to 'angle' position
     OCR1A = ((angle-239.95)/-0.05);
     // Convert angle to OCR1A (duty cycle) value
  }

❹ void servoAngleB(uint8_t angle)
  // Rotate servo on OCR1B to 'angle' position
  {
     OCR1B = ((angle-239.95)/-0.05);
     // Convert angle to OCR1A (duty cycle) value
  }

❺ void servoRange()
  {
     OCR1A=4799;                 // 0 degrees
     OCR1B=4799;
     _delay_ms(1000);
     OCR1A=1199;                 // 180 degrees
     OCR1B=1199;                 // 180 degrees
     _delay_ms(1000);
  }

  int main()
  {
     DDRB|=(1<<PB1)|(1<<PB2);  // Set PB1 and PB2 to outputs
     initPWM();
     uint8_t i;

     while(1)
     {
        servoRange();
        _delay_ms(1000);

        for (i=0; i<=180; i++)
        {
           servoAngleA(i);
           servoAngleB(i);
           _delay_ms(25);
        }
        for (i=180; i>0; --i)
        {
```

```
            servoAngleA(i);
            servoAngleB(i);
            _delay_ms(5);
        }
    }
}
```

In the `initPWM()` function ❶, after activating PWM we turn on the COM1B1 bit in TCCR1A to enable PWM for the second servo connected to PB2 ❷. Two `servoAngle()`-type functions, one for servo A ❸ and one for servo B ❹, allow for control by accepting the required rotational angle. I've modified the function `servoRange()` ❺ to control the first servo and the second servo by assigning the required values to OCR1A and OCR1B, respectively.

You could also experiment with the direction of both servos by altering the delays after the `servoAngleA/B()` functions or reversing the counting to go from higher values to lower values. Now that you can use two servos with ease, it's time to put them to work in the form of an analog clock.

## Project 59: Building an Analog Clock with Servo Hands

In this project you'll use two servos to display the time in the form of a dual-display analog clock. One servo will display the hour, and the other will display minutes.

### *The Hardware*

To build your circuit, you'll need the following hardware:

- USBasp programmer
- Solderless breadboard
- 5 V breadboard power supply
- ATmega328P-PU microcontroller
- One TMP36 temperature sensor
- Two 22 pF ceramic capacitors (C1–C2)
- 470 μF 16 V electrolytic capacitor (C3)
- 0.1 μF ceramic capacitor (C4)
- 16 MHz crystal oscillator
- DS3231 real-time clock module with backup battery
- Two SG90-compatible servos
- Jumper wires

Assemble your circuit as shown in Figure 14-10. Don't forget to connect the DS3231 board to 5 V and GND as well.

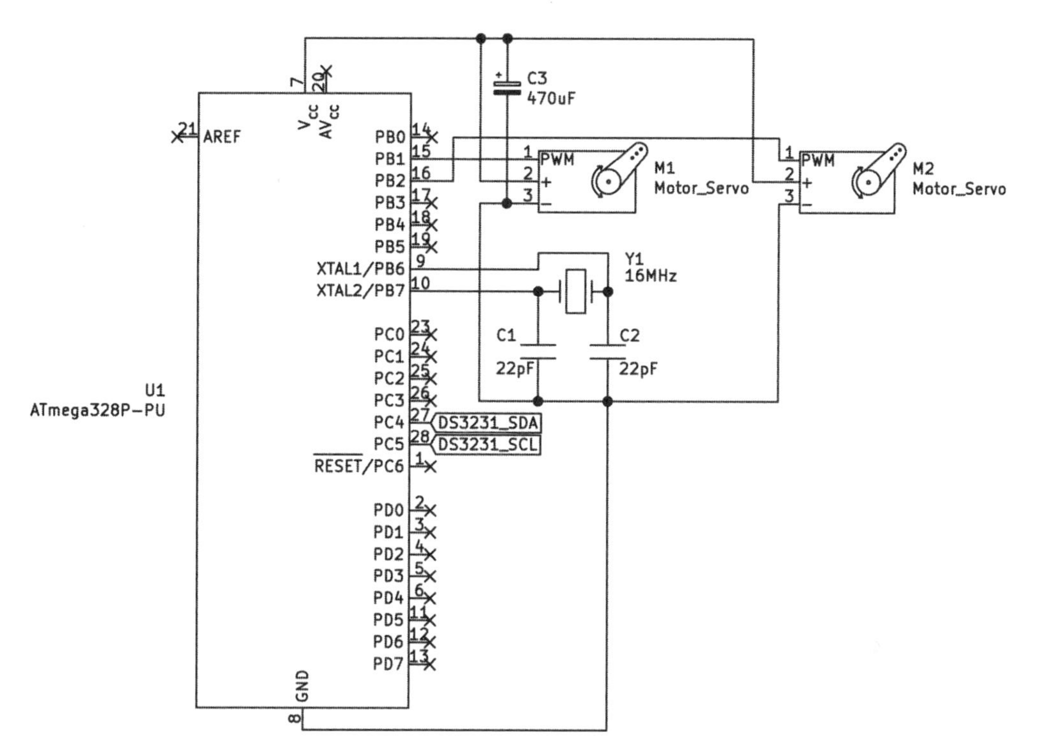


Figure 14-10: Schematic for Project 59

Before uploading the code, don't forget to set the time in the same manner as you did in previous projects that used the DS3231, such as Project 51. You may also want to create a backing display like the one used in Project 57, as shown in Figure 14-11—feel free to get creative. Note that servo M1 in the schematic is for hours, and M2 is for minutes.

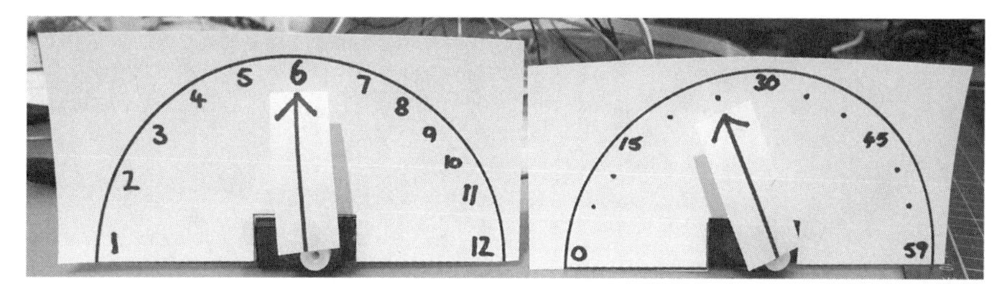


Figure 14-11: Example clock faces for Project 59

## The Code

Open a terminal window, navigate to the *Project 59* subfolder of this book's *Chapter 14* folder, and enter the command `make flash` as usual. Once you've flashed the code, you should see the current time displayed by way of the position of the servo horns.

Let's see how this works:

```
// Project 59 - Building an Analog Clock with Servo Hands

#include <avr/io.h>
#include <util/delay.h>

// Variables to store time and date
uint8_t hours, minutes, seconds, dow, dom, mo, years;

void I2Cenable()
// Enable I2C bus
{
   TWBR = 72;            // 100 kHz I2C bus
   TWCR |= (1 << TWEN); // Enable I2C on PORTC4 and 5
}

void I2Cwait()
// Wait until I2C finishes an operation
{
   // Wait until bit TWINT in TWCR is set to 1
   while (!(TWCR & (1<<TWINT)));
}

void I2CstartWait(unsigned char address)
{
   // Start I2C bus
   uint8_t status;
   while (1)
   {
      // Send START condition
      TWCR = (1<<TWINT) | (1<<TWSTA) | (1<<TWEN);

      // Wait until transmission completes
      I2Cwait();

      // Check value of TWSR, and mask out status bits
      status = TWSR & 0b11111000;
      if ((status != 0b00001000) && (status != 0b00010000)) continue;

      // Send device address
      TWDR = address;
      TWCR = (1<<TWINT) | (1<<TWEN);

      // Wait until transmission completes
      I2Cwait();

      // Check value of TWSR, and mask out status bits
      status = TWSR & 0b11111000;
      if ((status == 0b00100000 )||(status == 0b01011000))
      {
         TWCR = (1<<TWINT) | (1<<TWEN) | (1<<TWSTO);
         // Wait until stop condition is executed and I2C bus is released
         while(TWCR & (1<<TWSTO));
         continue;
      }
```

```
        break;
    }
}

void I2Cstop()
// Stop I2C bus and release GPIO pins
{
   // Clear interrupt, enable I2C, generate stop condition
   TWCR |= (1 << TWINT)|(1 << TWEN)|(1 << TWSTO);
}

void I2Cwrite(uint8_t data)
// Send 'data' to I2C bus
{
   TWDR = data;
   TWCR |= (1 << TWINT)|(1 << TWEN);
   I2Cwait();
}

uint8_t I2Cread()
// Read incoming byte of data from I2C bus
{
   TWCR |= (1 << TWINT)|(1 << TWEN);
   I2Cwait();
   return TWDR;
}

uint8_t I2CreadACK()
// Read incoming byte of data from I2C bus and ACK signal
{
   TWCR |= (1 << TWINT)|(1 << TWEN)|(1 << TWEA);
   I2Cwait();
   // Incoming byte is placed in TWDR register
   return TWDR;
}

uint8_t decimalToBcd(uint8_t val)
// Convert integer to BCD
{
   return((val/10*16)+(val%10));
}

uint8_t bcdToDec(uint8_t val)
// Convert BCD to integer
{
   return((val/16*10)+(val%16));
}

void setTimeDS3231(uint8_t hh, uint8_t mm, uint8_t ss, uint8_t dw,
                   uint8_t dd, uint8_t mo, uint8_t yy)
// Set time on DS3231
{
   I2CstartWait(0xD0);          // DS3231 write
   I2Cwrite(0x00);              // Start with hours register
   I2Cwrite(decimalToBcd(ss)); // Seconds
```

```
   I2Cwrite(decimalToBcd(mm)); // Minutes
   I2Cwrite(decimalToBcd(hh)); // Hours
   I2Cwrite(decimalToBcd(dw)); // Day of week
   I2Cwrite(decimalToBcd(dd)); // Date
   I2Cwrite(decimalToBcd(mo)); // Month
   I2Cwrite(decimalToBcd(yy)); // Year
   I2Cstop();
}

void readTimeDS3231()
// Retrieve time and date from DS3231
{
   I2CstartWait(0xD0);         // DS3231 write
   I2Cwrite(0x00);             // Seconds register
   I2CstartWait(0xD1);         // DS3231 read
   seconds = bcdToDec(I2CreadACK());
   minutes = bcdToDec(I2CreadACK());
   hours = bcdToDec(I2CreadACK());
   dow = bcdToDec(I2CreadACK());
   dom = bcdToDec(I2CreadACK());
   mo = bcdToDec(I2CreadACK());
   years = bcdToDec(I2CreadACK());
}

void initPWM()
// Activate PWM
{
   TCCR1A |= (1 << WGM11);
   TCCR1B |= (1 << WGM12)|(1 << WGM13)|(1 << CS11);

   // Connect PWM to PB1 and PB2
   TCCR1A |= (1 << COM1A1)|(1 << COM1B1);
   ICR1=39999;
}

void servoAngleA(uint8_t angle)        // Hours servo
// Rotate servo on OCR1A to 'angle' position
{
   OCR1A = ((angle-239.95)/-0.05);
   // Convert angle to OCR1A (duty cycle) value
}

void servoAngleB(uint8_t angle)        // Minutes servo
// Rotate servo on OCR1B to 'angle' position
{
   OCR1B = ((angle-239.95)/-0.05);
   // Convert angle to OCR1A (duty cycle) value
}

❶ void displayServoTime()
{  // Displays hours on servo A, minutes on servo B
   uint8_t _hours;
   uint8_t _minutes;

 ❷ _hours = hours * 15;
```

```
❸ servoAngleA(_hours);
❹ _minutes = minutes * 3;
❺ servoAngleB(_minutes);
}

int main()
{
   DDRB = 0b11111111;                   // Set PORTB as outputs
   I2Cenable();
   initPWM();

   // Uncomment to set time & date, then comment and reflash code
   // setTimeDS3231(9,13,0,5,29,4,21); // h,m,s,dow,dom,m,y
   while(1)
   {
      readTimeDS3231();
      displayServoTime();
      _delay_ms(1000);
   }
}
```

After reviewing the code, you should recognize the parts dedicated to enabling the $I^2C$ bus (as described in Chapter 12), along with setting and retrieving the time from the DS3231 RTC module (as described in Chapter 13) and controlling the servos via PWM (as discussed earlier in this chapter).

The new material in this project is in the `displayServoTime()` function ❶, which takes the values of the hours and minutes from the RTC and converts them into suitable angles to which the servos move. For the servo displaying hours, we divide the 180-degree servo range by 12 hours. That gives us 15, so we multiply the hours value by 15 to get the required servo angle ❷, then command the first servo to move to that position ❸. We use a similar process to convert minutes to angles: 180 divided by 60 is 3, so we multiply the minutes value by 3 ❹, then command the second servo to move to that position ❺.

For a final challenge, try altering the code so the hours display starts at 12 and finishes at 11 instead of going from 1 to 12, or making your own servo library. There are many ways you can expand on these clocks, and using servos in general: for example, you could try using two servos as the front arms of a crawling robot, or to control older mechanical light switches.

So where do you go from here? This book is only the beginning of your AVR journey. Check out the following epilogue for some next steps.

# EPILOGUE

At this point, after having read about (and hopefully built) the 59 projects in this book, you should have the understanding, knowledge, and confidence you need to create your own AVR microcontroller–based projects. I'm sure you'll be able to apply AVR technology to solve all sorts of problems, and have fun at the same time! (For a little inspiration, revisit the examples of more advanced projects in Chapter 1.)

I also hope this book has inspired you to dig deeper into the exciting world of electronics and electrical engineering. You won't be alone in doing so—you'll find a thriving community of AVR microcontroller users on the internet, in places like the following:

- AVR Freaks forum: *https://www.avrfreaks.net/*
- Instructables: *https://www.instructables.com/howto/AVR/*
- Reddit: *https://www.reddit.com/r/avr/new/*

You could even seek out a local hackerspace or club, either online or in person.

I'm always happy to receive feedback about this book via the contact details on the publisher's web page: *https://nostarch.com/contactus/*. Don't forget to check out my other books published by No Starch Press while you're there. And above all, don't just sit there—make something!

# INDEX

Page numbers followed by *f* and *t* refer to figures and tables, respectively.

## Numbers


0.25 W resistors, 16*f*, 18
1N4004 diodes, 20, 21*f*, 185
5 W resistors, 18, 18*f*
8-bit values, 80
10-bit numbers, 80
74HC595 shift register, 224–31, 224*f*
  connections, 225*t*
  schematic symbol, 224*f*
  timing diagram, 228*f*


## A


AA cells, 26, 26*f*, 172*f*
  testing, 76
AC (alternating current), 15, 35
acknowledge bit (ACK), 263
ADCH variable, 76, 78
ADC pins, 75, 79, 240
ADC register, 161
ADCs (analog-to-digital converters), 75–79, 159
  external, 240
  voltage range, 78
ADCSRA register, 75, 79–80, 306
`ADCvalue`, 80
ADMUX register, 75, 79
ADSC bit, 76
Allegro AS1107, 232
alternating current (AC), 15, 35
amperes (amps/A), 15
analog input pins, 6
analog inputs, 74–75
analog signals, 75, 75*f*
analog-to-digital converters (ADCs), 75–79, 159
  external, 240
  voltage range, 78
AND comparison operator (&&), 62, 84
AND operator (&), 54, 62, 247
anode pins, for power diodes, 21
anode side, of an LED, 19, 19*f*
ASCII code, 97–98, 114, 287–88
ATmega328P-PU
  ADCs, 75, 79–80, 82–84, 87
  compared to ATtiny85, 5*t*
  connecting to MCP3008, 241*t*
  connection to crystal circuit, 251*f*
  Controlling Two Servos (project), 323–26
  DDR registers, 40
  EEPROM, 190, 193, 263
  Experimenting with ATmega328P-PU Digital Outputs (project), 47
    physical layout diagram, 48*f*
    schematic diagram, 48*f*
  hardware interrupts, 116–35
  introduction to, 4–6, 4*f*
  LCDs and, 283
  MAX7219 connections, 233*t*
  pinout and port register diagram, 47
  pinout diagram, 127*f*
  pins, 96*t*
    interrupt, 116
    output, 39
    PWM, 161–63, 161*t*
  port register diagram, 47
    activating, 162–63
    PORTB (PB), 47
    PORTC, 75
    PORTD (PD), 47
  reset button, 238
    circuit, 238*f*
  SPCR, 222
  specifications, 5*t*

ATmega328P-PU (*continued*)
timers, 138
USART, 92
Atmel, 5
ATtiny85, 31*f*
ADCs, 75–79
compared to ATmega328P-PU, 5*t*
connected to USBasp, 34*f*
DDR registers, 39–40, 41–43
EEPROM, 190
Experimenting with the ATtiny85's EEPROM (project), 191–92
Experimenting with ATtiny85 Digital Outputs (project), 41–43
introduction to, 4–6, 4*f*
microcontroller symbol, 44, 44*f*
pins, 33*t*
alignment, 35
output, 39
PWM, 157–61
PORT registers, 40, 40*f*, 41–43
PWM, 154–61
activating, 157–58
deactivating, 158
Demonstrating PWM with the ATtiny85 (project), 155–57
reset button, 237
circuit, 237*f*
specifications, 5*t*
timers, 138
AVRDUDE, 34–35
AVR microcontrollers, xv. *See also* ATmega328P-PU; ATtiny85
arithmetic with, 84–85
internal pullup resistors, 70
introduction to, 1–27
specifications, 5*t*
speed, increasing, 250–52
voltage fluctuations, protecting from, 69–70
AVR programmers, 7

## B

backing sheet, with temperature range, 320, 321, 321*f*
batteries
AAA cells, 76
AA cells, 26, 26*f*, 76, 172*f*
C cells, 76
CR2032 coin cells, 271
D cells, 76
testing, 76–79
battery holders, 26, 26*f*
battery packs
black/negative lead, 172, 176, 181, 185
red/positive lead, 185
battery testers, 76–79
BCD (binary-coded decimal), 271
BCD-to-decimal conversions, 271
binary numbers, 41, 52, 54, 55, 56, 93, 147, 262, 307
bits, 92–93
bit shifting, 52–53, 244, 258, 270
Bit-Shifting Digital Outputs (project), 50–53
bitwise arithmetic, 53–56, 288. *See also* operators; AND operator; NOT operator; OR operator; XOR operator
bitwise operations, 76, 147–51, 244, 247
Experimenting with Overflow Timers Using Bitwise Operations (project), 150–51
individual bits, 147–49
toggling between high and low, 147–48
turning to high, 147
turning to low, 148
multiple bits, 149–51
toggling between high and low, 149
turning to high, 149
turning to low, 149–50
registers and, 147–51
`blankMAX7219()` function, 236
`blinkFast()` function, 206, 208
`blinko.c` library, 207, 208, 208*f*, 211–12, 212*f*
Using the `blinko2.c` Library (project), 212
`blinko.h` library, 207, 211–12
`blinkSlow()` function, 206, 208

`blinkType()` function, 210–11
breadboard power supplies, 25–26, 26*f*
breadboards, 24*f*, 25*f*
  power supply module, 85, 85*f*
  pushbutton insertions, 58–59, 58*f*
  solderless, 24–26, 30*f*, 31*f*, 32*f*, 34*f*
breakout boards, 270–71
brightness, of LEDs, 153, 154, 155, 157
  RGB, 164, 167
buttons. *See* pushbuttons
`byte _string[i]`, 288
bytes, 92–93
  represented on DSO, 92*f*, 220*f*
  sending and receiving, 110


## C


calculators, 111–14
capacitors, 21–23. *See also* ceramic capacitors; electrolytic capacitors
castor, 183–84
cathode pins, for power diodes, 21
cathode side, of an LED, 19, 19*f*
ceramic capacitors, 21–22, 22*f*, 250
character arrays, 98–100, 102, 114, 300
character generator RAM (CGRAM), 307–9, 312
char variables, 234
circuit boards, for breadboards, 25–26
circuits, 15, 21, 23–24, 25, 80–81
  assembling, 30
  building, 32*f*, 33*f*, 43, 59–61
circuit schematics. *See* schematic diagrams
`clearLCD()` function, 285, 286, 288, 292
Clear Timer on Compare Match (CTC timers), 142–45
  timing mode, 142
  Using a CTC Timer for Repetitive Actions (project), 142–45
  Using CTC Timers for Repetitive Actions with Longer Delays (project), 143–45
Clear Timer on Compare Match interrupts, 137
clock bit, 223
clock faces, 327*f*
clock line (CLK line), 220, 223
clock line (SCK line), 223, 252
clock-phase bit (CPHA bit), 223, 228, 244
clock pin (SCK pin), 33*t*, 49*t*, 221, 239
clocks, 270–79, 293
  Building an Analog Clock with Servo Hands (project), 326–31
clock signal, 220, 220*f*, 223, 234, 244, 252
clock source, 138
colors, of RGB LEDs, 163
  mixing, 164, 164*f*
COM1B1 bit, 326
`commandLCD()` function, 285, 308
comments, in programs, 37–38
common-anode configuration, 71
common-cathode configuration, 71
comparison operators, 62–63. *See also* AND comparison operator; OR comparison operator
compilers, 8
  error reports, 38–39
configuration registers, 258
connecting wires, 25, 25*f*
constant values (constants), 51
control codes, 98
  silent. *See* escape sequences
cooling fan, 175*f*
CoolTerm terminal emulator software, 93–95, 95*f*
  configuring, 94*f*
counter register, 138
counters, 131–35, 133*f*
  8-bit, 138
  16-bit, 138
  Building a Single-Digit Numerical Counter (project), 71–74
counting display routine, 290*f*
CPHA bit (clock-phase bit), 223, 228, 244
CPOL bit, 223, 228, 234, 244
C programming language, 36, 84
`createCC()` function, 308, 312
crystal circuit, 251*f*
crystal oscillators, 250–51, 250*f*
  schematic symbol, 250*f*

CTC timers (Clear Timer on Compare Match), 142–45
timing mode, 142
Using a CTC Timer for Repetitive Actions (project), 142–45
Using CTC Timers for Repetitive Actions with Longer Delays (project), 143–45
curly quotes, 100
current. *See also* alternating current; direct current
defined, 15
servos and, 314
of USBasp, 25
current-limiting resistors, 19
`cursorLCD()` function, 286–87, 288, 292, 299, 312
custom character data array (`ccdata[]`), 308
custom characters, LCD displays, 306, 309*f*
Displaying Custom LCD Characters (project), 309–12
smiley face, 307*f*

## D

data buses, 219–47
data-direction register (DDR), 39
data-logging device
Simple EEPROM Datalogger (project), 193–98
data sheets, 40, 47, 47*f*, 71, 148, 222
DC (direct current), 15
DC jack adaptors, 25, 25*f*
DDR (data-direction register), 39
DDRB register, 40–41
DDRx function, 223
decimal numbers, 93, 261
decision-making code, 61–63, 82
#define macro, 51
`delay.h` library, 206
`_delay_ms()` function, 69, 157, 208
`_delay_us()` function, 288
device signatures, 34
digital data, defined, 6
digital input pins, 59
voltage fluctuations, 69–70
digital inputs, 58–61
digital I/O pins, 58
digital output, controlling, 39
digital pins, defined, 6
digital signals, levels, 74, 74*f*
digital storage oscilloscope (DSO), 68, 68*f*, 92–93
data represented on, 92*f*, 220*f*, 252*f*
LCD Digital Thermometer with Min/Max Display (project), 301–6
output measurements, 146*f*
`digitData` array, 231
DIN line, 223
diodes, 191. *See also* light-emitting diodes; power diodes
LCD, 282
symbol, 45
direct current (DC), 15
direction registers, 258
discharging, 21
`displayNumber()` function, 73, 90, 214, 216
`displayServoTime()` function, 331
`displayTimeMAX7219()` function, 279
`dispMAX7219()` function, 236
`dispNumSR()` function, 228, 231
DORD bit, 222, 223, 234, 244
DS3231 real-time clock, 270–79
data registers, 271*f*
module, 271*f*, 293, 299, 331
DSO (digital storage oscilloscope), 68, 68*f*, 92–93
data represented on, 92*f*, 220*f*, 252*f*
output measurements, 146*f*
`dtostrf()` function, 102, 300
`dumpData()` function, 198, 204
duty cycles, 154, 154*f*, 159, 174, 315, 317
motor speed, 188
PWM, 161*t*
setting, 161

## E

EEPROM (electrically erasable programmable read-only memory), 5*t*, 189–204, 192, 198
defined, 6, 189
Experimenting with the ATtiny85's EEPROM (project), 191–92

introduction to, 190–91
library, 190
lifespan, 190
Temperature Logger with EEPROM (project), 199–204
Using an External IC EEPROM (project), 263–70
`eeprom.h` library, 190, 192, 198
`eeprom_read_float()` function, 199
`eeprom_update_byte()` function, 190
`eeprom_update_float()` function, 199
`eeprom_update_word()` function, 193
`eeprom_write_byte()` function, 190
`eeprom_write_float()` function, 198
`eeprom_write_word()` function, 193
EICRA register, 116–17, 117*t*
EIFR (external interrupt flag register), 118
eight-digit LED module, 232, 232*f*
EIMSK register, 117
electrically erasable programmable read-only memory. *See* EEPROM
electricity
basics of, 15
EEPROM and, 190
safety concerns, 35
electric motors, 313
electrolytic capacitors, 22–23, 22*f*
electronic components
fundamental, 15–27
suppliers for, 7
ENABLE pin, 180, 180*t*, 183, 188
`eraseData()` function, 198, 204
escape sequences, 99
example projects
AvrPhone, 4, 4*f*
AVR TV Game, 3–4, 3*f*
Digispark board, 3, 3*f*
logic analyzer, 2, 2*f*
external interrupt flag register (EIFR), 118
external interrupts, 115, 116–26

## F

Fahrenheit, converting to Celsius, 87, 217, 306
falling edge interrupts, 116
EICRA register, 117*t*
Experimenting with Falling Edge Interrupts (project), 121–23
Experimenting with Two Interrupts (project), 124–26
farads, 21
feedback controller, 315
first bit, 41
Fischl, Thomas, 8
flash memory, 6
floating-point math, 85, 90
floating-point numbers
converting, 300
displaying on LCDs, 300, 301
floating-point variables, 102
defined, 85
introduction to, 198–99
storing, 198–99
Temperature Logger with EEPROM (project), 199–204
`for` loops, 52–53, 192, 261, 292
frequency (*f*), in timer formula, 138
functions
custom, 63–68, 206, 208, 227, 228, 236, 246–47
Custom Functions That Return Values (project), 66–68
Custom Functions with Internal Variables (project), 64–65
EEPROM, 193
LCDs, 284–88
to pass values, 64–66
to perform a task, 64–65
to return values, 66–68
servos, 319
Simple Custom Function (project), 64–65
PORTx, 52–53

## G

global variables, 135, 136, 144, 145
GND (ground), 15, 33*t*, 49*t*, 175
battery pack leads, 172, 176, 181, 185
interrupt pins, 116

GND (ground) (*continued*)
    L293D motor driver IC, 179–80
    servo connections, 315
    symbol, 46*f*
    USB-to-serial converter, 95–96, 96*t*
    wire connected to ground, 46
GPIOA register, 258, 261
GPIOB register, 258, 261
GPIO pin, 221, 254, 256, 288
    adding, 256
ground. *See* GND

## H

hardware
    suppliers, 7
    testing, 30–39
    USART, 93–95
hardware interrupts, 115–16. *See also* external interrupts; pin-change interrupts
    ATmega328P-PUB, 116–35
hardware registers, 39–43
hardware timers, 137–45
H-BRIDGE IC, 178
heatsink, 180
Heinrich, Adam, 4
hexadecimal numbers, 234, 236, 253
high byte, 269, 270
horns, 313, 314*f*, 320
hysteresis, 178

## I

I$^2$C bus (Inter-Integrated Circuit bus), 249–79, 331
    functions, 261, 269, 279, 299
    introduction, 252–62
    pins, 254
    setting address, 257
    starting, 255
    stopping, 256
    wiring, 253*f*
    writing to, 256
`I2Cenable()` function, 254
`I2Cread()` function, 262
`I2CreadACK()` function, 262–63, 279
`I2CstartWait()` function, 255, 262
`I2Cstop()` function, 256
`I2Cwait()` function, 254
`I2Cwrite()` function, 256
IC extractors, 23–24, 23*f*
ICs (integrated circuits), 23–24
    maximum current, 39
    pins, 23, 23*f*
`if ... else` statements, 61, 62, 63
`if` statements, 61–62
`initCTC()` function, 143
`initLCD()` function, 286
`initMAX7219()` function, 236
`initOVI()` function, 141
`initPWM()` function, 157, 167, 316–17, 319, 320, 326
inline header pins, 232, 232*f*, 282*f*
integers, 51–52, 300
    unsigned, 52
integer variables, 102, 288
integrated circuits (ICs), 23–24
    maximum current, 39
    pins, 23, 23*f*
Inter-Integrated Circuit bus. *See* I$^2$C bus
interrupt functions, 128
`interrupt.h` library, 116, 117, 128
interrupt INT0, 116, 117, 126
interrupt INT1, 116, 117, 126
interrupt service routine (ISR), 115
    Creating an Up/Down Counter Using Interrupts (project), 131–35
    CTC and, 142–43
    defining, 117, 128
    functions, 127
    PCI bank codes, 127–28
    timers and, 139
    triggering, 142–43
`io.h` library, 38, 206
ISR. *See* interrupt service routine
`i` variable, 135

## J

jumper wires, 30, 32*f*
junction dot, 45, 46*f*

## K

Kettenburg, Erik, 2
KiCad package, for schematic diagramming, 47
kilohms (kΩ), 16

# L


L293D motor driver IC, 178–88, 178*f*, 180*t*
  block diagram, 179
  DC Motor Control with L293D (project), 180–83
  pinouts, 179*f*
L293D single motor control, 180*t*
last bit, 41
latch. *See* secondary select pin
LCDs (liquid crystal displays), 281
  16×2-character, 282*f*
    schematic symbol, 283*f*
  Building an AVR-Based LCD Digital Clock (project), 292–300
  characters, 307*f*
  clear display command, 284–85, 284*f*
  clearing, 286
  custom characters, 306–12, 309*f*
    smiley face, 307*f*
  Displaying Custom LCD Characters (project), 309–12
    temperature, 302*f*, 306
  functions, 292, 299
  initializing, 286
  introduction, 282–300
  pins, 283, 285
  printing to, 287–88
  sending commands, 285
  setting cursor, 286
  in solderless breadboard, 283
  text display, 289*f*
  Using a Character LCD with Your AVR (project), 288–92
least significant bit (LSB), 41, 222, 223, 244, 247
LEDs (light-emitting diodes), 19*f*, 30–32, 31*f*, 32*f*. *See also* RGB LEDs; seven-segment LED displays
  Blinking an LED (project), 36–39
  Blinking an LED on Command (project), 59–61
  controlling, 42
  current required, 16
  currents, 19–20
  Experimenting with Rising Edge Interrupts (project), 118–21
  introduction to, 19–20
  voltages, 19–20
LED symbol, 45, 45*f*
`leftDigit` variable, 231
level converter, 221–22, 221*f*, 254
libraries, 38
  accepting and acting on values, 210–12
  anatomy of, 207–8
  creating, 205–17
  example code, 206
  header files, 207, 211, 213
  installing, 208, 208*f*
  processing data and returning values, 212–17
  source, 213–14
  source files, 207, 211–12
`library.c` library, 207
`library.h` library, 207
light-emitting diodes. *See* LEDs
limit switches, 136
linear variable resistors, 80, 81*f*
Linux, installing software on, 10–11
liquid crystal displays. *See* LCDs
logarithmic variable resistors, 80
`logData()` function, 198, 204
`logDelay`, 198, 204
logging events, 198
logic change interrupts, 116
  EICRA register, 117*t*
loops, 52–53, 55, 174, 177–78, 183, 188, 198, 204, 247, 269. *See also* `for` loops
  stepping through, 90
low byte, 269, 270
low level interrupt, 116
  EICRA register, 117*t*
LSB (least significant bit), 41, 222, 223, 244, 247


# M


macOS, installing software on, 9
`main()` function, 114
`main.c` library, 36, 37
main device, 220

main in, secondary out (MISO pin), 33*t*, 49*t*, 221, 239
main out, secondary in (MOSI pin), 33*t*, 49*t*, 221, 239
main–secondary configuration, SPI bus, 220
mains plugpacks, 25
Makefile, 36, 252
  libraries and, 208, 208*f*, 212*f*, 214, 215*f*
  speed and, 251
`math.h` library, 85, 90
MAX7219 LED driver IC, 231–37, 242
  functions, 279
  module, 299
  surface-mount package type, 231, 231*f*, 232
  through-hole package type, 231–32, 231*f*
maximum processing speed, 6
MCUs. *See* AVR microcontrollers
megohms (MΩ), 16
Meier, Roger, 93
memory, 189–204. *See also* EEPROM; flash memory
metal-oxide-semiconductor field-effect transistors. *See* MOSFETs
Microchip 24LC512-E/P external EEPROM IC, 263–70, 263*f*
  $I^2C$ bus address configuration, 264*t*
  schematic symbol, 263*f*
Microchip MCP23017 16-bit I/O expander, 256–61, 257*f*
  address configuration, 258*t*
  schematic symbol, 257*f*
Microchip MCP3008 8-channel ADC IC, 240–47, 240*f*
  schematic symbol, 241*f*
  timing diagram, 243*f*
microcontrollers. *See* AVR microcontrollers
microfarads (µF), 21
milliamps (mA), 15
millivolts (mV), 247
miniature variable resistors. *See* trimpots
MISO pin (main in, secondary out), 33*t*, 49*t*, 221, 239
MOSFETs (metal-oxide-semiconductor field-effect transistors), 170*f*
  2N7000, 170
    schematic symbol, 170*f*
  DC Motor Control with PWM and MOSFET (project), 171–74
  defined, 169
  introduction to, 170–71
  operating, 170–71
  Temperature-Controlled Fan (project), 174–78
MOSI pin (main out, secondary in), 33*t*, 49*t*, 221, 239
most significant bit (MSB), 41, 222, 223, 244, 247
`motorBackward()` function, 183
`motorForward()` function, 183
`motorOff()` function, 174, 183
`motorOn()` function, 174
`motorPWM()` function, 174
motors, 169, 171*f*, 178. *See also* electronic motors; L293D motor driver IC
  Controlling a Two-Wheel-Drive Robot Vehicle (project), 183–88
  DC Motor Control with L293D (project), 180–83
  DC Motor Control with PWM and MOSFET (project), 171–74
  robot chassis, 184
  rotation, 188
  Temperature-Controlled Fan (project), 174–78
`motorsOff()` function, 188
`moveBackward()` function, 188
`moveForward()` function, 188
`moveLeft()` function, 188
`moveRight()` function, 188
MSB (most significant bit), 41, 222, 223, 244, 247
MSTR, 223
multimeters, 18, 18*f*, 85
multiplier band, 16
multitasking, 115, 144


## N


negative, 46
negative numbers, 52, 301

nibbles, 285, 288
nitPWM() function, 177
non-negative numbers, 38
NOT operator (~), 53–54
numberMAX7219() function, 236–37

## O

OCR0A register, 161
  PWM, 157, 161*t*
OCR1A register, 143, 317, 326
  duty cycle, 174
  PWM, 161*t*
OCR1B register, 326
  PWM, 157, 161*t*
ohms (Ω), 16
Ohm's law, 19–20
Ohm's law triangle, 20, 20*f*
operating voltage, 6
operators, in bitwise arithmetic, 53–56
OR comparison operator (||), 62
OR operator (|), 55, 244
oscillator circuit, 138
output pins, maximum current, 39
overflow timers. *See* timers: overflow

## P

PB (PORTB register), 40–41, 75, 140, 142, 143, 146
  duty cycle, 174
  EEPROM, 192, 193, 198
  PWM, 157, 161, 161*t*, 167
  setting as output, 147
  thermometer, 216
  toggling, 147
  toggling between high and low, 149
  turning on output, 147
  turning to high, 149
  turning to low, 148, 150
PCI banks (pin-change interrupt banks), 127–28
PCICR register, 128
PCIFR register, 128
PCINT (pin-change interrupts), 115, 116, 126–35
PCMSK*x* registers, 128
physical layout diagrams, 42*f*, 43, 48*f*
picofarads (pF), 21
piezo element, 159–61, 159*f*
  Experimenting with Piezo and PWM (project), 159–61
  schematic symbol, 159*f*
pin-change interrupt banks (PCI banks), 127–28
pin-change interrupts (PCINT), 115, 116, 126–35
pin-out interrupts, Creating an Up/Down Counter Using Interrupts (project), 131–35
pins, 39
PIN*x* register, 58, 59
  alignment, 35
  high state, 59
  low state, 59
PL2303TA-type USB-to-serial cable, 93, 93*f*
PMD Way module, 85
PORTB register (PB), 40–41, 75, 140, 142, 143, 146
  duty cycle, 174
  EEPROM, 192, 193, 198
  PWM, 157, 161, 161*t*, 167
  setting as output, 147
  thermometer, 216
  toggling, 147
  toggling between high and low, 149
  turning on output, 147
  turning to high, 149
  turning to low, 148, 150
PORTC register, 75
PORT registers, 40, 40*f*
PORTx function, 223
positive numbers, 301
potentiometers, 240. *See also* variable resistors
power, 25–27
  defined, 15
  external, 85, 171, 175
power diodes, 20–21
power, voltage, and current equation, 15, 18
prescalers, 75, 138, 306
prescaler value (*p*), in timer formula, 138
primary device, 219, 252, 253

printCCLCD() function, 308–9, 312
printLCD() function, 287–88, 292, 300, 309
processing speeds, 34
programming software, 8
pulldown resistors, 70, 70*f*, 116
pullup configuration, 238
pullup resistors, 69, 69*f*, 116, 121, 253
pulse-width modulation (PWM), 153–67, 154*f*, 320. *See also* duty cycles
  DC Motor Control with PWM and MOSFET (project), 171–74
  Demonstrating PWM with the ATtiny85 (project), 155–57
  Experimenting with Piezo and PWM (project), 159–61
  Experimenting with RGB LEDs and PWM (project), 164–67
  high-frequency operations, 157
    ATmega328P-PU, 161–63
    ATtiny85, 157–61
  MOSFETs, 170
  piezo element, 159–61
  servos, 326, 331
  signal, 154, 316*f*
pulse wires (PWM wires), 315
pushbuttons, 58–59, 58*f*, 116, 121, 124, 136
  reset, 237–39
  schematic symbol, 59*f*
  switch bounce, 68–70

## Q

quotes, straight vs. curly, 100

## R

readMCP3008() function, 246–47
read-only memory (ROM) ICs, 190
readTemperature() function, 204
readTimeDS3231() function, 279
readTMP36() function, 214, 216–17
real-time clock (RTC), 270–79
reduction gears, 315
reset buttons, 237–39
  circuits, 237, 238*f*
RESET pin, 238
resistance, 16–18, 19–20
  reading values, 16–18
  variable resistors and, 80–84
resistors, 16*f*, 31*f*, 32*f*. *See also individual resistor types*
  bands representing resistance, 16–17, 17*f*, 17*t*
  defined, 16
  diagrams, 17, 17*f*
  power ratings, 18
resistor symbol, 44, 45*f*
RGB LEDs, 163, 163*f*
  color mixing, 164*f*
  common-anode configuration, 163, 163*f*
  common-cathode configuration, 163, 163*f*
  Experimenting with RGB LEDs and PWM (project), 164–67
  schematic symbols, 163*f*
rightDigit variable, 231
rising edge interrupts, 116
  Controlling a Two-Wheel-Drive Robot Vehicle (project), 183–88
  Creating an Up/Down Counter Using Interrupts (project), 131–35
  EICRA register, 117*t*
  Experimenting with Rising Edge Interrupts (project), 118–21
  Experimenting with Two Interrupts (project), 124–26
rotational range, 314, 315*f*
round() function, 90
RST pin, 33*t*, 49*t*
RTC (real-time clock), 270–79
RX pin, USB-to-serial converter, 95–96, 96*t*
Ryves, Ben, 3

## S

safety, 35, 180
  batteries, 76
  electricity, 35
schematic diagrams, 43–47, 160*f*
  ATtiny85 digital outputs, 46*f*
  components, 44–45
  creating, 47

defined, 5
dissecting, 46–47
examples of, 44*f*
introduction to, 43–44
symbols. *See* schematic symbols
wires, 45–46
connected, 45–46, 46*f*
connected to ground, 46
non-connected, 45, 45*f*
schematic symbols
2N7000 MOSFET, 170*f*
16×2-character LCD, 283*f*
24LC512-E/P, 263*f*
74HC595 shift register, 224*f*
breadboard trimpots, 284*f*
crystal oscillators, 250*f*
MCP23017 microchip, 257*f*
MCP3008, 241*f*
piezo element, 159*f*
pushbutton, 59*f*
RGB LEDs, 163*f*
servos, 315*f*
seven-segment common-cathode LED display modules, 71*f*
TMP36 temperature sensor, 86*f*
variable resistor, 80*f*
SCK line (clock line), 223, 252
SCK pin (clock pin), 33*t*, 49*t*, 221, 239
SCL line (serial clock line), 252
SCL pin, 253, 254
SDA (serial data line), 252
SDA pin, 253, 254
secondary device, 219, 220, 222, 223, 252, 253
secondary select pin (SS pin), 221, 236, 239
`sei()` function, 117, 128
`sendString()` function, 100
Serasidis, Vassilis, 2
serial data, 92
serial peripheral interface bus. *See* SPI bus
`servoAngle()` function, 319, 326
servomechanisms (servos), 313, 314*f*
connecting to, 315
controlling, 315–17
multiple, 323–26
positioning, 317
rotational range, 314, 315*f*, 317, 320
schematic symbol, 315*f*
setting up, 314–17
speed, 314
`servoRange()` function, 319, 320, 326
`setTimeDS3231()` function, 273, 279, 293, 299
`setupSPI()` function, 227, 228
seven-segment LED display modules, 70–74, 70*f*
common-cathode configuration, 71
symbols, 71*f*
Creating a Digital Thermometer with the thermometer.c Library (project), 215–17
thermometer library, 213
SG90-type servos, 314, 315
shift registers, 223–24
silent control codes. *See* escape sequences
smoothing capacitors, 80–84
software
installing, on Linux, 10–11
installing, on macOS, 9, 9*f*
installing, on Windows, 11–15, 11*f*, 12*f*, 13*f*, 14*f*
required, 8–15
USART, 93–95
solderless breadboards, 24–26, 30*f*, 31*f*, 32*f*, 34*f*
heat safety, 180
power supplies, 25–26, 26*f*
sound waves, 159
SPCR register (SPI Control Register), 222–23, 234, 244
diagram, 222*f*
SPDR register, 236, 239
SPE, 222
SPI bus, 219–47
connection to AVR, 221*f*
data transfer, 239–40, 240*f*
implementing, 222–23
multiple SPI devices on, 239, 239*f*, 242
pins, 221
receiving data from, 239–47
represented on DSO, 220*f*
sending data, 223

SPI bus (*continued*)
timing diagram, 222–23, 222*f*
workings of, 220–23
SPIE, 222
spreadsheets, 102, 106, 107, 198, 199, 204, 307
temperature data, 108
SS pin (secondary select pin), 221, 236, 239
`startADC()` function, 177, 306
static random access memory (SRAM), 5*t*, 6
straight quotes, 100
`switch()` function, 114
switch bounce, 68–70, 120
measuring, 68, 68*f*
`switch ... case` statements, 61, 63, 74, 204, 299–300
symbols. *See* schematic symbols

## T

`targetCount` variable, 145
`targetFunction()` function, 145, 146
`target` variable, 145
TCCR1A register, 326
TCCR1B register, 141–42
prescaler values, 139*t*
temperature. *See* thermometers
temperature-controlled fan, 174–78
`temperature.h` library, 213
temperature sensor. *See* TMP36 temperature sensor
terminal emulators, 93–95, 94*f*, 95*f*
displaying data, 108*f*
raw mode, 108–9, 109*f*
recording data, 106–8, 106*f*, 107*f*
text delimiter, 107, 107*f*
text editors, 8, 36, 39, 106, 273, 293
auto-correct settings, 100
`thermometer.c` library, 213–14, 215*f*, 216–17
thermometer library, 212–17
thermometers
Creating a Digital Thermometer (project), 87–90
Creating a Digital Thermometer with the thermometer.c Library (project), 215–17
Creating an Analog Thermometer (project), 320–23
LCD Digital Thermometer with Min/Max Display (project), 301–6
thermostat, 177
time (*T*), in timer formulas, 138
TIMER0, 138
TIMER1, 138, 317
Experimenting with Timer Overflow and Interrupts (project), 139–42
prescaler values, 139*t*
Using a CTC Timer for Repetitive Actions (project), 142–43
TIMER2, 138
timer overflow interrupts, 137, 148, 151
timers. *See also* hardware timers
Experimenting with Timer Overflow and Interrupts (project), 139–42
internal, accuracy of, 146
overflow, 139
Experimenting with Overflow Timers Using Bitwise Operations (project), 150–51
PWM, 157
TIMSK1 register, 141, 148, 148*f*
TMP36 temperature sensor, 86–90, 86*f*, 102–4, 240, 306
Creating a Digital Thermometer with the thermometer.c Library (project), 215–17
Creating a Temperature Data Logger (project), 102–8
schematic symbol, 86
Temperature-Controlled Fan (project), 174–78
Temperature Logger with EEPROM (project), 199–204
thermometer library, 213
torque, 314
trimmers. *See* trimpots
trimpots, 81, 81*f*, 283
breadboard compatible, 283
schematic symbol, 284*f*
piezo element, 159–61

TWBR (TWI Bit Rate Register), 254
TWCR (Two-Wire Control Register), 254, 255, 256, 263
TWDR, 256
TWEA, 263
TWEN bit (Two-Wire Enable bit), 254, 255, 256, 263
TWI (two-wire serial interface), 252
TWI Bit Rate Register (TWBR), 254
TWINT bit (Two-Wire Interrupt bit), 254, 255, 256, 263
Two-Wire Control Register (TWCR), 254, 255, 256, 263
Two-Wire Enable bit (TWEN bit), 254, 255, 256, 263
two-wire interface start condition (TWSTA), 255
Two-Wire Interrupt bit (TWINT bit), 254, 255, 256, 263
two-wire serial interface (TWI), 252
TWSR, 255
TWSTA (two-wire interface start condition), 255
TWSTO bit, 256
TX pin, USB-to-serial converter, 95–96, 96*t*

## U

Ubuntu Linux, 10
universal synchronous and asynchronous receiver-transmitter (USART), 92
    Creating an Up/Down Counter Using Interrupts (project), 131–35
    defined, 91
    EEPROM, 193, 198, 204
    hardware, 93
    Receiving Data from Your Computer (project), 108
    Sending Numbers with the USART (project), 100–102
    Sending Text with the USART (project), 98–100
    Testing the USART (project), 95–98
unsigned integers, 52
`update` command, 190–91
USART (universal synchronous and asynchronous receiver-transmitter), 92
    Creating an Up/Down Counter Using Interrupts (project), 131–35
    defined, 91
    EEPROM, 193, 198, 204
    hardware, 93
    Receiving Data from Your Computer (project), 108
    Sending Numbers with the USART (project), 100–102
    Sending Text with the USART (project), 98–100
    Testing the USART (project), 95–98
`USARTInit()` function, 97
`USARTReceiveByte()` function, 110
`USARTSendByte()` function, 97, 110
USBasp drivers, installing, 14
USBasp programmers, 8*f*
    connected to ATtiny85, 34*f*
    connecting to breadboard, 33
    connection pins, 33*f*, 33*t*
    introduction to, 7–8
    as power method, 25
    testing, 30–39
USBasp to ATmega328P-PU Connections, 49*t*
USBasp to ATtiny85 Connections, 33*f*
USB-to-serial converter, 93, 93*f*
    microcontroller connections, 96*t*
    pins, 96*t*
    testing, 95–98
user interface, 198, 204

## V

variable resistors, 80–84. *See also* potentiometers; trimpots
    linear models, 80, 81*f*
    logarithmic models, 80
    symbol, 80*f*
variables, 51–52, 74
    `ADCvalue`, 80
    byte, 262
    floating point, 85, 102
    global, 135, 136, 144

variables (*continued*)
integer, 102, 288
internal, 65–66
$V_{CC}$ pin, 33*t*, 49*t*, 179–80
USB-to-serial converter, 95
vibration, 159
`void blinkFast()` function, 207
`void blinkSlow()` function, 207
`volatile` keyword, 135, 136
voltage
of AA cells, 26
defined, 15
dropping, with 1N4004 diodes, 21
levels, 92–93, 116
of mains plugpacks, 25
volts (V), 15
$V_{REF}$, 241, 244, 247

## W

watts (W), 15
`while()` function, 76
WinAVR software, 11
Windows, installing software on, 11–15, 11*f*, 12*f*, 13*f*, 14*f*
wires. *See* jumper wires; pulse wires
words of data, 192–98, 193, 198
A Simple EEPROM Datalogger (project), 193–98
`writeLCD()` function, 308
`writeMAX7219()` function, 236

## X

XOR operator (^), 56